Swarm Intelligence Algorithms

Swarm Intelligence Algorithms

Modifications and Applications

Edited by
Adam Slowik

CRC Press
Taylor & Francis Group
Boca Raton London New York

CRC Press is an imprint of the
Taylor & Francis Group, an **informa** business

First edition published 2020
by CRC Press
6000 Broken Sound Parkway NW, Suite 300, Boca Raton, FL 33487-2742

and by CRC Press
2 Park Square, Milton Park, Abingdon, Oxon, OX14 4RN

© 2021 Taylor & Francis Group, LLC

CRC Press is an imprint of Taylor & Francis Group, LLC

Library of Congress Cataloging-in-Publication Data
Names: Slowik, Adam, editor. Title: Swarm intelligence algorithms. Modifications and applications / edited by Adam Slowik. Description: First edition. | Boca Raton : Taylor and Francis, 2020. | Includes bibliographical references and index. Identifiers: LCCN 2020018734 (print) | LCCN 2020018735 (ebook) | ISBN 9781138391017 (hardback) | ISBN 9780429422607 (ebook) Subjects: LCSH: Swarm intelligence. | Algorithms. | Mathematical optimization. Classification: LCC Q337.3 .S9244 2020 (print) | LCC Q337.3 (ebook) | DDC 006.3/824--dc23 LC record available at https://lccn.loc.gov/2020018734 LC ebook record available at https://lccn.loc.gov/2020018735

ISBN: 978-1-138-39101-7 (hbk)
ISBN: 978-0-429-42260-7 (ebk)

**Visit the Taylor & Francis Web site at
http://www.taylorandfrancis.com**

**and the CRC Press Web site at
http://www.crcpress.com**

*— To my beloved wife Justyna,
and my beloved son Michal —*

*— My beloved late parents
Bronislawa and Ryszard,
I will always remember you —*

Contents

Preface

Swarm Intelligence (SI) is a relatively new area of artificial intelligence, in which the essential part is a cooperation between agents. In the literature we can find many algorithms which are in the family of SI, but as pioneers of swarm intelligence we can identify three algorithms such as: stochastic diffusion search, ant colony optimization (ACO), and particle swarm optimization (PSO). The swarm's intelligence family is constantly growing and newer algorithms based on the nature and cooperation between particular agents are created. Already existing SI algorithms have been strongly developed, and their new modifications are proposed. Due to these modifications, we can, for example, use the PSO algorithm for solving discrete problems (the original version of the PSO is adapted to the continuous optimization problems) or the ACO algorithm can be used for solving continuous problems (the original version of ACO is proposed for discrete optimization problems).

Currently, SI algorithms are widely used in various applications. In this book we present a short brief on 24 swarm intelligence algorithms which are chosen from the whole family of SI algorithms. In each chapter you will find a brief description of one selected algorithm together with a discussion of its modifications and a presentation of its practical application. Below you can see a list (in alphabetical order) of the SI algorithms which are presented in this book. In this list, for each SI algorithm, we present a short description of its exemplary application which is discussed in this book. The SI algorithms and their exemplary applications which are presented in this book are as follows:

- **Ant Colony Optimization**
Exemplary application: Optimal shunt capacitors allocation problem aiming to minimize power delivery loss and node voltage deviation subjected to various system and distribution network operator constraints,

- **Artificial Bee Colony**
Exemplary application: Software requirement selection in order to minimize the cost and maximize the customer satisfaction under resource constraints, budget bounds and requirement dependencies,

- **Bacterial Foraging**
Exemplary application: Simultaneous allocation problem of distributed generations (small generating units typically connected to the utility grid in parallel near load

centers) and shunt capacitors in distribution systems in order to reduce annual energy losses and to maintain better node voltage profiles,

- **Bat Algorithm**

Exemplary application: Problem of distribution network reconfiguration in order to find the best possible topology of the distribution network that can give minimum losses and a better node voltage profile in the system,

- **Cat Swarm Optimization**

Exemplary application: Generation of optimal diet which consists of four principal meals in a day namely, breakfast, lunch, dinner, supper, and which is based on the person's food preferences,

- **Chicken Swarm Optimization**

Exemplary application: Detection of falls in daily living activities,

- **Cockroach Swarm Optimization**

Exemplary application: Traveling salesman problem in which the main goal is to find the path of shortest length or minimum cost between all the requested points (cities), visiting each point exactly once and returning to the starting point,

- **Crow Search Algorithm**

Exemplary application: Tuning of the number of nodes of each layer of a deep neural network used for predicting the status of the jobs using the CIEMAT Euler log,

- **Cuckoo Search**

Exemplary application: Designing the power system stabilizer parameters of a multi-machine power system to enhance small-signal stability,

- **Dynamic Virtual Bats Algorithm**

Exemplary application: Identify/estimate the parameters of a quarter-car suspension system,

- **Dispersive Flies Optimisation**

Exemplary application: Neuroevolution-based approach for training a neural network which is applied to the problem of detecting false alarms in Intensive Care Units based on physiological data,

- **Elephant Herding**

Exemplary application: Optimal economic dispatch of microgrids composed of multiple distributed energy resources to minimize the daily operating cost of the systems,

- **FireFly Algorithm**

Exemplary application: Few areas of the potential applications of the firefly algorithm are highlighted,

- **Glowworm Swarm Optimization**

  Exemplary application: Multiple source localization and boundary mapping; clustering; sensor deployment scheme in wireless networks; signal source localization,

- **Grasshopper Optimisation**

  Exemplary application: Clustering problem - which represents a task of dataset division into a set of C disjoint clusters,

- **Grey Wolf Optimizer**

  Exemplary application: Five engineering optimization problems are shown: welded beam design problem, pressure vessel design problem, speed reducer design problem, three-bar truss design problem, tension compression spring problem,

- **Hunting Search**

  Exemplary application: Carbon steel rectangular cantilever beam design problem in order to carry a certain load acting at the free tip with minimum overall cost and fabrication,

- **Krill Herd**

  Exemplary application: Optimum design of retaining walls for minimizing their cost and weight,

- **Monarch Butterfly Optimization**

  Exemplary application: Optimal allocation of distributed generations (DGs) in distribution system; the objective is to determine the optimal sites (nodes) and sizes (DG capacities) of 3 DGs for minimum real power loss of 33-bus distribution system,

- **Particle Swarm Optimization**

  Exemplary application: Design of stable IIR (Infinite Impulse Response) digital filter with non-standard amplitude characteristic,

- **Salp Swarm Algorithm**

  Exemplary application: Welded beam design problem,

- **Social Spider Optimization**

  Exemplary application: Economic load dispatch problem - optimal combination of power generation where the total production cost of the system is minimized,

- **Stochastic Diffusion Search**

  Exemplary application: Identifying metastasis in bone scans,

- **Whale Optimization Algorithm**

  Exemplary application: Optimum design of shallow foundation.

Also, I would like to notice that in the first volume of this book – entitled: *Swarm Intelligence Algorithms: A Tutorial* – you can find a very detailed explanation for each SI algorithm (detailed description, pseudo-code, source-code in Matlab, source-code in C++, and detailed step-by-step numerical example).

At the end of this short preface, I would like to thank very much all the contributors for their hard work in preparation of the chapters for this book. I also would like to wish all readers enjoyment in reading this book.

Adam Slowik
Department of Electronics and Computer Science
Koszalin University of Technology, Koszalin, Poland

MATLAB® is a registered trademark of The MathWorks, Inc. For product information, please contact:

The MathWorks, Inc.
3 Apple Hill Drive
Natick, MA, 01760-2098 USA
Tel: 508-647-7000
Fax: 508-647-7001
E-mail: info@mathworks.com
Web: www.mathworks.com

Editor

ADAM SLOWIK (IEEE Member 2007; IEEE Senior Member 2012) received the B.Sc. and M.Sc. degrees in computer engineering and electronics in 2001 and the Ph.D. degree with distinction in 2007 from the Department of Electronics and Computer Science, Koszalin University of Technology, Koszalin, Poland. He received the Dr. habil. degree in computer science (intelligent systems) in 2013 from Department of Mechanical Engineering and Computer Science, Czestochowa University of Technology, Czestochowa, Poland. Since October 2013, he has been an Associate Professor in the Department of Electronics and Computer Science, Koszalin University of Technology. His research interests include soft computing, computational intelligence, and, particularly, bio-inspired optimization algorithms and their engineering applications. He is a reviewer for many international scientific journals. He is an author or coauthor of over 80 refereed articles in international journals, two books, and conference proceedings, including one invited talk. Dr. Slowik is an Associate Editor of the IEEE TRANSACTIONS ON INDUSTRIAL INFORMATICS. He is a member of the program committees of several important international conferences in the area of artificial intelligence and evolutionary computation. He was a recipient of one Best Paper Award (IEEE Conference on Human System Interaction - HSI 2008).

Contributors

Adam Slowik
Department of Electronics and Computer Science
Koszalin University of Technology, Koszalin, Poland
e-mail: aslowik@ie.tu.koszalin.pl, adam.slowik@tu.koszalin.pl

Pushpendra Singh
Department of Electrical Engineering
Govt. Women Engineering College, Ajmer, India
e-mail: pushpendragweca@gmail.com

Nand K. Meena
School of Engineering and Applied Science
Aston University, Birmingham, B4 7ET, United Kingdom
e-mail: nkmeena@ieee.org

Jin Yang
School of Engineering and Applied Science
Aston University, Birmingham, B4 7ET, United Kingdom
e-mail: j.yang8@aston.ac.uk
James Watt School of Engineering
University of Glasgow, Glasgow, G12 8LT, United Kingdom
e-mail: jin.yang@glasgow.ac.uk

Bahriye Akay
Department of Computer Engineering
Erciyes University, 38039, Melikgazi, Kayseri, Turkey
e-mail: bahriye@erciyes.edu.tr

Neeraj Kanwar
Department of Electrical Engineering
Manipal University Jaipur, Jaipur, India
e-mail: nk12.mnit@gmail.com

Sonam Parashar
Department of Electrical Engineering
Malaviya National Institute of Technology, Jaipur, 302017, India
e-mail: sonam_ee@yahoo.com

Dorin Moldovan
Department of Computer Science
Technical University of Cluj-Napoca, Romania
e-mail: dorin.moldovan@cs.utcluj.ro

Viorica Chifu
Department of Computer Science
Technical University of Cluj-Napoca, Romania
e-mail: viorica.chifu@cs.utcluj.ro

Ioan Salomie
Department of Computer Science
Technical University of Cluj-Napoca, Romania
e-mail: ioan.salomie@cs.utcluj.ro

Joanna Kwiecien
Department of Automatics and Robotics
AGH University of Science and Technology, Krakow, Poland
e-mail: kwiecien@agh.edu.pl

Dhanraj Chitara
Department of Electrical Engineering
Swami Keshvanand Institute of Technology (SKIT), Jaipur, India
e-mail: dhanraj.chitara@gmail.com

Ali Osman Topal
Department of Computer Engineering
Epoka University, Tirana, Albania
e-mail: aotopal@epoka.edu.al

Mohammad Majid al-Rifaie
School of Computing and Mathematical Sciences, University of Greenwich
Old Royal Naval College, Park Row, London SE10 9LS, United Kingdom
e-mail: m.alrifaie@greenwich.ac.uk

Hooman Oroojeni M. J.
Department of Computing
Goldsmiths, University of London, London SE14 6NW, United Kingdom
e-mail: h.oroojeni@gold.ac.uk

Mihalis Nicolaou
Computation-based Science and Technology Research Center
The Cyprus Institute, Nicosia, Cyprus
e-mail: m.nicolaou@cyi.ac.cy

Xin-She Yang
School of Science and Technology
Middlesex University, London NW4 4BT, United Kingdom
e-mail: x.yang@mdx.ac.uk

Krishnanand Kaipa
Department of Mechanical and Aerospace Engineering
Old Dominion University, Norfolk, Virginia, United States
e-mail: kkaipa@odu.edu

Debasish Ghose
Department of Aerospace Engineering
Indian Institute of Science, Bangalore, India
e-mail: dghose@iisc.ac.in

Szymon Lukasik
Faculty of Physics and Applied Computer Science
AGH University of Science and Technology, Krakow, Poland
e-mail: slukasik@agh.edu.pl

Ahmed F. Ali
Department of Computer Science
Suez Canal University, Ismaillia, Egypt
e-mail: ahmed_fouad@ci.suez.edu.eg

Mohamed A. Tawhid
Department of Mathematics and Statistics
Thompson Rivers University, Kamloops, BC, Canada V2C 0C8
e-mail: mtawhid@tru.ca

Ferhat Erdal
Department of Civil Engineering
Akdeniz University, Turkey
e-mail: eferhat@akdeniz.edu.tr

Osman Tunca
Department of Civil Engineering
Karamanoglu Mehmetbey University, Turkey
e-mail: osmantunca@kmu.edu.tr

Erkan Dogan
Department of Civil Engineering
Celal Bayar University, Turkey
e-mail: erkan.dogan@cbu.edu.tr

Ali R. Kashani
Department of Civil Engineering
University of Memphis, Memphis, TN 38152, United States
e-mail: `kashani.alireza@ymail.com`, `akashani@memphis.edu`

Charles V. Camp
Department of Civil Engineering
University of Memphis, Memphis, TN 38152, United States
e-mail: `cvcamp@memphis.edu`

Hamed Tohidi
Department of Civil Engineering
University of Memphis, Memphis, TN 38152, United States
e-mail: `htohidi@memphis.edu`

Essam H. Houssein
Faculty of Computers and Information
Minia University, Minya, Egypt
e-mail: `essam.halim@mu.edu.eg`

Ibrahim E. Mohamed
Faculty of Computers and Information
South Valley University, Luxor, Egypt
e-mail: `ibrahim.elsayed.ibrahim@gmail.com`

Aboul Ella Hassanien
Faculty of Computers and Information
Cairo University, Cairo, Egypt
e-mail: `aboitcairo@gmail.com`

J. Mark Bishop
Department of Computing
Goldsmiths, University of London, SE14 6NW, United Kingdom
e-mail: `m.bishop@gold.ac.uk`

Moein Armanfar
Department of Civil Engineering
Arak University, Arak, Iran
e-mail: `m.armanfar.ce@gmail.com`

1

Ant Colony Optimization, Modifications, and Application

Pushpendra Singh

Department of Electrical Engineering
Govt. Women Engineering College, Ajmer, India

Nand K. Meena

School of Engineering and Applied Science
Aston University, Birmingham, United Kingdom

Jin Yang

James Watt School of Engineering
University of Glasgow, Glasgow, United Kingdom

CONTENTS

1.1   Introduction

Ant colony optimization (ACO) is a robust and very effective optimization technique, proposed by Marco Dorigo in the year 1992 [1]. This is a population based meta-heuristic technique, inspired by the biological characteristic of ants. Like humans, ants too prefer to live in colonies and work together. They exhibit very interesting behaviors; for instance while a single ant has very limited capabilities, the whole ant colony system is highly organized [2]. The ant colony travels through the shortest path between their nest and food source. Ants basically have low visibility but good senses. They communicate with each other by the help of organic substances known as *'pheromone trails'*. This is a chemical released by an ant while traveling on the ground, to provide clues to fellow ants especially when traveling with food. The neighboring ants sense this chemical and follow the path with a high level of pheromone concentration.

Basically, ACO is one of the class of model based search (MBS) techniques [3, 4], with a characteristic to employ a specified probabilistic model and without reforming the model configuration during the run. The chapter briefly describes the fundamental characteristics of artificial ants and their mathematical representations adopted in standard ACO. This method iteratively simulates the behavior of ants with an aim to find the shortest path between food and nest. Furthermore, some improved variants of ACO are discussed which overcome some of the limitations observed in its standard variant.

1.2   Standard ant system

1.2.1   Brief of ant colony optimization

The most popular, successful and standard version of the ACO technique is based on the ant system [1]. It has been observed that the capabilities of a single ant are very limited but an ant system has very complex behavior. Each ant is collectively cooperating in this ant system without knowing their cooperative behavior [2]. By laying the pheromone on the ground, they are helping their fellows to take that path. The purpose of the ACO technique is not to simulate the complete ant colony but instead to use the collective behavior of ants to develop an optimization tool for complex real-life optimization problems. Fig. 1.1 demonstrates the shortest path finding behavior of ants [5].

In Fig. 1.1(a), ants follow the shortest path between nest and food source. In order to demonstrate the path seeking behavior of ants, a block is placed in this path, as shown in Fig. 1.1(b). Abruptly, ants scatter to find the path to their food source or nest. Now they have two paths to follow: paths through

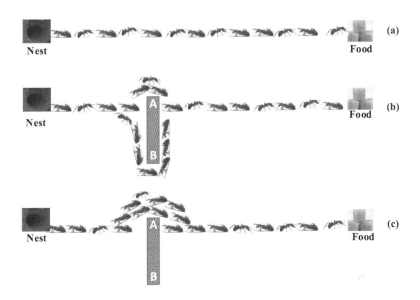

FIGURE 1.1
Ant system: a) ants follow the shortest path between nest and food source, b) a block interrupts the ant path and ants again seek the shortest path, c) ants again follow the shortest path after some time.

edge A and B. The ants which are going through edge A will take less time to return than ants following edge B. Therefore, more pheromone concentration will be deposited on the path through edge A. Due to this increasing pheromone concentration all ants will follow this shortest path, as shown in Fig. 1.1(c). This discussed behavior of the ant system is adopted in the ACO algorithm.

The brief description based on this working mechanism of the ant system in the ACO algorithm is as follows:

1. initially, a frame is designed with edges and vertexes;

2. n_a artificial ants of the ant colony system are randomly initialized;

3. ants are arbitrarily set on vertexes of the frame; and

4. considering the fact that initial concentration of pheromone trails on all the edges is set to a small value, i.e., $\tau_0 \in [0, 1]$.

At each solution building stage, every ant cumulatively contributes to the pheromone trail by adding her share of organic substance on the partially constructed frame of solution. Suppose the kth ant is starting its journey from vertex i during the tth building phase, as shown in Fig. 1.2.

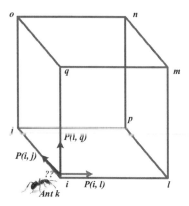

FIGURE 1.2
Decision making on ant system frame.

The ant k pursues a random walk from the vertex i to the next or nearby possible vertexes (j, l, q) for constructing an expedient solution cumulatively. The next vertex to travel is probabilistically chosen according to the transition probabilities of other vertexes such as $P_k(i, j)$, $P_k(i, l)$, and $P_k(i, q)$, with respect to the present vertex i of ant k. This transition probability of ant k to choose the next vertex j from its original vertex i is determined by using the random proportional state transition law [1], mathematically expressed as

$$P_k(i,j) = \begin{cases} \frac{[\tau(i,j)]^\alpha \cdot [\eta(i,j)]^\beta}{\sum_{j \in allowed}[\tau(i,j)]^\alpha \cdot [\eta(i,j)]^\beta}, & \text{if } j \in allowed \\ 0, & \text{othewise} \end{cases} \tag{1.1}$$

where, $\tau(i, j)$ represents the quantity of pheromone trail on the edge that links the vertex i and j. Further, $\eta(i, j)$ is a heuristic value also called the *desirability* or *visibility* of the ant to the building solution on the edge connecting vertexes i and j. It is set as the inverse of connection cost or distance between these vertexes i.e. $\eta(i, j) = 1/d(i, j)$. It is generally suggested for promoting the cost effective vertex of the frame which has a large quantity of pheromone concentration.

$\alpha \in (0, 1]$ and $\beta \in (0, 1]$ are known as regulating parameters that help to control the relative significance of pheromone versus heuristic values.

When all the ants of the ant system have completed their journey, the pheromone concentration is updated on the edges by the pheromone global updating rule, defined as

$$\tau(i,j) \leftarrow (1 - \rho) \cdot \tau(i,j) + \sum_{k=1}^{n_a} \Delta\tau_k(i,j) \tag{1.2}$$

where, $\rho \in (0, 1]$ represents the fractional amount of evaporated pheromone between two steps or iterations of the ACO algorithm. The value of ρ is set to be very small to avoid unlimited pheromone deposition on the path. $\Delta\tau_k(i,j)$ represents the amount of pheromone deposited by ant k on path (i,j), measured in per unit length and expressed as

$$\Delta\tau_k(i,j) = \begin{cases} \frac{Q}{L_k}, & \text{if ant } k \text{ travels through path } (i,j) \\ 0, & \text{otherwise} \end{cases} \tag{1.3}$$

here, Q is a constant and L_k represents the total travel length of ant k. The value of constant Q is generally set to zero.

1.2.2 How does the artificial ant select the edge to travel?

In this section, we demonstrate how an artificial ant selects the shortest path to travel. Figure 1.3 shows the mathematical expressions and explains the phenomena of an ant selecting its traveling edge. In this figure, a frame has been designed with four vertexes labeled *star, rectangle, circle, triangle*. The cost and pheromone graph of the edges are given in sub-figure (A) which can be treated as input data. Suppose an ant is starting its journey from the star vertex. Sub-figure (B) shows that currently the ant is standing at vertex *star*. It is willing to travel in any one direction out of three available vertexes, i.e. *rectangle, circle, triangle*. To make the decision, the transition probability explained in (1.1) should be calculated for each possible traveling

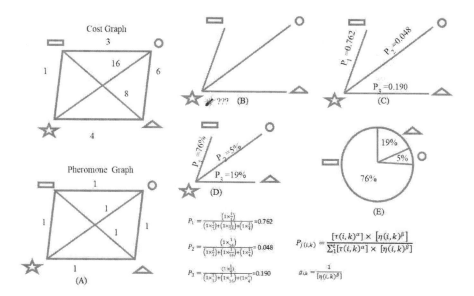

FIGURE 1.3
Mathematical and graphical representations of rules adopted by an artificial ant to select path in ACO algorithm.

path available to this ant. Now, the transition probability of each possible traveling edge is determined and presented in sub-figure (C). Similarly sub-figures (D) and (E) represent the percentage chances of an ant to travel on these edge. It can be observed from this mathematical analysis that the ant on edge *star* has the highest probability to travel towards the *rectangle* vertex. Similarly, the same approach will be used to select the future path.

1.2.3   Pseudo-code of standard ACO algorithm

The pseudo-code of the standard ACO is presented in Algorithm 1.

Algorithm 1 Pseudo-code of standard ACO.

1: determine the objective function $OF(.)$
2: set the ant population n_a, maximum number of iterations $iter_{max}$, constant and algorithm parameters such Q, τ_0, α, β, ρ, etc.
3: determine the distances $d(i,j)$ between vertexes from where ants start their journey and visibility $\eta(i,j) = 1/d(i,j) \; \forall \; i,j$
4: set iteration $t = 0$
5: **while** $t < iter_{max}$ **do**
6:      $t = t + 1$
7:      **for** each k-th ant **do**
8:          select a random vertex or point i to start the journey of kth ant.
9:          determine transition probability $P_k(i,j)$ of ant k to move from vertex i to j, by using (1.1)
10:         apply roulette wheel selection criteria on $P_k(i,j) \; \forall \; i,j$ to select the next vertex or point of travel for ant k ▷ make sure ant does not revisit any node or vertex
11:         determine the fitness value $OF(.)$ for ant k
12:         preserve the best ant, *best_ant* and its fitness, *best_fit*
13:      **end for**
14:      update the pheromone concentration $\tau(i,j) \; \forall \; i,j$ visited by all ants, by using (1.2)
15:      return the best ant, *best_ant* and best fitness, *best_fit*
16: **end while**

1.3   Modified variants of ant colony optimization

In this section, popularly known variants of ACO algorithms are discussed.

1.3.1   Elitist ant systems

The first improved variants of the ACO algorithm were proposed in [6]. One of these adopted an elitist strategy known as elitist ant system (EAS). It provides reinforcement to the solution components belonging to the best ant found since the start of the algorithm. In EAS, the global best solution pheromone, in each iteration, is deposited on its trail for all the ants of the colony, irrespective of their visit to this trail. As explained in [8], suppose the global best solution is represented by s^{gb}; then a substance quality $n_e/OF(s^{gb})$ will be added on this trail, where n_e and $OF(s^{gb})$ denote the number of elitist ants and solution cost of s^{gb}. In [6], it has been investigated that an adequate number of elitist ants should be chosen in the elitist ant system to determine the best path in fewer numbers of iterations. Alternatively, the ACO technique can converge to the suboptimal solution with premature convergence if an excessive number of elitist ants are present in the ant system.

1.3.2 Ant colony system

Another improved variant of ACO is known as ant colony system (ACS), introduced in [7]. The ACS has the following additional features which make it different from the ant system.

1. It has a state transition rule which straightway balances the exploration of new edges and exploitation of the superior solution and accumulated information of the problem.

2. In this technique, the global updating rule is applied only to the path traveled by the best ant of the colony.

3. while ants construct the solution, a local pheromone updating rule is also employed.

In ACS, n_a ants are initially placed on randomly selected vertexes of a constructional graph or cities. Repeatedly, each ant starts its tour by using the state transition rule, i.e., stochastic greedy rule. A local pheromone updating rule is also employed for each ant, on visited edges, to modify the concentration of its deposited pheromone. Like in the ant system, the ant will choose the next path by using the visibility heuristic and pheromone concentration. The edge with the highest pheromone intensity will always be preferred by the ants in ACS as well.

Suppose the kth ant is currently at vertex or city i in the tth step of construction; then the probability of this ant to travel towards the next edge or city j is determined by using the state transition rule known as the random-proportional rule [7], expressed as

$$j = \begin{cases} \arg\max_{l \in Z_k(i)}\{[\tau(i,l)] \cdot [\eta(i,l)]^{\beta}\}, & \text{if } q \leq q_0 \text{ (exploitation)} \\ B, & \text{othewise} \quad \text{(biased exploitation)} \end{cases} \quad (1.4)$$

where, $Z_k(i)$ represents the set of the node/vertexes/cities that remain to visit by ant k currently placed at vertex i. Further, q and q_0 are the uniformly distributed random number and a parameter $0 \leq q_0 \leq 1$ respectively. B is a random variable selected according to the probability expressed in (1.1). The state transition rule consequent from (1.1) and (1.4) is known as the *pseudo-random-proportional rule*. It helps to achieve the most cost effective vertexes with high concentration of pheromone trail. The adequate value of q_0 provides the balance between exploitation and exploration to the ACO algorithm.

In ACS, the best ant will only be allowed to deposit the pheromone on shorted visited edges to make the search faster. This rule is applied when all ants complete their journey. In this rule, the pheromone intensity on the edges is updated as

$$\tau(i,j) \leftarrow (1 - \rho) \cdot \tau(i,j) + \rho \cdot \Delta\tau(i,j) \tag{1.5}$$

where,

$$\Delta\tau(i,j) = \begin{cases} 1/J_{gb}, & \text{if } (i,j) \in S_{best} \\ 0, & \text{othewise} \end{cases} \tag{1.6}$$

where, S_{best} and J_{gb} represent the global best tour and its length respectively, from the start of the trail.

While building the solution for every step, ants change the pheromone concentration by applying the local updating rule, defined as

$$\tau(i,j) \leftarrow (1 - \zeta) \times \tau(i,j) + \zeta \cdot \Delta\tau(i,j) \tag{1.7}$$

where, ζ is a parameter varying between 0 to 1.

1.3.3   Max-min ant system

Fundamentally, the ACO method inspired by the food seeking behavior of an ant colony has shown poor performance for large and complex benchmark problems. In order to overcome some of the limitations of the standard ant system, Stutzle T. and Hoos HH. [8] have proposed a new model of ACO technique known as max-min ant system (MMAS). It is inspired by the conventional ant system model that claims to be an effective method to solve the quadratic assignment problem (QAP), the traveling salesman problem [7]. The MMAS has some unique advantages over AS, given below.

- The MMAS deploys a strategy in which only the best solution found during the iteration will construct the pheromone trail. It means only a single ant which is having the best solution can add the organic substance on the path; it can be the best ant of the current iteration or the best ant found so far from the beginning of the trial.

- Initially, the highest possible intensity of pheromone, τ_{max} is considered on all the edges/paths, for better exploration of the solution.

- the range of possible pheromone trails for each solution element is limited between τ_{min} and τ_{max} to avoid search stagnation of the algorithm.

As discussed in the conventional ACO, suppose n_a number of ants are randomly placed at vertexes of a construction framework. Initially, the value of the pheromone trail on all the connecting edges is set at upper limit τ_{max}. Similar to the traditional model of ACO, every ant of the colony travels randomly to build its solution. From its current vertex i, each ant k will select the next suitable vertex j by using the transition probability, $P_k(i,j)$, defined in (1.1).

In MMAS, when all the ants of the colony have obtained their respective solution or the program has completed one iteration, the best solution of pheromone trail is upgraded by applying the rule of global update, expressed below

$$[\tau(i,j) \leftarrow (1-\rho) \cdot \tau(i,j) + \Delta\tau^{best}(i,j)]_{\tau^{min}}^{\tau^{max}} \tag{1.8}$$

The mathematical operator $[z]_b^a$ is defined as

$$[z]_b^a = \begin{cases} a, & \text{if } z > a \\ b, & \text{if } z < b \\ z, & \text{otherwise} \end{cases} \tag{1.9}$$

Moreover, the $\Delta\tau^{best}(i,j)$ is determined as

$$\Delta\tau^{best}(i,j) = \begin{cases} 1/g_{best}, & \text{if } (i,j) \text{ belongs to the best tour} \\ 0, & \text{otherwise} \end{cases} \tag{1.10}$$

here, g_{best} is representing the tour length of the best ant. This may be the best tour solution obtained in the current iteration or the best solution sought from the beginning of the trial. Furthermore, the value of upper and lower boundaries, i.e. τ^{max} and τ^{min}, are problem specific [8].

1.3.4 Rank based ant systems

Bullnheimer *et al.* [9], proposed a rank based ACO algorithm. In this system, all the obtained solutions are ranked based on their length and only a fixed number of ants which are having the best solutions in that iteration are allowed to update the *pheromone trial*. Furthermore, the quantity of deposited pheromone is weighted for every obtained solution. This system helps to achieve a higher concentration level of pheromone on the shortest path as compared to other corresponding paths.

1.3.5 Continuous orthogonal ant systems

A continuous orthogonal ant system was also developed by Hu *et al.* [10]. In this model, a new mechanism of pheromone deposition was suggested that enables ants to seek a possible solution by an effective and collaborative manner. By the help of an orthogonal framework, the ants in a feasible region can effectively explore their selected area rapidly while improving the global solution searching capability of the algorithm. The elitist strategy is also adopted to preserve the most valuable solutions.

1.4 Application of ACO to solve real-life engineering optimization problem

1.4.1 Problem description

Reactive power compensation has always been an important concern for power system operators to maintain the desired node voltage profile and stability in small to large-scale power systems. Shunt capacitors are traditionally deployed in power distribution networks. The optimal integration of shunt capacitors (SCs) can provide reduced real and reactive power losses while improving the node voltage profile in distribution systems. The optimal SC allocation is a complex combinatorial optimization problem aiming to minimize power delivery loss and node voltage deviation subjected to various system and distribution network operator (DNO) constraints.

1.4.2 Problem formulation

In this section, we formulate a optimal SC allocation problem for a benchmark 33-bus test distribution system. The bus and line data of this system is obtained from [11]. It is a 12.66 kV network with total real and reactive power loads of 3715 kW and 2300 kVar respectively. In this problem, the objective function for power loss minimization is expressed as

$$F = \sum_{i=1}^{N} \sum_{j=1}^{N} \alpha_{ij} \left(P_i P_j + Q_i Q_j \right) + \beta_{ij} \left(Q_i P_j - P_i Q_j \right) \tag{1.11}$$

where,
$$\alpha_{ij} = \frac{R_{ij} \cos(\delta_i - \delta_j)}{V_i V_j} \text{ and } \beta_{ij} = \frac{R_{ij} \sin(\delta_i - \delta_j)}{V_i V_j} \tag{1.12}$$

subjected to:
$$P_i = V_i \sum_{j=1}^{N} V_j Y_{ij} \cos(\theta_{ij} + \delta_j - \delta_i) \quad \forall \ i \tag{1.13}$$

$$Q_i = -V_i \sum_{j=1}^{N} V_j Y_{ij} \sin(\theta_{ij} + \delta_j - \delta_i) \quad \forall \ i \tag{1.14}$$

$$V^{\min} \leq V_i \leq V^{\max} \qquad \forall \ i \tag{1.15}$$

$$I_{ij} \leq I_{ij}^{\max} \qquad \forall \ i, j \tag{1.16}$$

$$n_i^{sc} Q_{bank}^{SC} \leq Q^{\max} \qquad \forall \ i \tag{1.17}$$

$$\sum_{i=1}^{N} \sigma_i n_i^{sc} Q_{bank}^{SC} \leq \chi \sum_{i=1}^{N} Q_{D_i} \tag{1.18}$$

Equations (1.13) to (1.18) express the constraints known as nodal real and reactive power balances, node voltage, feeder current, allowed SC capacity at a single node and total system constraints respectively. Here, P_i, Q_i, Q_{D_i}, V_i, δ_i, n_i^{sc}, and σ_i denote the real power injection, reactive power injection and demand, voltage magnitude and angle, number of SC banks and binary decision variable of SC deployment respectively, all at node i. The R_{ij}, Y_{ij}, θ_{ij}, I_{ij} and $I_{ij}^{\max}$ are respectively representing the resistance, impedance matrix element, impedance angle, current and maximum allowed current limit of branch connecting nodes i and j. Furthermore, the constants N, χ, $V^{\max}$, $V^{\min}$, Q_{bank}^{SC}, $Q^{\max}$ represent the total number of nodes in the system, nominal to peak demand conversion factor (usually, $\chi = 1.6$), maximum and minimum allowed node voltage limits in per units, VAr capacity of a capacitor bank and maximum allowed reactive power compensation at any single node respectively.

1.4.3   How can ACO help to solve this optimization problem?

Before solving this optimization problem, we apply some engineering knowledge to make it simple. In a 33-bus system, the peak reactive power load of this system would be equal to $\chi \sum_{i=1}^{N} Q_{D_i} = 1.6 \times 2300 = 3680$ kVAr. In this problem, a $Q_{bank}^{SC} = 300$ kVAr capacitor bank is considered. Generally, three nodes are found to be optimal for this system; therefore each node can accommodate a maximum of $\left(\chi \sum_{i=1}^{N} Q_{D_i}\right) / (3.Q_{bank}^{SC}) = 3680/(300 \times 3) \approx 4$ banks.

Now, we use the graph representation of search space for ants. One of the suitable graphical structures for the optimal SC allocation problem is shown in Fig. 1.4. This graph is designed for N number of SCs and corresponding number of capacitor banks to be deployed at these nodes. In C_1, '1' represents

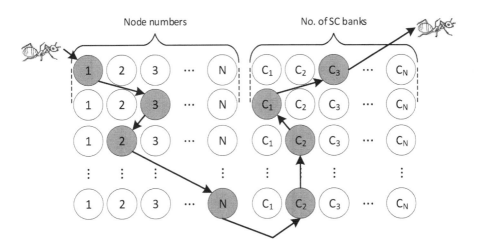

FIGURE 1.4
Proposed graph for SC allocation problem.

the one capacitor bank, i.e. 300 kVAr. In this problem, we deploy three capacitors on three different sites. Therefore, no ant should revisit any of the nodes in site selection; however the same number of capacitor banks can be obtained for two different nodes while selecting sites. A max-min ant system is adopted to solve this problem and simulation results are presented in the following section.

1.4.4   Simulation results

The optimal sites and sizes of the SCs obtained by the ACO algorithm based on max-min ant system are presented in Table 1.1. The convergence characteristic of the ACO method is also shown in Fig. 1.5.

TABLE 1.1
Simulation results of optimal SC deployment in 33-bus distribution system.

Cases	Optimal sites of SCs	Optimal capacity of SCs (kVAr)	Real power loss (kW)
Uncompensated network	–		202.6650
Optimally compensated network	15, 23, 30	300, 300, 1200	134.7107

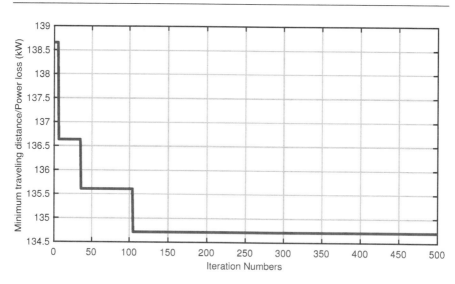

FIGURE 1.5
Convergence characteristic of ACO for best solution search.

1.5   Conclusion

In this chapter, the basic ACO technique and its popularly known modified variants are discussed. The standard ACO algorithm is also explained for better understanding of this meta-heuristic approach. Furthermore, some of the limitations of the standard ACO method and advantages of modified variants are discussed. The improved versions overcome some of the limitations observed in the standard ACO variant and help to build better solution frameworks. At the end, the application of the ACO technique is explained by solving a real-life optimal SC allocation problem of a 33-bus distribution system for real power loss minimization. A MMAS based ACO method has been used to solve this optimization problem.

Acknowledgment

This project has received funding from the European Union's Horizon 2020 research and innovation programme under the Marie Sklodowska-Curie grant agreement No 713694. We would like to thank Dr. Srikanth Sugavanam for valuable discussions and comments.

References

1.   Dorigo M, "Optimization, Learning and Natural Algorithms", *Ph.D. dissertation, Dipartimento di Elettronica*, Politecnico di Milano, Italy, 1992.

2.   Colorni A, Dorigo M, Maniezzo V, "Distributed optimization by ant colonies,". In Proceedings of the First European Conference on Artificial Life, 1992, Vol. 142, pp. 134-142.

3.   Dorigo M, Di Caro G, "Ant colony optimization: a new meta-heuristic," *Proceedings of IEEE Congress on Evolutionary Computation", (CEC-99)*, vol. 2. 6-9 July 1999, pp. 1470-1477

4.   Lee K.Y, El-Sharkawi, M.A. eds., Modern Heuristic Optimization Techniques: Theory and Applications to Power Systems, John Wiley & Sons , Vol 39, 2008.

5.   Deneubourg JL, Goss S, "Collective patterns and decision-making". *Ethology Ecology & Evolution*, 1989, vol. 1, no. 4, pp. 295-311.

6. Dorigo M, Maniezzo V, Colorni, "Ant system: optimization by a colony of cooperating agents," *IEEE Trans Syst Man Cybern B Cybern*, 1996, doi:10.1109/3477.484436

7. Dorigo M, Gambardella LM, "Ant colony system: A cooperative learning approach to the traveling salesman problem," *IEEE Trans Evol Computat.*, 1997 vol. 1, no. 1, pp. 53-66.

8. Stutzle T, Hoos HH., "MAX-MIN ant system," *Future Generation Computer Systems*, 2000, vol. 16, no. 8, pp. 889-914.

9. Bullnheimer B, Hartl RF, Strauss C, "A new rank based version of the Ant System – A computational study," *Central European Journal for Operations Research and Economics*, vol. 7, pp. 25-38, 1997.

10. Hu XM, Zhang J, Li Y, "Orthogonal methods based ant colony search for solving continuous optimization problems," *Journal of Computer Science and Technology*, 2008, vol. 23, no. 1, pp. 2-18.

11. Baran ME, Wu FF. "Network reconfiguration in distribution systems for loss reduction and load balancing," *IEEE Transactions on Power Delivery*, 1989, 4(2), pp. 1401-7.

2

Artificial Bee Colony – Modifications and An Application to Software Requirements Selection

Bahriye Akay

Department of Computer Engineering
Erciyes University, Melikgazi, Kayseri, Turkey

CONTENTS

2.1 Introduction

A honey bee colony exhibits collective intelligence in some of their daily activities. One of them is the foraging task which is crucial for survival of the colony. In the foraging task, bees are distributed into three roles: employed foragers, onlooker bees and scouts. The employed foragers are responsible for exploiting the nectar of discovered food sources. When she arrives at the hive, after unloading the nectar, she gives information about her food source by dancing. The onlooker bees watch the dances of some bees in the dance area and select food source toward which to fly. After having been exploited by the bees, an exhausted food source is abandoned and its bee starts to search for a new source as a scout bee. In 2005, Karaboga was inspired by honey bee foraging and proposed the Artificial Bee Colony (ABC) algorithm [1]. In ABC, a food source location corresponds to a solution which is a vector of design parameters to be optimized and its nectar corresponds to the solution's fitness to be maximized. The ABC algorithm has three phases: the employed bee phase which exploits the information of discovered solutions, the onlooker bee phase which has a stochastic selection scheme to select high quality solutions and performs local search around them and the scout bee phase to detect exhausted solutions. The ABC algorithm combines division of labour, positive and negative feedback, multiple interaction and fluctuation properties leading swarm intelligence.

The original ABC algorithm has been proposed for solving continuous and unconstrained optimization problems. Each design parameter, $x_i \in \mathbb{R}$ and for a n-dimensional problem, the search space, $\mathbb{S}$, is defined as a n-dimensional rectangle in $\mathbb{R}^n$ ($\mathbb{S} \subseteq \mathbb{R}^n$). Therefore, a real-valued representation is adopted and local search operators are able to produce solutions in $\mathbb{R}^n$. Since there is no additional constraint and a single objective function is considered in the problem, a greedy selection is employed to favour the solutions with less objective value. The basic ABC algorithm gained success on especially high dimensional and multi-modal unconstrained problems [2].

When the algorithms proposed for unconstrained optimization are intended to solve combinatorial, constrained or multi-objective problems, they are required to be modified to adopt the search space discretion and/or to handle with feasible solutions and multiple objectives. This chapter aims to give information about problem-related modifications in the ABC algorithm related to representation, local search strategy and selection strategy to solve different kinds of problems.

The rest of the chapter is organized as follows: in Section 2.2 a brief description of the ABC algorithm is presented. In Section 2.3, the modified versions of the ABC algorithm are provided and explained in detail. In Section 2.4, an application of the ABC algorithm to a software requirements selection problem is demonstrated. Finally, Section 2.5 is devoted to conclusions of the chapter.

2.2   The Original ABC algorithm in brief

The ABC algorithm has three phases: employed bees, onlooker bees and scout bees as in the foraging task performed by a real honey bee colony. Operations performed in each phase are presented as pseudo-code in Alg. 2.

Algorithm 2 Main steps of the ABC algorithm.

1: Set values for the control parameters:
2: SN: The number of food sources,
3: MCN: The maximum number of cycles,
4: $limit$: The maximum number of exploitations for a solution, ▷ Set problem-specific
   parameters:
5: f: Objective function to be minimized,
6: D: Dimension of the problem,
7: x_j^{lb} : Lower bound of jth parameter,
8: x_j^{ub}: Upper bound of $j$the parameter
9: **for** $i = 1$ to SN **do** ▷ //Initialization
10: $\quad$ $\vec{x}_i = x_j^{lb} + rand(0, 1)(x_j^{ub} - x_j^{lb})$, $j = 1 \dots D$
11: $\quad$ $f_i = f(\vec{x}_i)$
12: $\quad$ $trial_i = 0$
13: **end for**
14: $cyc = 1$
15: **while** $cyc < MCN$ **do** ▷ //Employed Bees' Phase
16: $\quad$ **for** $i = 1$ to SN **do**
17: $\quad\quad$ $\vec{\hat{x}} = \vec{x}_i$
18: $\quad\quad$ $k \longleftarrow randint[1, CS], k \neq i$
19: $\quad\quad$ $j \longleftarrow randint[1, D]$
20: $\quad\quad$ $\phi_{ij} \longleftarrow rand[-1, 1]$
21: $\quad\quad$ $\hat{x}_{ij} = x_{ij} + \phi_{ij}(x_{ij} - x_{kj})$
22: $\quad\quad$ **if** $f(\vec{\hat{x}}) < f_i$ **then**
23: $\quad\quad\quad$ $\vec{x}_i = \vec{\hat{x}}$
24: $\quad\quad\quad$ $f_i = f(\vec{\hat{x}})$
25: $\quad\quad\quad$ $trial_i = 0$
26: $\quad\quad$ **else**
27: $\quad\quad\quad$ $trial_i = trial_i + 1$
28: $\quad\quad$ **end if**
29: $\quad$ **end for**
30: $\quad$ **for** $i = 1$ to SN **do** ▷ //Calculate probabilities
31: $\quad\quad$ $p_i = 0.1 + 0.9 * \frac{fitness_i}{max(\vec{fitness})}$
32: $\quad$ **end for**
33: $\quad$ $i = 0$ ▷ //Onlooker Bees' Phase
34: $\quad$ $t = 0$
35: $\quad$ **while** $t < SN$ **do**
36: $\quad\quad$ **if** $rand(0, 1) < p(i)$ **then**
37: $\quad\quad\quad$ $t = t + 1$
38: $\quad\quad\quad$ $\vec{\hat{x}} = \vec{x}_i$
39: $\quad\quad\quad$ $k \longleftarrow randint[1, CS], k \neq i$
40: $\quad\quad\quad$ $j \longleftarrow randint[1, D]$
41: $\quad\quad\quad$ $\phi_{ij} \longleftarrow rand[-1, 1]$
42: $\quad\quad\quad$ $\hat{x}_{ij} = x_{ij} + \phi_{ij}(x_{ij} - x_{kj})$
43: $\quad\quad\quad$ **if** $f(\vec{\hat{x}}) < f_i$ **then**
44: $\quad\quad\quad\quad$ $\vec{x}_i = \vec{\hat{x}}$
45: $\quad\quad\quad\quad$ $f_i = f(\vec{\hat{x}})$
46: $\quad\quad\quad\quad$ $trial_i = 0$
47: $\quad\quad\quad$ **else**
48: $\quad\quad\quad\quad$ $trial_i = trial_i + 1$
49: $\quad\quad\quad$ **end if**
50: $\quad\quad$ **end if**
51: $\quad\quad$ $i = (i + 1) \ mod \ (SN - 1)$
52: $\quad$ **end while**
53: $\quad$ Memorize the best solution
54: $\quad$ $si = \{i : trial_i = max(trial)\}$ ▷ //Scout bee phase
55: $\quad$ **if** $trial_{si} > limit$ **then**
56: $\quad\quad$ $\vec{x}_{si} = x_j^{lb} + rand(0, 1)(x_j^{ub} - x_j^{lb})$, $j = 1 \dots D$
57: $\quad\quad$ $f_{si} = f(\vec{x}(si))$
58: $\quad\quad$ $trial_{si} = 0$
59: $\quad$ **end if**
60: $\quad$ $cyc + +$
61: **end while**

Before starting the algorithm, values are set to the control parameters of ABC and problem-specific parameters. The control parameters of ABC are the number of food sources (SN), the maximum number of cycles (MCN) and the parameter *limit* to determine the exhausted sources. In steps 9-13, each solution vector (food source location, $\vec{x}_i$) is initialized within the range $[x_j^{lb}, x_j^{ub}]$ and evaluated in the cost function (f). *trial* counters holding the number of exploitations are set to 0. Steps 16-29 correspond to the employed bee phase. A local search is carried out in the vicinity of each solution. In step 21, a new solution ($\vec{\tilde{x}}$) is produced by using the information of $\vec{x}_i$ and $\vec{x}_k$. This local search is an analogy of food source exploitation. Steps 22-28 perform greedy selection. If the new solution, $\vec{\tilde{x}}$ is better than the current solution, $\vec{x}$, the new solution is kept in the population. Otherwise, the current solution is retained and the counter (*trial*) holding the number of exploitations is incremented by 1 (Step 27). Once the employed bee phase is completed, each solution is assigned a probability value (Steps 30-32) proportional to its fitness which can be calculated by $1/f_i$. Lines 33-52 are the steps of the onlooker bee phase in which the solutions to be exploited are selected based on a stochastic scheme. If the solution's probability is higher than a random number drawn within range [0,1] (Step 36), the solution is exploited which means a local search is conducted around the selected solution (Step 42) and a greedy selection is applied between the new solution and the selected one. In the employed bee phase, each solution is involved in the local search by turn while in the onlooker bee phase, only selected solutions are involved. High quality solutions are likely to be selected (positive feedback) but low quality solutions also have a chance proportional to their fitness. Before the scout bee phase, the best solution is memorized. In the scout bee phase, the counters, $\vec{trial}$, are checked to determine whether an exhausted source exits (Lines 54-59). In each cycle of the basic algorithm, only one scout bee is allowed to occur to avoid losing the population's previous knowledge. If there is an exhausted source, this source is abandoned and a new random source is produced instead of it (Line 56).

2.3   Modifications of the ABC algorithm

In this section, some different versions of ABC are presented to give insight into how an algorithm can be modified to handle different problem characteristics. The combinatorial, constrained and multi-objective versions of the ABC algorithm are explained in the following subsections.

2.3.1   ABC with modified local search

In the local search operator of the basic ABC algorithm, information only in one dimension is perturbed to form a new solution vector and the rest are copied from the current solution. Although this operator is good at fine tuning, it may slow down the convergence on some problems. Therefore, some new local operators are integrated to the algorithm to gain a speed up in the convergence.

One modification in the local search operator is increasing the number of dimensions to be changed by Eq. 2.1 [3].

$$v_{ij} = \begin{cases} x_{ij} + \phi_{ij}(x_{ij} - x_{kj}) & , \quad if \quad R_{ij} < MR \\ x_{ij} & , \quad \text{otherwise} \end{cases} \qquad (2.1)$$

where R_{ij} is a random number drawn from uniform distribution within the range (0,1) and MR is the modification rate controls the number of dimensions to be perturbed. For each dimension j, if $R_{ij} < MR$, then the information in the jth dimension is changed, otherwise the previous information is kept. In [3], additionally, ϕ_{ij} is drawn at random within the range $[-SF, SF]$ where SF is a variable scaling factor based on Rechenberg's 1/5 rule.

Another modification in the ABC algorithm is superimposing the best solution in a neighbourhood topology defined in a certain radius (Eq. 2.2). This operation is proposed to be used in the onlooker bee phase in [4].

$$v_{ij} = x_{bestN,j} + \phi_{ij}(x_{bestN,j} - x_{kj}) \qquad (2.2)$$

where $x_{bestN,j}$ is the best solution in a neighbourhood. In [4], for each solution, mean Euclidian distance md_i is calculated and the solutions x_j are assumed to be neighbours if $d_{ij} \leq r \times md_i$. It means that two solutions are neighbours if the distance between them is less then the mean distance weighted by r which is a predefined control parameter.

2.3.2 Combinatorial version of ABC

Combinatorial optimization finds the minimum cost set, I, among all subsets of the power set of a finite set E. Hence, the algorithms should be able to search finite set E space and some efficient operators are needed to find a good approximate solution in polynomial time. When a combinatorial problem is intended to be solved by the ABC algorithm, first, a discrete representation should be adopted. In the initialization phase, random orders of the nodes are generated unlike basic ABC generating random points in continuous space. In the employed bee and onlooker bee phases, a local search operator suitable for the problem such as 2-opt, greedy sub tour mutation (GSTM), etc., can be employed to produce mutant solutions based on the order representation [5-6]. In [5], ABC is integrated with 2-opt which aims to find a different edge by re-linking the ith and jth nodes of the current solution and a randomly chosen neighbour. In [6], a GSTM operator is used to generate new solutions. The main steps of GSTM operator used in [6] are given below.

1. Select solution x_k randomly in the population, $k \neq i$.
2. Select a city x_{ij} randomly.
3. Select the value of searching way parameter $\phi \in \{-1, 1\}$ randomly.
4. if (ϕ=1) then

    - The city visited before x_{kj} is set as the previous city of x_{ij}.

5. else

    - The city visited after x_{kj} is set as the next city of x_{ij}.

6. endif
7. A new closed tour $\hat{T}$ is generated.
8. By this new connection, an open sub tour T^* is generated. R_1: first city of T^* and R_2: last city of T^*
9. if (random < P_{RC}) then

    - Subtract $[R_1$-$R_2]$ tour from tour T ($T^{\#} \leftarrow T - [R_1 - R_2]; T^* \leftarrow [R_1 - R_2]$)
    - Hold on T^* sub-tour $T^{\#}$ so that there will be minimum extension.

10. else

    - if (random < P_{CP}) then
        - Copy $[R_1 - R_2]$ sub-tour in the T tour, $T^* \leftarrow [R_1 - R_2]$
        - Add each city of T^* to the T starting from the position R_1 by rolling or mixing with probability P_L.
    - else
        - Select randomly one neighbor from neighbor lists for the points R_1 and R_2 (NL_{R_1} and NL_{R_2}).
        - Invert the points NL_{R_1} or NL_{R_2} that provide maximum gain in such a way that these points will be neighbors to the points R_1 or R_2.
        - Repeat if the inversion does not take place.
    - endif

11. endif

Here, P_{CP} is correction and perturbation probability, P_{RC} is reconnection probability and P_L is linearity probability.

2.3.3 Constraint handling ABC

Constrained optimization aims to find a parameter vector $\vec{x}$ that minimizes an objective function $f(\vec{x})$ subject to inequality and/or equality constraints (2.3):

$$
\begin{aligned}
&\text{minimize } f(\vec{x}), \quad \vec{x} = (x_1, \ldots, x_n) \in \mathbb{R}^n \\
&\qquad\qquad\qquad l_i \leq x_i \leq u_i, \qquad\qquad i = 1, \ldots, n \\
&\text{subject to :} \quad g_j(\vec{x}) \leq 0, \qquad\qquad \text{for } j = 1, \ldots, q \\
&\qquad\qquad\qquad h_j(\vec{x}) = 0, \qquad\qquad \text{for } j = q+1, \ldots, m
\end{aligned}
\tag{2.3}
$$

The objective function f is defined on a search space, $\mathbb{S}$, which is defined as a n-dimensional rectangle in $\mathbb{R}^n$ ($\mathbb{S} \subseteq \mathbb{R}^n$). All variables are bounded by their lower and upper values. Feasible region $\mathbb{F} \subseteq \mathbb{S}$ is constructed by a set of m additional constraints ($m > 0$) and the global optimum is located in feasible space ($\vec{x} \in \mathbb{F} \in \mathbb{S}$). Because infeasible solutions violating the constraints may have lower cost values, the local search and/or selection operators are modified to favour the search through the feasible space. The local search operator can pass only feasible solutions into the population or infeasible solutions can be repaired to form feasible solutions. In this case, the population includes only feasible solutions. Repairing solutions or trying to generate only feasible solutions can be a time consuming task. Alternatively, the population can hold both feasible and infeasible solutions. However, when a selection is performed, a feasible solution is preferred against an infeasible one. Deb defined the precedency rules to be used in selection phase [7] and in the constraint handling ABC [8], the greedy selection mechanism is replaced with Deb's rules to make a distinction between feasible and infeasible solutions. Deb's rules are given below:

- Any feasible solution is preferred to any infeasible solution ($violation_j > 0$) (solution i is dominant),

- Among two feasible solutions, the one having better objective function value is preferred ($f_i < f_j$, solution i is dominant),

- Among two infeasible solutions ($violation_i > 0$, $violation_j > 0$), the one having smaller constraint violation is preferred ($violation_i < violation_j$, solution i is dominant).

In the constraint handling ABC, the algorithm does not need the initial population holding only feasible solutions and the population can include both feasible and infeasible solutions. Since infeasible solutions are accepted to provide diversity, ABC does not employ a reconstruction mechanism. In the onlooker bee phase, the probability assignment scheme based on only cost function should also be replaced to assign probability to infeasible solutions based on their violation values (Eq. 2.4). In this scheme, the probability values of feasible solutions are higher than those of infeasible solutions and proportional to the cost values while the probability values of infeasible solutions are proportional to their violation values.

$$
p_i =
\begin{cases}
0.5 + \left(\dfrac{\text{fitness}_i}{\sum\limits_{j=1}^{\text{sn}} \text{fitness}_j} \right) * 0.5 & if\ solution\ is\ feasible \\[3em]
\left(1 - \dfrac{violation_i}{\sum\limits_{j=1}^{sn} violation_j} \right) * 0.5 & if\ solution\ is\ infeasible
\end{cases}
\tag{2.4}
$$

where $violation_i$ is the summation of constraint values going beyond the feasible region. A feasible solution's probability is within the range [0.5,1] while it is in the range [0, 0.5] for an infeasible solution.

2.3.4 Multi-objective ABC

An optimization problem with more than one criterion to be optimized simultaneously is called a multi-objective optimization problem (MOP) and the solutions that do not have dominance on all objectives (especially on conflicting objectives) are called Pareto-optimal solutions (PS) or non-dominated solutions. A MOP can be defined by Eq. 2.5:

$$
\begin{aligned}
&\min/\max\ F(x) = (f_1(x), \ldots, f_m(x))^T \\
&\text{subject to } x \in \Omega
\end{aligned}
\tag{2.5}
$$

where Ω is decision variable space, $F : \Omega \to R^m$ is the objective vector and R^m is the objective space.

In single optimization, the ABC algorithm employs a greedy selection mechanism which compares the values of two solutions evaluated in the single objective function. A higher fitness value is assigned to a solution with better objective function. A multi-objective ABC algorithm should be able to make a comparison between solutions evaluated in multi-objective cost functions and should rank the solutions by assigning a fitness value considering all objective values.

One of the multi-objective ABC algorithms is asynchronous MOABC which replaces greedy selection with a Pareto-dominance-based selection scheme (A-MOABC/PD) [9]. In Pareto-dominance, a solution vector x is partially less than another solution y ($x \prec y$), when none of the objectives of y is less than those of x, and at least one objective of y is strictly greater than that of x. If x is partially less than y, the solution x dominates y or y is inferior to x. Any solution which is not dominated by an other solution is said to be non-dominated or non-inferior [10]. A-MOABC/PD assigns cost values by Eq. 2.6 based on Pareto rank, distance and Gibbs distribution probability of a solution.

$$
f_i = R(i) - TS(i) - d(i)
\tag{2.6}
$$

where $R(i)$ is the Pareto rank value of the individual i, $S(i) = -p_T(i)logp_T(i)$, where $T > 0$ is temperature. $p_T(i) = (1/Z)exp(-R(i)/T)$ is the Gibbs distribution, $Z = \sum_{i=1}^{N} exp(-R(i)/T)$ is called the partition function, and N is the population size [11]. $d(i)$ is the crowding distance calculated by a density estimation technique [12].

Algorithm 3 S-MOABC/NS Algorithm.

1: Set the control parameters of the ABC algorithm.
2: CS: The Number of Food Sources,
3: MCN: The Maximum Cycle Number,
4: $limit$: The Maximum number of trial for abandoning a source,
5: M: The Number of Objectives   ▷ //Initialization
6: **for** $s = 1$ to CS **do**
7:     $X(s) = x_j^{lb} + rand(0,1)(x_j^{ub} - x_j^{lb})$, $j = 1 \ldots D$;
8:     $f_{si} = f_i(X(s))$,$i = 1 \ldots M$;
9:     $trial(s) \longleftarrow \emptyset$
10: **end for**
11: $cycle = 1$
12: **while** $cycle < MCN$ **do**
                                                                    ▷ //Employed Bees' Phase
13:     **for** $s = 1$ *to* CS **do**
14:         $x' \longleftarrow$ a new solution produced by Eq. 2.1
15:         $f_i(x') \longleftarrow$ evaluate objectives of new solution, $i = 1 \ldots M$
16:         $X(CS + s) = x'$
17:     **end for**
18:     $\mathcal{F} \longleftarrow$ Non-dominated sorting(X)
19:     Select best CS solutions based on rank and crowding distance to form new population
20:     Assign fitness values to solutions depending on the cost values defined by Eq. 2.6
21:     Calculate probabilities for onlookers by (2.4)
                                                                    ▷ //Onlooker Bees' Phase
22:     $s = 0$
23:     $t = 0$
24:     **while** $t < CS$ **do**
25:         $r \longleftarrow rand(0,1)$ ▷ //Stochastic sampling
26:         **if** $r < p(s)$ **then**
27:             $t = t + 1$
28:             $x' \longleftarrow$ a new solution produced by Eq. 2.1
29:             $f_i(x') \longleftarrow$ evaluate objectives of new solution, $i = 1 \ldots M$
30:             $X(CS + s) = x'$
31:         **end if**$s = (s + 1) \mod (CS - 1)$
32:     **end while**
33:     $\mathcal{F} \longleftarrow$ Non-dominated sorting(X)
                                                                    ▷ //Scout bee phase
34:     $mi = \{s : trial(s) = max(trial) \wedge X(s) \notin \mathcal{F}\}$
35:     **if** $trial(mi) > limit$ **then**
36:         $X(mi) = x_j^{lb} + rand(0,1)(x_j^{ub} - x_j^{lb})$, $j = 1 \ldots D$
37:         $f_{i,mi} = f_i(X(mi))$
38:         $trial(mi) \longleftarrow \emptyset$
39:     **end if**
40:     $cycle + +$
41: **end while**

Another multi-objective ABC applies non-dominated sorting (S-MOABC/NS) on the pool of solutions generated after the employed bee phase and again after the onlooker bee phase [9]. (S-MOABC/NS) employs a ranking selection method [12] and a niche method. Steps of S-MOABC/NS are given in Alg. 3. S-MOABC/NS can generate efficient, well distributed and high quality Pareto-front solutions.

2.4   Application of ABC algorithm for software requirement selection

2.4.1   Problem description

For a large scale software project, developing the software in an incremental manner may reduce the effects of a moving target problem due to changing requirements by time. In incremental development, it is essential to assign requirements to the next release, which is called a Next Release Problem (NRP) [13, 14]. The main goal in selecting the requirements is minimizing the cost and maximizing the customer satisfaction under resource constraints, budget bound and requirement dependencies. It is an $\mathcal{NP} - hard$ problem which means an exact solution cannot be produced in a polynomial time. Therefore, approximation algorithms can be used to produce sub-optimal solutions in polynomial time without violating the budget constraint.

2.4.2   How can the ABC algorithm be used for this problem?

The next release problem is a high dimensional, binary and constrained optimization problem. Therefore, the basic ABC algorithm may not be applied directly. Some modifications in decoding solutions and local search operator are required and a constraint handling method should be adopted to favour the feasible solutions whose satisfied requirements' costs comply with the budget limit.

2.4.2.1   Objective function and constraints

Let R be the set of requirements where $|R| = m$ and C be set of customers where $|C| = n$. Each customer $c_i \in C$ has an importance $\omega_i \in Z^+$ and has a set of requirements $R_i \subset R$. Each $r_i \in R$ has a cost $cost_i$ and dependencies in R can be represented by an acyclic graph $G = (R, E)$ where edge $(r, r') \in E$ means r is a prerequisite of r'. Since $(r, r') \in E \& (r', r'') \in E \Rightarrow (r, r'') \in E$, G is transitive. Parents of R_i are the set of all prerequisites of each requirement in R_i, $parents(R_i) = \{r \in R | (r, r') \in E, r' \in R_i\}$. Therefore, in order to meet the requirements of the customer, c_i, in addition to R_i, $parents(R_i)$ should also be satisfied. $\hat{R}_i = R_i + parents(R_i)$. The company needs to select a subset of customers, $S \subset C$, in order to maximize total profit without violating the budget constraint. This constrained optimization problem can be expressed as follows:

$$\begin{aligned} \max \quad & \sum_{i \in S} \omega_i \\ subject\ to \quad & cost(\bigcup_{i \in S} \hat{R}_i) \leq b \end{aligned} \tag{2.7}$$

When the requirements of customers are independent, it means $R_i \cap R_j = \emptyset, i \neq j$ and $\hat{R}_i = R_i$. In this case the problem is called a basic next release problem. In order to formulate the problem, we define a boolean variable $x_i \in 0, 1, i = 1 \ldots n$ for each customer to indicate whether the customer c_i will be a selected customer ($x_i = 1$) whose requirements are satisfied or not ($x_i = 0$).

$$\max \quad \sum_{i=1}^{n} \omega_i x_i$$
$$subject\ to \quad \sum_{i=1}^{n} cost(\hat{R}_i)x_i \leq b \tag{2.8}$$

The main goal is to find $\vec{x}$ that maximizes the profit under the budget constraint.

2.4.2.2 Representation

Because each x_i is in the set $\{0, 1\}$, the next release problem is a constrained binary problem. In this application, in order to exploit the search space efficiently by numeric local search operators, the solutions in the population are encoded with continuous representation and they are decoded to binary values in the objective value calculation by rounding to 0 or 1.

2.4.2.3 Local search

The dimensions of next release problems are high due to the high number of customers. Hence, for an efficient and fast convergence, we used the local search defined by Eq. 2.1.

2.4.2.4 Constraint handling and selection operator

In the study, we adopted a penalty approach to convert the constrained problem into an unconstrained one. The budget values over the upper limit (b) are added to the cost function to avoid infeasible solutions violating the constraints. Since the resulting problem is unconstrained, a greedy selection scheme is applied as in the basic ABC algorithm.

2.4.3 Description of the experiments

In the study, realistic next release problem instances collected from open source bug repositories in [14, 15] are used. We abide by the instance names in the web source. There are 12 realistic next release problem instances in the set. The number of customers, the number of requirements and the total cost of each problem are given in Table 2.1. The budget upper limit is 30 and 50 percent of the total cost of the problems.

Control parameters of the ABC algorithm are the number of food sources, maximum cycle number, *limit* and modification rate (MR). We set 100 for the number of food sources, 100 for *limit*, 0.25 for MR and 5000 for the maximum cycle number.

TABLE 2.1

Properties of the Realistic NRP instances used in the experiments.

Problem	Number of customers(D)	Number of Req.s	Total Cost
nrp-e1	536	3502	13150
nrp-e2	491	4254	15928
nrp-e3	456	2844	10399
nrp-e4	399	3186	11699
nrp-g1	445	2690	13277
nrp-g2	315	2650	12626
nrp-g3	423	2512	12258
nrp-g4	294	2246	10700
nrp-m1	768	4060	15741
nrp-m2	617	4368	16997
nrp-m3	765	3566	13800
nrp-m4	568	3643	14194

2.4.4   Results obtained

The results obtained by the ABC algorithm for the software requirements selection problem are presented in Table 2.2. Each instance is analysed with the budget upper limits which are 30 and 50 percent of the total cost and each line in Table 2.2 gives the statistics of 30 runs. Satisfied costs and profit values

TABLE 2.2

Results obtained by the ABC algorithm on realistic NRP instances.

Percentage	Problem	Budget	Satisfied Costs (Best)	Profit Best	Mean	Std. Dev.
	nrp-e1	3945	3944	7140	7101.8	25.7026
	nrp-e2	4778.4	4775	6603	6565.6	24.6901
	nrp-e3	3119.7	3118	6166	6142.8	13.2061
	nrp-e4	3509.7	3506	5365	5347.4	11.0975
	nrp-g1	3983.1	3982	5879	5858.8	15.2009
0.3	nrp-g2	3787.8	3787	4349	4336.6	9.0086
	nrp-g3	3677.4	3673	5645	5621.3	14.0479
	nrp-g4	3210	3210	4040	4028.7	6.8807
	nrp-m1	4722.3	4714	8904	8768.7	81.2076
	nrp-m2	5099.1	5081	7272	7214.3	23.8097
	nrp-m3	4140	4136	8311	8216.5	50.6387
	nrp-m4	4258.2	4253	6517	6466.2	36.8203
	nrp-e1	6575	6569	10053	10025.8	11.9331
	nrp-e2	7964	7956	9231	9204.3	11.3827
	nrp-e3	5199.5	5199	8656	8648.3	6.7831
	nrp-e4	5849.5	5842	7495	7485.5	7.0435
	nrp-g1	6638.5	6637	8537	8508.8	12.4615
0.5	nrp-g2	6313	6309	6212	6203.5	6.2937
	nrp-g3	6129	6121	8156	8142.5	7.4722
	nrp-g4	5350	5345	5787	5777.2	8.804
	nrp-m1	7870.5	7871	12976	12936.6	28.1512
	nrp-m2	8498.5	8489	10510	10463.6	25.1891
	nrp-m3	6900	6894	12447	12389.1	27.2782
	nrp-m4	7097	7094	9670	9638.7	18.8034

of the best of 30 runs and the mean and standard deviation of the profit values of 30 runs are also presented in Table 2.2. As seen from the results, the ABC algorithm can find high profitable and feasible solutions which do not violate budget constraints on a large-scale realistic software requirements selection problem. When the budget upper limit constraint is relaxed to 50% of total budget, more profitable solutions are obtained in terms of both the best and mean values. On all problems except for nrp-g4, the standard deviation is less and the algorithm is more stable when the budget is relaxed.

From the results, it can be concluded that, the ABC algorithm can determine software requirements resulting in the maximum profit without violating the budget constraint.

2.5 Conclusions

In this chapter, the ABC algorithm and its modifications according to the problem characteristics are presented. Local search modifications to enhance the convergence rate are described. Combinatorial local search operators integrated in the ABC algorithm to solve combinatorial problems are summarized. The nature of the constrained problems and the adaptations of the ABC algorithm to generate feasible solutions or to prefer the solutions in the feasible region are explained. ABC algorithm variants that can handle problems with more than one objective function are presented. In Section 2.4, the ABC algorithm is applied to a binary and constrained software requirement selection problem in which the aim is to minimize the cost and maximize the customer satisfaction under resource constraints, budget bound and requirement dependencies. The representation, objective function and constraint definition, local search and selection operators that can be used for this problem are mentioned in the chapter. Some large-scale realistic problems collected from the repositories are employed in the experiments. Based on the experiments performed, it can be concluded that with minor modifications in the basic ABC algorithm, it can be easily adapted to solve the problem and to select requirements that yield high profit and do not violate budget limits.

References

1. D. Karaboga. An idea based on honey bee swarm for numerical optimization. Technical Report TR06, Erciyes University, Engineering Faculty, Computer Engineering Department, 2005.

2. D. Karaboga, B. Akay. A comparative study of artificial bee colony algorithm. *Applied Mathematics and Computation*, 214:108-132, 2008.

3. B. Akay, D. Karaboga. A modified artificial bee colony algorithm for real-parameter optimization. *Information Sciences,* 192: 120-142, 2012.

4. D. Karaboga, B. Gorkemli. A quick artificial bee colony (Qabc) algorithm and its performance on optimization problems, 2012 International Symposium on Innovations in Intelligent Systems and Applications (INISTA), pp.227-238, 2012.

5. B Akay, E. Aydogan, and L. Karacan. 2-opt based artificial bee colony algorithm for solving traveling salesman problem. Proceedings of 2nd World Conference on Information Technology (WCIT-2011), pp. 666-672, 2011.

6. D. Karaboga and B. Gorkemli. A combinatorial artificial bee colony algorithm for traveling salesman problem. In *Innovations in Intelligent Systems and Applications (INISTA), 2011 International Symposium on,* pages 50-53, June 2011.

7. K. Deb. An efficient constraint handling method for genetic algorithms. *Computer Methods in Applied Mechanics and Engineering,* 186(2- 4), 311-338, 2000.

8. D. Karaboga and B. Akay. A modified artificial bee colony (abc) algorithm for constrained optimization problems. *Applied Soft Computing,* 11(3):3021-3031, 2011.

9. B. Akay. Synchronous and asynchronous pareto-based multi-objective artificial bee colony algorithms. *Journal of Global Optimization,* 57(2):415-445, October 2013.

10. N. Srinivas, K. Deb. Muiltiobjective optimization using nondominated sorting in genetic algorithms. *Evolutionary Computation,* 2: 221-248, 1994.

11. X. Zou, Y. Chen, M. Liu, L. Kang. A new evolutionary algorithm for solving many-objective optimization problems. *Systems, Man, and Cybernetics, Part B: Cybernetics, IEEE Transactions on,* 38(5): 1402-1412, 2008.

12. K. Deb, A. Pratap, S. Agarwal, T. Meyarivan. A fast and elitist multiobjective genetic algorithm: NSGA-II. *IEEE Transactions on Evolutionary Computation,* 6(2): 182-197, 2002.

13. A. Bagnall, V. Rayward-Smith, and I. Whittley. The next release problem, *Information and Software Technology,* 883-890, 2001.

14. J. Xuan, H. Jiang, Z. Ren, Z. Luo. Solving the large scale next release problem with a backbone based multilevel algorithm, *IEEE Transactions on Software Engineering,* 38(5): 1195-1212, 2011.

15. http://researchers.lille.inria.fr/~xuan/page/project/nrp/

3

Modified Bacterial Foraging Optimization and Application

Neeraj Kanwar

Department of Electrical Engineering
Manipal University Jaipur, Jaipur, India

Nand K. Meena

School of Engineering and Applied Science
Aston University, Birmingham, United Kingdom

Jin Yang

School of Engineering and Applied Science
Aston University, Birmingham, United Kingdom

Sonam Parashar

Department of Electrical Engineering
Malaviya National Institute of Technology, Jaipur, India

CONTENTS

3.1   Introduction

The bacterial foraging optimization (BFO) algorithm is a recently developed nature-inspired meta-heuristic optimization method that mimics the foraging behaviour of Escherichia coli (i.e. *E. coli*) bacteria. It was introduced by Passino [1] in 2002 by observing the foraging behaviour of *E. coli* bacteria. The foraging behaviour should happen in such a way that energy spent in search of food will remain less than energy gained after eating it. Bacteria obtain their nutrients optimally to get maximum intake of energy per unit time. Foraging behaviour varies from one species of animal to an other. Bacteria that have a poor foraging strategy are eliminated from nature or move to some other places. On the other hand, nature always favours the bacteria that have a worthy foraging strategy. A bacterium has 8-10 thin flagella that help them in searching of food. By sensing the nutrient present in environment, flagella can move in a clockwise or anticlockwise direction. If favourable nutrients are sensed, bacteria swim in that direction. And if a noxious environment is sensed, they move away.

The food searching behaviour found in bacteria are modelled in BFO in four steps, i.e. chemotaxis, swarming, reproduction and elimination–dispersal. In chemotaxis, bacteria move in a nutrient-rich site and avoid a harmful site. This is accomplished by two types of actions, i.e., swimming and tumbling. In swimming, bacteria will trail in the same direction as in the previous stage and in tumbling they will travel in a different direction than the former one. Swarming modelled bacteria move towards the favourable or non-favourable environment. This is a group behaviour where other bacteria can be attracted toward or repelled from a bacteria by cell-to-cell signalling. Reproduction shows the depletion of less healthy bacteria from the environment. If temperature, water flow and nutrient concentration is changed, it affects the bacteria population. The elimination-dispersal step models the elimination and dispersal of bacteria due to sudden changes in environment. It introduces some diversity in population.

This chapter is organized as follows. In Section 3.2, the standard BFO algorithm is discussed along with its algorithm. All steps of the BFO algorithm are presented. In Section 3.3, some modifications suggested in literature for improving BFO performance are presented as non-uniform elimination-dispersal probability distribution, adaptive chemotaxis step, varying population. In Section 3.4, the application of the BFO algorithm for a simultaneous DGs and SCs allocation problem of distribution system is demonstrated.

3.2 Original BFO algorithm in brief

The functioning of the BFO algorithm is based on cumulative efforts of all bacteria in a group searching for food. Each bacteria should be responsive to inducements in its foraging environment. It should be capable of exchanging information with other bacteria in the group. Each bacteria searches their food by chemotaxis procedure and this mimicking of chemotactic movement forms the basic working of this algorithm. This chemotaxis movement is a complex mixture of swimming and tumbling that retains bacteria in places where the chance of finding nutrients in large concentrations is high. It shows the foraging capabilities of bacteria in a group. Like other nature-inspired optimization techniques, this technique also follows the philosophy of "Survival of the fittest". Foraging animals change their foraging behaviour. These changes basically simulate the action and response to reduce energy consumption per unit time by following all types of constraints. Bacteria that have poor foraging behaviour are eliminated in the successive generation or it may improve to a better one. The four basic steps of the BFO algorithm namely chemotaxis, swarming, reproduction, and elimination-dispersal, are discussed in following sections.

3.2.1 Chemotaxis

The strategies by which any bacteria will move to search of food are called chemotaxis. The movement of bacteria is modelled based on the movement of flagella that may be of two types i.e. tumbling and swimming. Random movement of bacteria is called tumbling whereas directed progression is called swimming. In tumbling, flagella turn clockwise; this shows independent movement and leads to erratic displacement. In swimming, flagella turn counterclockwise and show rotation in the same direction, thus pushing the bacterium steadily in a forward direction characterized as a 'run'. A decision system of bacteria enables it to avoid noxious substances in searching of food. Bacteria always search for an area having a greater concentration of nutrients. By following these two steps bacteria can achieve a high nutrient area and show an improvement in fitness value. If x shows the position of a bacteria, $J(x)$ is the objective function to be minimized. The position after the j-th chemotactic step is updated as

$$x_i(j+1, k, l) = x_i(j, k, l) + C(i) \frac{\Delta(i)}{\Delta^T(i)\Delta(i)} \tag{3.1}$$

where, $x_i(j, k, l)$ is the i-th bacteria at the j-th chemotactic step, k-th reproductive and l-th elimination-dispersal step. $C(i)$ is a scalar called run length unit that specifies the step size for tumble in the random directions. Δ shows a random directed unit length vector between -1 and 1.

For a maximization function, $J(x) < 0$ shows a nutrient rich area, $J(x) = 0$ shows a neutral area and $J(x) > 0$ shows a negative environment with noxious substances. Fitness value is calculated at this new position of bacteria given by (3.1). If it is less than the previous fitness of that bacteria then the bacteria will move in that particular direction until the number of chemotactic steps are exhausted.

3.2.2   Swarming

It is always desired that if a bacteria finds a nutrient-rich area, it should send a signal to other bacteria to concentrate them in this nutrient-rich area. In BFO, this process is modelled by using swarming. Bacteria send attracting or repelling signals to other bacteria depending upon environment. The cell-to-cell attraction and repelling is represented as

$$J_{CC}(x, x^i(j, k, l)) = \sum_{i=1}^{S} \left[-d_{attractant} \cdot \exp\left(-\omega_{attractant} \sum_{m=1}^{d} (x_m - x_m^i)^2 \right) \right]$$
$$+ \sum_{i=1}^{S} \left[h_{repellent} \cdot \exp\left(-\omega_{repellent} \sum_{m=1}^{d} (x_m - x_m^i)^2 \right) \right]$$

$$(3.2)$$

where $d_{attractant}$, $\omega_{attractant}$, $h_{repellent}$ and $\omega_{repellent}$ are different coefficients that represent the attraction and repulsion strength of the signal. d defines the problem dimension, S represent the total swarm size, x_m is a point in d-dimensional search space, x_m^i is the position of the ith bacteria. The obtained fitness value after swarming is added in the existing fitness value, expressed as

$$J(i, j, k, l) = J(i, j, k, l) + J_{CC}(x, x^i(j, k, l)) \qquad (3.3)$$

3.2.3   Reproduction

After the chemotaxis stage, it is assumed that bacteria present in a swarm having sufficient nutrients. The fitness value obtained for each bacteria represents its health. If a bacteria has good fitness value this means it got sufficient nutrients in its life cycle. This healthy bacteria can reproduce itself under a favourable environment and temperature. They increase their length and split into two parts. These parts are the replica of each other. On the other hand, if it has poor fitness value, it shows that the bacteria did not get sufficient nutrients in its life cycle and is not healthy to reproduce itself; therefore, it cannot survive further. To simulate this step, the bacterial population is sorted according to their fitness values. Half of population with poor fitness values will be liquidated while the other better bacteria split into two parts and start exploring the search space from the same location. This phenomenon is similar to the elitist-selection mechanism of the other classical evolutionary algorithms.The newly generated bacteria will replace the dead bacteria of the swarm, so swarm size remains constant after reproduction.

3.2.4 Elimination and dispersal

The environment where bacteria live may change suddenly or gradually. Due to this noxious environment some bacteria may die or some may disperse to some other locations. In general, dispersion probability is very low. Dispersed bacteria follow different paths, so this process generates a swarm having a new set of bacteria.

3.2.5 Pseudo-codes of the original BFO algorithm

The pseudo-code of modified BFO (MBFO) is presented in Algorithm 4. The BFO algorithm parameters are: d=Total number of parameters to be optimized or dimension of the search space; S=Total number of bacteria in the population; Nc=total number of chemotactic steps; Ns=Maximum swimming length before tumbling; Nre=total number of reproduction steps; Ned=total number of elimination-dispersal events; Ped=probability for elimination-dispersal; $C(i)$=Step size in random direction that is specified by the tumbling process.

Algorithm 4 Pseudo-code of MBFO.

1: define the objective function $OF(.)$ and initialize BFO parameters
2: set initial value of j,k and l to zero, calculate $OF(.)$ for each bacteria x_i
3: Start elimination-dispersal loop
4: **for** each l-th elimination-dispersal **do**
5:    Start reproduction loop
6:    **for** each k-th reproduction **do**
7:       Start chemotactic loop
8:       **for** each j-th chemotactic step **do**
9:          Start bacterial loop
10:          **for** each i-th bacteria **do**
11:             calculate objective function, $J(x^i(j,k,l))$
12:             $Jlast = J(x^i(j,k,l))$
13:             generate a tumble angle for bacteria i, as $\frac{\Delta(i)}{\Delta^T(i)\Delta(i)}$
14:             evaluate new position of bacteria i, using (3.1)
15:             for new position, recalculate $J(x^i(j+1,k,l))$
16:             Start counter for swimming process
17:             $m = 0$
18:             **while** $m \leq Ns$ **do**
19:                **if** $J(x^i(j+1,k,l)) < Jlast$ **then**
20:                   $Jlast = J(x^i(j+1,k,l))$;
21:                   calculate new position as
22:                   $x_i(j+1,k,l) = x_i(j,k,l) + C(i)\frac{\Delta(i)}{\Delta^T(i)\Delta(i)}$
23:                   for new position, recalculate $J(x^i(j+1,k,l))$
24:                **else**
25:                   $m = Ns$

26:               **end if**
27:                   $m = m + 1$
28:               **end while**
29:           **end for**
30:           Reproduction
31:               sort the bacteria according to fitness values, $J(x^i)$
32:               split the top $S_r = S/2$ fittest bacteria and replace poor ones
33:       **end for**
34:       Elimination-dispersal
35:       **for** each i-th bacteria **do**
36:           $rr = rand$
37:           **if** $rr < P_{ed}$ **then**
38:               disperse the bacterium at random location
39:           **end if**
40:       **end for**
41:   **end for**
42: **end for**
43: return the best result

3.3   Modifications in bacterial foraging optimization

3.3.1   Non-uniform elimination-dispersal probability distribution

In elimination-dispersal step of standard BFO (see steps 35–40 in Algorithm 4), the probability of elimination for each bacteria is taken constant. It is completely independent from any ranking of bacteria in population. This constant probability may affect the convergence speed of the algorithm as in the iterative process the bacteria that are near to the optimal solution may be replaced by some other non-optimal solution. In other words, both types of fitness, best and worst, will have the same probability of elimination. This will ultimately affect the performance of the algorithm.

An improved version of BFO is proposed in [2] by suggesting a non-uniform probability distribution for the elimination and dispersal phase. It improves convergence speed due to implementation of a linear and nonlinear probability distribution to replace the constant distribution of standard BFO.

The linear elimination-dispersal probability is defined as

$$p_e(i) = \frac{(i-1)}{\left(\frac{S}{2} - 1\right)} \qquad \forall i = 1, 2, \ldots, \frac{S}{2} \tag{3.4}$$

The non-uniform probability distribution is defined as

$$p_e(i) = \left[\frac{(i-1)}{\left(\frac{S}{2} - 1\right)} \right]^2 \qquad \forall i = 1, 2, \ldots, \frac{S}{2} \tag{3.5}$$

Equations (3.4) and (3.5) ensure that best bacteria will be assigned a lower probability than the worst one. However, in these equations, the probability is non-linearly proportional to the bacterium index.

A large value of elimination probability may avoid local trapping by relocating the bacteria in a better position and give exploration of the search space. On the contrary, a small value of elimination probability strengthens the search in local area and may stuck in local optima. Therefore, elimination probability should be selected in a proper fashion. During initial iterations, the whole search space should be explored by the algorithm and during anaphase the exploitation rate should be high. It can be achieved by assigning a high value of p_e during exploration and a low value during exploitation. Therefore, (3.4) and (3.5) are proposed in a more generalised way than (3.6) and (3.7) for linear and non-linear distributions [2].

Linear:

$$p_e(i,l) = P_e + \left[i - \left(\frac{S}{4} + 0.5 \right) \right] \frac{(l-1)}{(N_{ed} - 1)(\frac{S}{2} - 1)} \qquad \forall i, l \qquad (3.6)$$

Non-linear:

$$p_e(i,l) = P_e \left(\frac{l-1}{N_{ed} - 1} \right) \left[2 \times \left(\frac{i-1}{\frac{S}{2} - 1} \right)^2 - 1 \right] + P_e \qquad \forall i, l \qquad (3.7)$$

where, i and l are population and elimination index respectively. P_e shows constant probability.

3.3.2 Adaptive chemotaxis step

Run-length unit parameter $C(i)$ plays a crucial role in deciding the performance of BFO. A large value can increase exploration of the search space. It benefited the BFO to escape local optima very easily but sometimes global optima may be missed due to this large step size. On the contrary, a bacteria with small run-length parameter can only search the region which is near the starting point. So there is a fair chance that it may be stuck in the local region. Therefore, run-length unit parameter can be made adaptive with the chemotaxis step. In [3], an adaptive BFO algorithm is proposed to maintain a proper balance between exploration and exploitation of the search space using the self-adaptive chemotaxis step. Two factors are considered that guide the algorithm for switching in exploration or exploitation. These factors depend on the improvement in fitness value. When there is an improvement in fitness value beyond a certain limit from the last to the current chemotactic generation, it means that bacteria have searched a new promising region and now there is a requirement to exploit this area further to obtain a better solution. Therefore run-length unit parameter should be small compared to the previous generation. So bacteria should be self-adaptive with the exploitation state. If there is no improvement in fitness value from the previous generation to the current generation continuously for a predefined time, it shows that it is not fruitful to search that area any more; it requires a large change in run-length unit so that bacteria can jump to a new search area. It helps to escape local trapping. So bacteria should be self-adaptive with the exploration state under such situation.

3.3.3   Varying population

The iterative process of all evolutionary algorithms (EAs) in general starts with a fixed population size. So search space is not fully explored and it also increases the computational time. An improved version of BFO i.e. bacterial foraging algorithm with varying population (BFAVP) is proposed in [4]. Bacterial energy and bacterial age indexes have been introduced to measure the search ability and life of each bacteria. It simulates three behaviours of each bacteria, i.e. metabolism, proliferation and elimination. The better position of bacteria than the previous step is assured by incorporating these. Diversity is controlled by combining chemotaxis behavior with the phenomena of quorum sensing. It also speeds up the convergence and avoids local trapping. In [5], BFAVP is used to solve an optimal power flow problem.

3.4   Application of BFO for optimal DER allocation in distribution systems

3.4.1   Problem description

Reactive power compensation in distribution systems using SCs is typically an old classical problem of power systems. The capacitors must be allocating optimally; otherwise line losses may increase and develop over-voltages during light load hours. The optimal capacitor placement problem involves the determination of their optimal number, location and capacity. In an electric distribution system with specified structure, the value of the active component of losses cannot be reduced by the use of the SCs alone. Therefore, to supply active loads of the distribution system, local generation should be privileged by DGs. DGs refer to small generating units typically connected to the utility grid in parallel near load centres. The DG allocation problem also involves the determination of the optimal number, location and capacity of DG units to achieve certain objectives under specified constraints. The optimal generation of active and reactive power from these devices reduces power import from the substation and thus regulates feeder power flows. Optimal capacitor placement achieves this goal by regulating reactive power flow, whereas optimal DG placement does the same by regulating active power flow in the system. These key technologies may be coordinated together to get better solutions so that distribution systems can achieve optimum performance. The simultaneous allocation problem of DGs and SCs in distribution systems is modelled to reduce annual energy losses and to maintain better node voltage profiles while considering a piece-wise multi-level annual load profile.

3.4.2   Individual bacteria structure for this problem

For a simultaneous DG and SC placement problem, each bacteria is structured as shown in Fig. 3.1 which is composed of candidate location and capacity for the respective candidate DGs and SCs. Here, l^{DG} and l^{SC} represents location, and, P^{DG} and Q^{SC} represents capacity of these components, respectively. The candidate nodes are allocated randomly between defined limits. Repetition of nodes is avoided using non-repeatability constraints as given in the next section. The capacity of the candidate DGs and SCs is selected randomly within their respective predefined bounds. However, the number of locations for DGs and SCs is set more than what should be required and the lower bound of the capacity for DGs and SCs is set equal to zero.

$$\underbrace{l_1^{DG}, l_2^{DG}, \ldots, l_{T_{DG}}^{DG}}_{\text{DG Locations}} \ \underbrace{P_1^{DG}, P_2^{DG}, \ldots, P_{T_{DG}}^{DG}}_{\text{DG Capacities}} \ \underbrace{l_1^{SC}, l_2^{SC}, \ldots, l_{T_{SC}}^{SC}}_{\text{SC Locations}} \ \underbrace{Q_1^{SC}, Q_2^{SC}, \ldots, Q_{T_{SC}}^{SC}}_{\text{SC Capacities}}$$

FIGURE 3.1
Individual structure of each bacteria.

3.4.3   How can the BFO algorithm be used for this problem?

We want to generate an optimization method to give optimal allocation of DGs and SCs in a distribution system. The optimal allocation is represented by optimal number, location and capacity of these components. Here, the BFO algorithm consists of mainly five steps to find out the optimal solution for any optimization problem. Initially we define the different control parameters of the BFO algorithm, i.e. dimension of the search space (d), total number of bacteria in the population (S), total number of chemotactic steps (N_c), maximum swimming length before tumbling (N_s), total number of reproduction steps (N_{re}), total number of elimination–dispersal events (N_{ed}), probability for elimination-dispersal (Ped), step size for tumbling process $(C(i))$, maximum number of iterations, maximum and minimum limit of DGs and SCs at each candidate node, total compensation limit at all candidate nodes, minimum and maximum voltage limit at each node.

After defining all problem and algorithm specific control parameters of BFO, the first step is the initialization of the initial swarm. It consists of S bacteria (each bacteria is a d-dimensional vector). The dimensions depend upon design variables considered for optimal allocation of DGs and SCs in the distribution system. Simultaneous optimal allocation of DGs and SCs is a complex, combinatorial problem as so many combinations of siting and sizing of DGs and SCs are possible. The initial swarm of bacteria is randomly generated. Each bacteria in this swarm has a defined position in the swarm i.e. x^i and structured as in Fig. 3.1. Each individual bacteria represents an allocation of DGs and SCs simultaneously and corresponds to some fitness value of objective function. Here objective function is formulated for annual energy loss reduction. So each bacteria will represent annual energy loss for that allocation. The fitness value of each bacteria in the swarm is calculated and saved.

The second step is chemotaxis operation of bacteria. Bacteria can move in one direction or multiple directions in searching fore food. These movements basically depend upon environmental condition. A bacteria starts tumbling to change its present location with small steps. If they find sufficient nutrients at this new position, they continuously swim in the same direction using (3.1), until the maximum swimming length (N_s) is exhausted. This process is done for all bacteria in the swarm until the maximum chemotactic steps (N_c) are reached. After the chemotaxis step, bacteria will have a different combination of siting and sizing of DGs and SCs. Fitness value is calculated by placing the given capacity at a given location for each bacteria.

The third step of BFO is swarming. It basically simulates the singling process of bacteria from one cell to other. Bacteria inform other bacteria in the swarm about favorable (nutrient-rich) or non-favorable (noxious) environment. For that they release some chemicals for attracting or repelling the other. This step is completed using (3.2). This equation gives a cell-to-cell communication value (J_{cc}) for each bacteria. This value is added in the obtained fitness value of each bacteria after the chemotaxis step using (3.3).

The fourth step of BFO is reproduction. This step shows that unhealthy bacteria cannot survive in nature while a healthy bacteria can reproduce itself in a suitable environment. To simulate it all bacteria present in swarm are sorted according to their fitness value. Generally, half the bacteria of high fitness values (means unhealthy) are removed and the other half of the bacteria are split into two parts that are replicas of each other. They replace the position of unhealthy bacteria in the swarm. By following this step all individuals having poor allocation for DGs and SCs are removed from the swarm.

The fifth step is elimination and dispersal of bacteria. It generates a new set of swarm having inclusion of some new bacteria and deletion of some bacteria. After this last step, the best/healthiest bacteria having minimum fitness value (J_{cc}) and its position are saved. For the simultaneous optimal allocation of DGs and SCs, this bacteria will generate optimal siting and sizing of these components. When this solution is implemented on the network, it will give maximum energy loss reduction ($1/J_{cc}$). This solution shows optimal allocation of these components after the first iteration. The iterative process continues until maximum iteration counts are exhausted.

3.4.4   Description of experiments

The integration of DGs and SCs alters power flow in distribution feeders. This causes reduction in annual energy losses and node voltage deviations. Therefore, the simultaneous DGs and SCs allocation problem is formulated to maximize the annual energy loss reduction while maintaining better node voltage profiles. In order to limit the voltage deviation at different nodes, a hard voltage constraint is used as desired. Similarly, to check the current carrying capacities of distribution feeders, a feeder ampacity constraint is essential. A multi-level piece-wise linearized annual load duration profile of the system is considered to evaluate annual energy losses of distribution feeders. The DR allocation problem is structured as a single-objective constrained optimization problem where optimal number, size and location of DGs and SCs are determined simultaneously. The problem objective is formulated as

$$\max F = \sum_{ll=1}^{N_{Total}} \left(ELOSS_{bj} - ELOSS_{aj} \right); \qquad \forall \; ll \epsilon L \qquad (3.8)$$

OR

$$\min J_{cc} = \frac{1}{(1+F)} \qquad (3.9)$$

This objective function is solved using the BFO technique, subject to the power flow and node voltage limit constraints. Node voltages $V_{n,ll}$ of all system buses must be kept within the minimum and maximum permissible limits, i.e., V_{min} and V_{max}, respectively, during the optimization process as defined by (3.10). The current flow in each branch must satisfy the rated ampacity of each branch as defined by (3.11). The active power injected by DG at each bus must be within their permissible range as given by (3.12), where, P_{min}^{DG} and P_{max}^{DG} are the minimum and maximum active power generation limit at a bus, respectively. The reactive power injected by SC at each bus must be within their permissible range as defined by (3.13), where, Q_{min}^{SC} and Q_{max}^{SC} are the minimum and maximum reactive power generation limit at a bus, respectively. The sum of active power injected by DGs at all candidate nodes should be less than nominal active power demand (P_D) of the distribution system as given in (3.14). Similarly, the sum of reactive power injected by SCs at all candidate nodes should be less than nominal reactive power demand (Q_D) of the distribution system as given in (3.15). Equations (3.16) and (3.17) prohibit the repetition of candidate sites for DGs and SCs, respectively.

$$V_{min} \leq V_{n,ll} \leq V_{max} \qquad \forall \; n, ll \qquad (3.10)$$

$$I_{n,ll} \leq I_n^{max} \qquad \forall \; n, ll \qquad (3.11)$$

$$P_{min}^{DG} \leq P_{n,ll}^{DG} \leq P_{max}^{DG} \qquad \forall \; n \qquad (3.12)$$

$$Q_{min}^{SC} \leq Q_{n,ll}^{SC} \leq Q_{max}^{SC} \qquad \forall \; n \qquad (3.13)$$

$$\sum_{n=1}^{T_{DG}} P_n^{DG} \leq P_D \qquad \forall \; n \qquad (3.14)$$

$$\sum_{n=1}^{T_{SC}} Q_n^{SC} \leq Q_D \qquad \forall \; n \qquad (3.15)$$

$$l_a^{DG} \neq l_b^{DG} \qquad \forall \; a, b \epsilon N \qquad (3.16)$$

$$l_a^{SC} \neq l_b^{SC} \qquad \forall \; a, b \epsilon N \qquad (3.17)$$

3.4.5   Results obtained

The simultaneous DGs and SCs allocation problem is solved using standard BFO algorithm for benchmark 33-bus [6] test distribution systems. The population size and maximum iterations are set at 10 and 200, respectively. The best result obtained after 100 independent trials of MBFO is presented in Table 3.1. The table shows the optimal DG and SC capacity as well as their obtained optimal locations. The annual load profile is segmented into three different load levels, i.e., light, nominal and peak to show 50%, 100% and 160% of the nominal system loading, respectively. The corresponding load durations are taken as 2000, 5260 and 1500 hours, respectively.

TABLE 3.1
Optimal solution of DGs and SCs using MBFO.

Parameter	DGs	SCs
Location (nodes)	14, 25, 32	14, 24, 30
Individual node capacity (kW/kVAr)	898, 944, 934	600, 300, 900
Total capacity (kW/kVAr)	2776	1800

TABLE 3.2
Comparison of power and energy loss with and without DRs.

Particular	Power loss (kW)			Annual energy loss (kWh)	% loss reduction
	Light	Nominal	Peak		
Without compensation	47.07	202.50	575.39	2022398.44	–
With compensation	3.53	15.12	93.40	226700.42	88.79

The improvement in network performance is evaluated with and without compensation. The comparison results are presented in Table 3.2. The table shows the power loss obtained at each load level. The table also shows that an annual energy loss reduction of about 89% is achieved after allocating the optimal solution of Table 3.1 in the distribution network.

3.5   Conclusions

In this chapter, the BFO algorithm was presented in detail. Some modifications of the BFO algorithm were demonstrated and discussed. Finally, the application of the BFO algorithm to a complex combinatorial problem was shown. The BFO algorithm was used for a simultaneous optimal allocation problem of DGs and SCs in a distribution system. The results of the BFO algorithm show that there is a significant improvement in the desired objectives.

Acknowledgement

This work was supported by the Engineering and Physical Sciences Research Council (EPSRC) of United Kingdom (Reference Nos.: EP/R001456/1 and EP/S001778/1).

References

1. K.M. Passino, "Biomimicry of bacterial foraging for distributed optimization and control", IEEE Control Systems Magazine (2002) pp. 52-67.

2. Mouayad A. Sahib, Ahmed R. Abdulnabi, and Marwan A. Mohammed, "Improving bacterial foraging algorithm using non-uniform elimination-dispersal probability distribution". Alexandria Engineering Journal (2018) 57, pp. 3341-3349.

3. Nandita Sanyal, Amitava Chatterjee, Sugata Munshi. "An adaptive bacterial foraging algorithm for fuzzy entropy based image segmentation". Expert Systems with Applications 38 (2011) pp. 15489-15498.

4. M.S. Li, T.Y. Ji, W.J. Tang, Q.H. Wu, J.R. Saunders. "Bacterial foraging algorithm with varying population". BioSystems 100 (2010) pp. 185-197.

5. M. S. Li, W. J. Tang, W. H. Tang, Q. H. Wu, and J. R. Saunders. "Bacteria foraging algorithm with varying population for optimal power flow" in Proc. Evol. Workshops 2007, LNCS vol. 4448. pp. 32-41.

6. M. E. Baran and F. F. Wu. "Network reconfiguration in distribution systems for loss reduction and load balancing". IEEE Transactions on Power Delivery, vol. 4(2), 1989, pp. 1401-1407.

4

Bat Algorithm – Modifications and Application

Neeraj Kanwar

Department of Electrical Engineering
Manipal University Jaipur, Jaipur, India

Nand K. Meena

School of Engineering and Applied Science
Aston University, Birmingham, United Kingdom

Jin Yang

School of Engineering and Applied Science
Aston University, Birmingham, United Kingdom

CONTENTS

4.1   Introduction

With the beginning of fast computational services, a large number of evolutionary or swarm-based artificial intelligence (AI) techniques gained the attraction to solve real-life complex combinatorial optimization problems. However, these techniques have their own qualities and shortcomings. BA is a recently developed optimization technique proposed by Xin-She Yang [1] in 2010. This is a bio-inspired search technique which has shown an advanced competency to reach into a promising region. It is inspired by the social behavior of bats and the phenomenon of echolocation to sense distance between the bat's current location and prey. It is a simple, easy to implement, significantly faster than other algorithms, and robust numerical optimization approach [2]. It has been used in literature to solve several optimization problems of diverse nature such as: distributed resources allocation, feature selection, economic load dispatch, load frequency control, distribution network reconfiguration (DNR), etc. It is very suitable for highly nonlinear problems and can generate optimal solutions with good accuracy for such type of problem. So, it has good exploitation potential, but its exploration is insufficient to reach the promising region. It happens due to its local trapping, insufficient diversity or slow movement of the algorithm. Therefore, several attempts have been made to improve the performance of the standard BA to overcome its inherent flaws.

DNR is a frequently required process in a distribution system. It is done by managing the open and closed status of sectionalizing and tie-switches. This switch changing process basically alters the topological structure of the distribution network. It reallocates load from one distribution feeder to another and balances the load on the system. It also improves the performance of the distribution network by reducing losses, improving voltage profile, etc. Thus, this is a very important operational strategy that can enhance distribution network performance. In this chapter, the application of BA is shown to solve a complex optimal DNR problem of a distribution network.

The chapter is organized as follows. In Section 4.2 an introduction of the standard BA algorithm is given. The working of the standard BA is also presented in brief. In Section 4.3, some modifications suggested in literature to improve the performance of BA are presented as improved BA, bat algorithm with centroid strategy, self-adaptive BA, chaotic mapping based BA, self-adaptive BA with step-control and mutation mechanisms, adaptive position update, smart BA, adaptive weighting function and velocity. In Section 4.4, the application of the BA is presented for the optimal DNR problem of a distribution network.

4.2 Original bat algorithm in brief

BA is a new bio-inspired algorithm that mimics the bats' behavior for searching their prey. Bats use the echolocation phenomenon to sense proximity of prey. In BA, all bats randomly fly in different directions with some definite velocity and frequency at a certain position. Two control parameters are used to guide the bats' movement in random directions i.e. loudness and PER. Bats update their velocity and position by assigning time varying values to these parameters. Bats emit pulses and adjust PER depending on the vicinity of the prey. To characterize the behavior of foraging bats the optimization model is developed. Foraging behavior of bats is modeled in two steps i.e. random fly and local random walk (LRW) which are briefly explained below [3].

4.2.1 Random fly

Each bat is defined by its position x_b, velocity v_b, frequency f_b, loudness A_b, and PER r_b in a D-dimensional problem search space. The velocity and position of bth bat at the tth iteration are updated using (4.1)–(4.3). These equations show the global searching process of BA.

$$f_b = f_{\min} + (f_{\max} - f_{\min})\beta \tag{4.1}$$

$$v_b^t = v_b^{t-1} + (x_b^t - x^*)f_b \tag{4.2}$$

$$x_b = x_b^{t-1} + v_b^t \tag{4.3}$$

Here β denotes a random number in range [0, 1] which is obtained from uniform distribution, x^* is current best solution, $f_{\max}$ and $f_{\min}$ are maximum and minimum frequency range.

4.2.2 Local random walk

A new solution is generated for each bat by using LRW, expressed as

$$x_{new,b} = x_{old,b} + \epsilon\langle A_b^t \rangle \tag{4.4}$$

where $\epsilon \in$ [-1, 1] is a uniformly distributed random number, $\langle A_b^t \rangle$ is the average loudness value of all bats and r_b^t is the PER of the bth bat. Loudness and PER of each bat are updated with iteration using (4.5) and (4.6).

$$A_b^{t+1} = \alpha A_b^t \tag{4.5}$$

$$r_b^{t+1} = r_b^0\left[1 - exp(-\gamma t)\right] \tag{4.6}$$

where α and γ are constants having value of 0.9. r_b^0 shows the maximum possible PER. As the algorithm proceeds, loudness tends to zero and PER become maximum. The pseudo-code of standard BA is presented in Algorithm 5.

Algorithm 5 Pseudo-code of the original bat algorithm.

1: Determine the objective function $OF(.)$
2: Initialize the bat population x_b, v_b and $f_{b,d}$, $\forall$ $b = 1, 2, \ldots, N_b$, $d = 1, 2, 3, \ldots, D$
3: Initialize pulse rate r_b and loudness A_b
4: Set $t = 0$
5: **while** $t < itr_{\max}$ **do**
6:     Generate new solutions by adjusting frequency, updating velocities, and positions by using (4.1)–(4.3).
7:     **if** $rand > r_b$ **then**
8:         Select the current best solution x^*
9:         Generate a local solution $OF(x_b)$ around x^*
10:    **end if**
11:    Generate a new solution by flying randomly
12:    **if** $rand < Ab$ and $OF(x_b) < OF(x^*)$ **then**   ▷ for minimization
13:        Accept the new solutions
14:        Increase r_b and reduce A_b
15:    **end if**
16:    Rank the bats and find the current best
17:    $t = t + 1$
18: **end while**
19: return the best bat and its fitness as a result

4.3   Modifications of the bat algorithm

Standard BA shows distinct performance while dealing with lower-dimensional optimization problems but performance degrades as the dimension of the problem increases. It all happens due to lack of diversity which ultimately drags BA to local optima and thus degrades convergence. Therefore, different modifications are suggested in literature to refine its exploration and exploitation potentials. Some of the modifications suggested by different authors are described in the following sections.

4.3.1   Improved bat algorithm

In [4] an improved BA (IBA) is proposed to adjust the loudness and PER in a better way and to enhance LRW of bats to refine its exploration and exploitation potentials, respectively. Furthermore, additional diversity in population is suggested to cope with the intense exploitation capability of BA. The first modification is self-adapted loudness and PER. Each bat is allowed to vary in accordance to the loudness assigned to it and are not allowed to vary too quickly. The gradual switching of exploration to exploitation is governed by the values that are assigned to loudness and PER. In this way, these two parameters become self-adapted for each bat. The suggested modeling for loudness and PER is given in (4.7) and (4.8).

$$A_b^t = \alpha^t A_b \qquad (4.7)$$

$$r_b^t = 1 - A_b^t \qquad (4.8)$$

where, $\alpha = \left(\frac{1}{2A_b}\right)^{\frac{1}{k*iter_{\max}}}$, and k is the desired fraction of maximum iteration at which loudness and PER equalize. In this manner both parameters become self-adaptive to each other.

The second modification is improved local random walk (ILRW) which is suggested to improve LRW of bats. First M_c number of replicas of the current best bat are generated and then each of them is mutated in the range [-1, 1] at randomly selected dimension. ILRW initiates only when the selected random number is less than PER, otherwise LRW will be performed as in the standard BA. Thus ILRW confirms the current best bat LRW during the evolutionary process and helps to maintain an appropriate balance between exploitation and exploration of the search space. If a better solution is found by ILRW, it replaces the current best bat.

BA performance also degrades when its intense exploitation capability is merged with its inherent poor exploration potential. It happens especially for large-scale optimization problems. It occurs because bats may remain busy to exploit the undesirable search space and be unable to identify the promising region. This causes local trapping of the algorithm. The mutation operator of GA has intense potential to discover a new area in the problem search space. As BA has very poor diversity, a very high mutation rate is required. Therefore the population is reinitialized in each iteration using mutation. Fitness is evaluated for all mutated bats and the current best bat is updated by the better mutated bat if obtained.

4.3.2 Bat algorithm with centroid strategy

In standard BA, the global and local search patterns are such that it blocks some neighboring area and limits the search ability. To overcome these flaws of BA, a new centroid strategy is proposed in [5]. Three different centroid strategies are proposed with six different designs. The main idea is to consider all the effects from the best position and neighbor bats. The local search capability is improved using the proposed strategy. A detailed comparison of six different centroid strategies is provided to give deep insight. These include arithmetic centroid, geometric centroid, harmonic centroid, weighted arithmetic centroid, weighted geometric centroid and weighted harmonic centroid.

4.3.3 Self-adaptive bat algorithm (SABA)

Fister *et al.* [6] presented the self-adaptive bat algorithm (SABA) in which the parameters (loudness and PER) at the beginning of the search process change as the process reaches maturity. It overcomes the difficulties of setting parameter values at the beginning and in different searching phases. A self-adaptive feature is also employed in [7] to improve algorithm performance. It enables a self-adaptation of its control parameters and gives the advantages that

1. the control parameters need not to be set before starting the algorithm.

2. the control parameters are self-adapted during the fitness evaluation

4.3.4   Chaotic mapping based BA

The chaotic mapping mechanism is a new optimization method to extend the searching range of some evolutionary algorithms to eliminate their premature convergence [8]. Three search algorithms are proposed in [8] for chaotic mapping. The first sub-algorithm is the global chaos traversal disturbance algorithm. In this algorithm, traversal search is performed in the region of feasible solutions. It helps to overcome the declining diversity of the BA population. The second sub-algorithm is the local niche accelerate search algorithm. It is suggested to improve local search in the anaphase of the algorithm. It increases the searching speed of BA. First the optimal solution is set as the center of a local niche acceleration searching area. Then, each component is mapped in interval [0, 1]. Search radius is calculated to set the local niche acceleration. Then, a traversal search is conducted. At the end, the optimal solution, loudness, PER and frequency are updated. The third sub-algorithm is adaptive speed control algorithm. It is suggested to modify conversion of bats between global and local search. It first defines the population gathering rate and then based on it updates the weight of the speed between global and local search.

4.3.5   Self-adaptive BA with step-control and mutation mechanisms

PER increases and loudness decreases with the distance from the prey. It is done to observe the movement of prey and to remain undetected by prey. In the early stage of BA, the search process is mainly affected by loudness and help in global search, whereas during anaphase of the algorithm, the search process is more influenced by PER. Therefore, the global and local search probability of BA should be guided by loudness and PER. These characteristics inspired Shilei Lyu et al. to propose a self-adaptive BA (SABA) [9]. Two modifications are proposed i.e. step-control and mutation mechanisms.

In SABA, step sizes are controlled by step-control during the search process. It is different from standard BA. It uses two frequencies f_1 and f_2 to control step sizes during global and local search. New modeling is suggested for velocity, frequency and position as given by (4.9)–(4.12).

$$v_b^t = \omega v_b^{t-1} + f_1 r_1 \left(h_b^* - x_b^{t-1} \right) \times f_2 r_2 \left(x^* - x_b^{t-1} \right) \tag{4.9}$$

$$f_1 = \alpha\Big(1 - exp\big(-|F_{avg} - F_{best}|\big)\Big) + \gamma\big(1 - k\big) + f_{\min} \qquad (4.10)$$

$$C_w = f_1 + f_2 \qquad (4.11)$$

$$x_b^t = x_b^{t-1} + \mu v_b^t \qquad (4.12)$$

where ω is the weight coefficient, h_b^* is the optimal solution for bth bat, x^* is the current best bat, f_1 and f_2 are the frequencies, and r_1 and r_2 are uniform random numbers in the range of [0.5, 1.5]. F_{avg} is the average fitness value for current optimal solutions and F_{best} is the current best fitness. $F_{\min}$ is a constant value to represent minimum value of f_1. k is the evaluation index. α, γ, and μ are weight coefficients.

The mutation mechanism is proposed to avoid local trapping. In this mechanism, PER and loudness are improved by suggesting (4.13).

$$A_b^{t+1} = \frac{f_1}{f_{\max}}; \quad \text{and} \quad R_b^{t+1} = \frac{f_2}{f_{\max}} \qquad (4.13)$$

SABA is effective to avoid local trapping during the early stage of the algorithm and has ability to improve accuracy during anaphase of the algorithm.

4.3.6 Adaptive position update

An adaptive position update based BA (APU-BA) is proposed [10] to improve the search performance and accuracy of the BA. The local search part is upgraded by assigning values for the BA position update index. It is assigned to each bat per dimension. Also a discrete layer peeling algorithm is combined with APU-BA to improve its convergence. This modification improves the performance of APU-BA compared to standard BA.

4.3.7 Smart bat algorithm

In [11], smart BA is proposed with an idea to make an artificial bat that regulates its searching process using fuzzy logic and decision theory. It provides better search direction, frequency and velocity to the artificial bat. A utility function is defined for each bat to decide its search direction. It is made up of two parts; the first part consists of difference between this artificially generated bat and the best bat, whereas the second one shows the ratio of the bat number threshold of a cluster to the number of bats in the cluster. It helps to avoid local trapping.

4.3.8 Adaptive weighting function and velocity

A modified version of (4.1) is proposed in [12] to improve the exploration and exploitation potential of BA.

$$v_b^t = \omega v_b^{t-1} + f_b^t\big(x_b^t - x^*\big)k_1 + f_b^t\big(x_b^t - x_{cbest}\big)k_2 \qquad (4.14)$$

$$k_1 + k_2 = 1 \qquad\qquad (4.15)$$

$$k_1 = 1 + (k_{init} - 1)\frac{(iter_{\max} - iter)}{iter_{\max}^n} \qquad\qquad (4.16)$$

where k_1 and k_2 are weighting factors, k_{init} is the initial value of k_1, ω shows positive weighting factor. This modification shows a balance between global and local search. An adaptive weighting function is also proposed as given by (4.17)

$$\Omega_b^t = \omega_b^0 e^{-\psi t} \qquad\qquad (4.17)$$

Here, ψ represents a positive value of fixed quantity inspired by the Lyapunov stability theorem. Weighting function controls the gap between obtained solution and expected solution.

State error between expected solution and current bth bat is defined by (4.18). The velocity equation is modified using the state error equation. When bats come near the final solution, velocity reduces and bats move slowly towards the final solution. It ultimately improves convergence of BA by reducing fluctuation around the final solution.

$$error_b^t = x_{\exp} - x_b^{t-1} \qquad\qquad (4.18)$$

4.4 Application of BA for optimal DNR problem of distribution system

In this section, the BA is used to solve the real-life engineering optimization problem.

4.4.1 Problem description

DNR is a well-known and effective operational strategy used to improve performance of modern automated radial distribution systems. The problem of DNR is characterized as finding the best possible topology of the distribution network that can give minimum losses and better node voltage profile in the system. Here, the DNR problem is formulated to reduce line losses while improving node voltages. Different operating constraints should be satisfied while solving the objective function such as voltage limits, feeder current capacity and feasible radial arrangement of the distribution system.

4.4.2 How can the BA algorithm be used for this problem?

We want to generate an optimization method to give optimal radial topology in the distribution system. The optimal topology is represented by optimal switches that are to be open to generate the optimal configuration of a given distribution network. In this chapter, the standard version of BA is used to solve this problem. All optimization techniques require some algorithm-specific and some common control parameters. Algorithm-specific parameters are different for each algorithm. The common control parameters are number of individuals and maximum number of iterations to solve an optimization problem. However, optimum number of these control parameters are found out by trade-off.

BA is a nature inspired algorithm. All bats in a swarm use echolocation in the searching process for their food and to avoid obstacles. They emit pulses in their surroundings and based on return echoes they are able to locate their prey. In addition, bats also identify the most nutritious area to move on. DNR is a complex, combinatorial optimization problem, therefore, so many combinations of switches are possible. Each bat has a defined position i.e. x_b. Initially we define the different control parameters i.e. dimension of the search space (d), total number of bats in the population (N_b), maximum number of iterations (N_{iter}), design constants (α, γ), pulse loudness (A), maximum and minimum pulse frequencies $(f_{\max}/f_{\min})$, initial pulse emission rate (r_b^0).

The initial swarm of bats is randomly generated using a defined problem and algorithm specific control parameters. It consists of N_b bats (each bat is a d-dimensional vector). The dimensions depend upon design variables considered for the optimal DNR problem in a distribution system. The design variables in each bat is switches that are randomly selected in a given range. The maximum range is the system size for which the DNR problem is solved. Each bat is generated using an initially defined frequency, velocity, loudness and pulse rate. Each bat in the swarm represents a possible solution of the objective function. After initialization, fitness is evaluated for each bat. Each individual represents a combination of switches corresponding to some fitness value of objective function. Here objective function is formulated for power loss reduction. So the combination of switches in each bat has some value of power loss. The fitness value of each bat is calculated and the best one is saved.

Initially bats fly in random directions with some initial velocity, fixed frequency and loudness. The wavelength of emitted pulses is automatically adjusted using control equations. The frequency and rate of emitted pulses are selected in a defined range. Frequency is generated in between minimum and maximum defined frequency. The pulses emission rate is selected in the range of [0, 1]. Loudness varies from large to minimum value. Bats movement in the random direction is modeled by (4.1) to (4.3). For better exploration and exploitation of search space both loudness and pulse emission rate are varied during each iteration using (4.5) and (4.6). The loudness of each bat is decreased and pulse emission rate is increased after each bat updates its current position using equation (4.4). Rank the bats and update the current best solution. If the termination criterion is satisfied, it generates the final results. When this solution is implemented on network, it will give minimum power loss. This solution shows the optimal topology of the distribution network after the first iteration. The iterative process continues until the maximum iteration counts are exhausted.

4.4.3   Description of experiments

By interchanging sectionalizing and tie switches of any distribution network, a new topology can be obtained. In BA, the final optimal solution shows an optimal configuration of switches. This configuration causes reduction in power losses and node voltage deviations. Therefore, the optimal DNR problem is formulated to minimize power losses while maintaining better node voltage profiles in the distribution system. In order to limit the voltage deviation at different nodes, a hard node voltage constraint is used as desired. Similarly to check the current carrying capacities of distribution feeders, a feeder ampacity constraint is essential. The problem is structured as a single-objective constrained optimization problem where optimal radial topology is obtained. The objective function $f(x)$ is formulated as:

$$\text{Min } P_{Loss} = \sum_{j=1}^{N_L} R_j \frac{P_{j,k}^2 + Q_{j,k}^2}{V_{j,k}^2} \qquad (4.19)$$

This objective function is solved using the standard BA technique, subject to different network and operational constraints as defined by (4.20)–(4.23). Power flow constraints are defined by (4.20) that represents a set of power flow equations. Equation (4.21) shows a node voltage limit constraint. Node voltages V_n of all system buses must be kept within the minimum and maximum permitted limits i.e. $V_{\min}$ and $V_{\max}$, respectively, during the optimization process. The current flow in each branch must satisfy the rated ampacity of each branch as defined by the branch current constraint of (4.22). In addition to these operational constraints, the radial topology constraint is also defined to solve the optimal DNR problem. The reconfigured network topology must always be radial. It means it should not have any closed path or loop. The radiality constraint for the Yth radial topology is defined and shown by (4.23).

$$g(h) = 0 \qquad (4.20)$$

$$V_{\min} \leq V_n \leq V_{\max} \qquad \forall\, n \in N \qquad (4.21)$$

$$I_{j,k} \leq I_j^{\max} \qquad \forall\, j, k \qquad (4.22)$$

$$\Phi_j(Y) = 0 \qquad (4.23)$$

Here, N, N_L, $P_{j,k}$, $Q_{j,k}$, $V_{j,k}$, $I_{j,k}$, $I_j^{\max}$ denote the total number of nodes and branches in the system, real and reactive power flows in line j, node voltage at the sending end of line j, present and maximum current carrying capacity of line j respectively, all at load level k.

4.4.4 Results

The benchmark 33-bus [13] test distribution system is used to solve this optimal DNR problem. It is a 12.66 kV system with 32 sectionalizing and 5 tie-switches (i.e 33-37). The nominal active and reactive demand of this system is 3.175 MW and 2.3 MVAr, respectively. The swarm size and maximum iterations in BA are set at 10 and 200, respectively. The best result obtained after 100 independent trials of BA is presented in Table 4.1. The annual load profile is segmented into three different load levels, i.e., light, nominal and peak to show 50%, 100% and 160% of the nominal system loading, respectively. The corresponding load durations are taken as 2000, 5260 and 1500 hours, respectively. The objective function value is calculated at each load level separately. The table shows the optimal configuration obtained at all three load levels. The network performance is improved at each load level when compared with base configuration. After applying this solution an annual loss reduction of about 32.24% can be obtained. This shows that a significant reduction in annual energy losses can be obtained by optimal DNR.

TABLE 4.1

Optimal solution for DNR using standard bat algorithm.

Cases	Particulars	Load Levels			Annual energy loss reduction
		L	N	P	
Case 1	Base configuration	33, 34, 35, 36, 37	33, 34, 35, 36, 37	33, 34, 35, 36, 37	—
	P_{Loss}	47.06	202.67	575.39	
	$V_{min}(p.u.)\,\forall\,n$	0.9583	0.9131	0.8528	
Case 2	Optimal configuration	7, 9, 14, 32, 37	7, 9, 14, 32, 37	7, 9, 14, 32, 37	32.24%
	P_{Loss}	33.27	139.52	379.95	
	$\Delta P_{Loss}(\%)$	29.30	31.16	33.95	
	$V_{min}(p.u.)\,\forall\,n$	0.9698	0.9378	0.8968	

4.5 Conclusion

In this chapter the Bat Algorithm is presented in detail. Some modifications suggested by different authors to improve the performance of BA are demonstrated and discussed. Finally, the application of the algorithm to a complex combinatorial problem of optimal DNR is shown. Power losses at different load levels are calculated and compared with the base case. The application results of the algorithm show that there is a significant improvement in the desired objectives.

Acknowledgement

This work was supported by the Engineering and Physical Sciences Research Council (EPSRC) of United Kingdom (Reference Nos.: EP/R001456/1 and EP/S001778/1).

References

1. X. S. Yang, Bat algorithm for multi-objective optimization, International Journal of Bio-Inspired Computation (2011) vol. 3(5), pp. 267-274.

2. T. Niknam, R. A. Abarghooee, M. Zare, and B. B. Firouzi, Reserve constrained dynamic environmental/economic dispatch: a new multiobjective self-adaptive learning Bat Algorithm, IEEE Systems Journal (2013) vol. 7(4) pp. 763-776.

3. N. Kanwar, N. Gupta, K. R. Niazi, A. Swarnkar, and R. C. Bansal, Multi-objective optimal DG allocation in distribution networks using bat algorithm, 3rd Southern African Solar Energy Conference (SASEC2015), Kruger National Park, South Africa, 11th-13th May 2015.

4. N. Kanwar, N. Gupta, K.R. Niazi, and A. Swarnkar, Optimal distributed generation allocation in radial distribution systems considering customer-wise dedicated feeders and load patterns, Journal of Modern Power and Clean Energy (2015) vol. 3(4), pp. 475-484.

5. Z. Cui, Y. Cao, X. Cai, J. Cai, and J. Chen, Optimal LEACH protocol with modified bat algorithm for big data sensing systems in Internet of Things, Journal of Parallel and Distributed Computing (2019) vol. 132, pp. 217-229.

6. I. Fister Jr., S. Fong, J. Brest, and I. Fister, Towards the self-adaptation in the bat algorithm, in: Proceedings of the 13th IASTED International Conference on Artificial Intelligence and Applications, Innsbruck, Austria, pp. 400-406, 2014.

7. S. Das, S.S. Mullick, and P.N. Suganthan, Recent advances in differential evolution- An updated survey, Swarm and Evolutionary Computation. (2016) vol. 27 pp. 1-30.

8. W. C. Hong, M. W. Li, J. Geng, and Y. Zhang, Novel chaotic bat algorithm for forecasting complex motion of floating platforms, Applied Mathematical Modelling (2019) vol. 72 pp. 425-443.

9. S. Lyu, Z. Li, Y. Huang, J. Wang, and J. Hu, Improved self-adaptive bat algorithm with step-control and mutation mechanisms, Journal of Computational Science (2019) vol. 30, pp. 65-78.

10. A. Al-Muraeb and H. A. Aty-Zohdy, Optimal design of short fiber bragg grating using bat algorithm with adaptive position update, IEEE Photonics Journal (2016), vol. 8(1).

11. C. K. Ng, C. H. Wu, W. H. Ip, and K. L. Yung, A smart bat algorithm for wireless sensor network deployment in 3-D environment, IEEE Communications Letters (2018) vol. 22(10), pp. 2120-2123.

12. D. Igrec, A. Chowdhury, B. Stumberger, and A. Sarjas, Robust tracking system design for a synchronous reluctance motor-SynRM based on a new modified bat optimization algorithm, Applied Soft Computing (2018) vol. 69, pp. 568-584.

13. M.E. Baran and F.F. Wu, Network reconfiguration in distribution system for loss reduction and load balancing. IEEE Transaction on Power Delivery (1989) vol. 4(2), pp. 1401-1407.

5

Cat Swarm Optimization - Modifications and Application

Dorin Moldovan
Department of Computer Science
Technical University of Cluj-Napoca, Romania

Adam Slowik
Department of Electronics and Computer Science
Koszalin University of Technology, Koszalin, Poland

Viorica Chifu
Department of Computer Science
Technical University of Cluj-Napoca, Romania

Ioan Salomie
Department of Computer Science
Technical University of Cluj-Napoca, Romania

CONTENTS

5.1   Introduction

Cat Swarm Optimization (CSO) was introduced in [1] and it is a part of a larger family of algorithms that are nature inspired. This family of algorithms includes algorithms such as Particle Swarm Optimization (PSO) [2], Cuckoo Search (CS) [3], Bat Algorithm (BA) [4], Ant Colony Optimization (ACO) [5], Elephant Herding Optimization (EHO) [6] and Lion Optimization Algorithm (LOA) [7]. The CSO algorithm is inspired by the behavior of the cats in nature and it is an algorithm that can be used for solving complex engineering problems in which the search space has many dimensions. In the case of the CSO algorithm each solution of the optimization problem is represented by a cat that has a velocity and a position and the cats are characterized by two types of behavior namely, seeking mode and tracing mode. The objective of the algorithm is to determine the best cat according to a fitness function that depends on the optimization problem that will be solved. The main idea of optimization in the case of the CSO algorithm is the movement of the cats towards a local best *Lbest* in the case of the local version of the algorithm or towards a global best *Gbest* in the case of the global version of the algorithm. In literature there are many varieties of CSO algorithm such as: Crazy Cat Swarm Optimization (CCSO) [8], Harmonious Cat Swarm Optimization (HCSO) [9], Binary Cat Swarm Optimization (BCSO) [10] and Parallel Cat Swarm Optimization (PCSO) [11]. The chapter is organized as follows: in Section 5.2 is given a short presentation of the CSO algorithm, of the global version of CSO (GCSO) and of the local version of CSO (LCSO), in Section 5.3 are illustrated modifications of CSO such as velocity clamping, inertia weight, mutation operators, acceleration coefficient c_1 and adaptation for diets recommendation, in Section 5.4 is presented the application of the CSO algorithm for generation of diets and in Section 5.5 are presented the main conclusions.

5.2   Original CSO algorithm in brief

The original version of the CSO algorithm in the global version (GCSO) and in the local version (LCSO) can be represented using the pseudo-code from Algorithm 6. The code that is used only for the local version is underlined with a dotted line, while the code that is used only for the global version is underlined with a solid line.

Algorithm 6 Pseudo-code of the original CSO.

Input M - the number of dimensions, $\left[P_{i,j}^{min}, P_{i,j}^{max}\right]$ - range of variability for i-th cat and j-th dimension, $MR, SMP, SRD, CDC, SPC, N$ - the total number of cats, max - maximum number of iterations, c_1 - a constant for updating the velocity of the cats in tracing mode, number of returned results $Re \in [1, N]$, the topology and the number of nearest neighbors Ne

Output *Gbest* - the best position achieved by a cat; select Re the best cats among the *Lbest* cats and return these cats as results

1: determine the M-th dimensional objective function $OF(.)$
2: randomly create the population of cats $X_i(i = 1, 2, ..., N)$
3: create the M-dimensional *Gbest* vector
4: **while** stopping criterion not satisfied or $I < I_{max}$ **do**
5:　　set the cats in tracing mode or seeking mode according to MR
6:　　**for** each i-th cat X_i **do**
7:　　　create a $Xbest_i$ cat equal to i-th cat X_i
8:　　　create a $Lbest_i$ cat for each cat X_i
9:　　　create a velocity V_i for each cat X_i
10:　　　evaluate the cat X_i using the $OF(.)$ function
11:　　**end for**
12:　　assign the best cat X_i to the *Gbest*; for each cat $Xbest_i$ assign the best cat among Ne nearest neighbors of cat X_i
13:　　**for** $i = 1 : N$ **do**
14:　　　**if** X_i is in seeking mode **then**
15:　　　　create SMP copies
16:　　　　update the position of each copy using the formula
17:　　　　$X_{cn} = X_c \times (1 \pm SRD \times R)$
18:　　　　pick randomly a position to move to from the set of SMP copies
19:　　　**else**
20:　　　　update the velocity of the cat using the formula
21:　　　　$v_{i,d} = v_{i,d} + R \times c_1 \times (Gbest_d - X_{i,d})$
22:　　　　$v_{i,d} = v_{i,d} + R \times c_1 \times (Lbest_{i,d} - X_{i,d})$
23:　　　　update the position of the cat using the formula
24:　　　　$X_{i,d,new} = X_{i,d,old} + v_{i,d}$
25:　　　**end if**
26:　　　select the best cat among Ne nearest neighbors of cat X_i and assign it to the cat T
27:　　　　**if** $OF(T)$ better than $OF(Lbest_i)$ **then**
28:　　　　　assign the cat T to the cat $Lbest_i$
29:　　　　**end if**
30:　　**end for**
31:　　select the best cat and assign it to cat T
32:　　**if** $OF(T)$ better than $OF(Gbest)$ **then**
33:　　　assign the cat T to the cat *Gbest*
34:　　**end if**
35: **end while**
36: return the *Gbest* as a result; the best Re cats among the *Lbest* cats

---

5.2.1 Description of the original CSO algorithm

The pseudo code of the original CSO algorithm is presented in Algorithm 6 and in this subsection the algorithm is described in more detail. The inputs of the algorithm are M - the number of dimensions, $\left[P_{i,j}^{min}, P_{i,j}^{max}\right]$ - the range of variability for the i-th cat for the j-th dimension, MR - the mixture ratio, a numerical value that indicates how many cats are in tracing mode and how many cats are in seeking mode, SMP - the seeking memory pool, SRD - the seeking range of the selected dimension, CDC - the count of dimensions to change, SPC - the self-position consideration, N - the total number of cats, max - the maximum number of iterations and c_1 - a constant that is used for updating the velocity of the cats. In addition to these inputs, the local version of the algorithm uses as inputs the number of returned results $Re \in [1, N]$, the topology and Ne which is the number of nearest neighbors. The output of the algorithm in the case of the global version is represented by $Gbest$, while the output of the algorithm in the case of the local version is represented by the best Re cats among the $Lbest$ cats.

In step 1 of the algorithm is created the M-th dimensional objective function $OF(.)$ and in step 2 of the algorithm the initial population of N cats is created randomly. For the GCSO algorithm in step 3 is created the M-dimensional $Gbest$ vector. While the stopping criterion that is set initially is not satisfied or the number of iterations is less than a threshold max, a number of steps is repeated. In step 5 of the algorithm the cats are set either in tracing mode or in seeking mode according to a numerical value MR which is the mixture ratio. Then for each cat X_i the following operations are performed: an equal copy $Xbest_i$ is created, in the LCSO version of the algorithm a $Lbest_i$ cat is created, a velocity vector V_i is initialized and the cat X_i is evaluated using the objective function $OF(.)$. In step 12 of the algorithm, in the GCSO version the best cat X_i according to the value returned by $OF(.)$ is assigned to $Gbest$ while in the LCSO version of the algorithm for each cat $Xbest_i$ the best cat among the Ne nearest neighbors of cat X_i is assigned.

The steps from step 13 to step 30 are executed for each cat X_i. If the cat X_i is in seeking mode then in step 15 of the algorithm SMP copies of that cat are created, the position of each copy is updated using a formula that considers the SRD - the seeking range of the selected dimension and R a random numerical value from the interval $[0, 1]$ and in step 18 a position to move to is picked randomly from the set of SMP copies. Otherwise, if the cat is in tracing mode, in step 20 the velocity of the cat is updated using a different formula depending on whether the algorithm is GCSO or LCSO and in step 23 the position of the cat is updated using a formula that considers the computed velocity. If the algorithm is LCSO then steps 26-29 are performed: the best cat from the Ne nearest neighbors is selected and assigned to cat T and if the value of $OF(T)$ is better than the value of $OF(Lbest_i)$, the new value of $Lbest_i$ becomes T. The steps 31-34 are performed only in the GCSO version of the algorithm: the best cat is assigned to T and if $OF(T)$ is better than $OF(Gbest)$ then cat T is assigned to $Gbest$. Finally in step 36 of the algorithm, in the GCSO version the $Gbest$ is returned as a result and in the $LCSO$ version the best Re cats among the $Lbest$ cats are returned.

5.3 Modifications of the CSO algorithm

5.3.1 Velocity clamping

In the original version of the CSO algorithm the velocity explodes quickly to large values. One solution for this problem is represented by the introduction of limited step sizes as shown in the following equation:

$$V_{i,j} = \begin{cases} V_{i,j} & \text{if } |V_{i,j}| < V_j^{max} \\ V_j^{max} & \text{if } |V_{i,j}| \geq V_j^{max} \end{cases} \tag{5.1}$$

where V_j^{max} is the maximum value of velocity for the j-th decision variable. This value can be computed using the formula:

$$V_j^{max} = \delta \times \left(P_j^{max} - P_j^{min} \right) \tag{5.2}$$

where $\delta \in [0, 1]$ is a numerical value that can be chosen by trial-and-error and it usually depends on the optimization problem [12], P_j^{max} is the maximum value from the search domain for the j-th decision variable and P_j^{min} is the minimum value from the search domain for the j-th decision variable.

5.3.2 Inertia weight

The application of inertia weight in the case of CSO is inspired by the application of inertia weight in the case of Particle Swarm Optimization (PSO) [13]. In literature there are various types of inertia weights such as the constant inertia weight [14], the random inertia weight [15] and the adaptive inertia weight [16]. The equation for velocity update in the case of GCSO when inertia weight is used is:

$$V_{i,j} = \omega \times V_{i,j} + R \times c_1 \times (Gbest_j - X_{i,j}) \tag{5.3}$$

and the equation for velocity update in the case of LCSO when inertia weight is used is:

$$V_{i,j} = \omega \times V_{i,j} + R \times c_1 \times (Lbest_{i,j} - X_{i,j}) \tag{5.4}$$

where $V_{i,j}$ is the value of velocity of the i-th cat for the j-th dimension, ω is the inertia weight, R is a random numerical value from the interval $[0, 1]$, $Gbest_j$ is the position of the global best cat for the j-th dimension, $Lbest_{i,j}$ is the position of the local best for the i-th cat and the j-th dimension and $X_{i,j}$ is the position of the i-th cat for the j-th dimension. The value of the inertia factor ω is usually in the interval $[0.4, 0.9]$ and there are two major cases: when $\omega \geq 1$ the swarm of cats diverges and when $0 < \omega < 1$ the cats decelerate. The value of ω depends on the optimization problem that is solved.

5.3.3 Mutation operators

The mutation operators [17] are used in order to improve the performance of the CSO algorithm and to escape from the local minima. In some cases the global best cat is mutated while in other cases the local best cat is mutated. The mutation operator that is presented in this chapter is adapted after the one presented in [18]. First, a weight vector is computed using the following formula:

$$W_i = \frac{\sum_{j=1}^{N} V_{j,i}}{N} \tag{5.5}$$

where N is the size of the population, $V_{j,i}$ is the i-th velocity value of the j-th cat from the swarm and W_i is the value at position i of the weights vector W. The value of W_i is from an interval $[-W_{max}, W_{max}]$ and the value of W_{max} is usually equal with 1. Second, the value of the global best cat is mutated using the following formula:

$$Gbest'_i = Gbest_i + W_i \times \mathcal{N}(X_{min}, X_{max}) \tag{5.6}$$

where $Gbest_i$ is the i-th value of the $Gbest$ vector, $Gbest'_i$ is the i-th value of the mutated $Gbest_i$ vector, W_i is the value at the i position of the weights vector and $\mathcal{N}(X_{min}, X_{max})$ is a random number from a Cauchy distribution that has the scale parameter equal with 1 where X_{min} and X_{max} are the minimum and the maximum values that can be taken by the positions of the cats for each dimension.

Using a similar approach, the Cauchy operator can be used in order to mutate the value of the local best cat using the following formula:

$$Lbest'_i = Lbest_i + W_i \times \mathcal{N}(X_{min}, X_{max}) \tag{5.7}$$

where $Lbest_i$ is the i-th value of the $Lbest$ vector and $Lbest'_i$ is the i-th value of the mutated $Lbest$ vector.

5.3.4 Acceleration coefficient c_1

The coefficient c_1 is named acceleration coefficient because it is responsible for the social part of the algorithm. We can introduce the adaptive change of the acceleration coefficient using the following formula:

$$c_1 = \left(c_1^{min} - c_1^{max}\right) \times \frac{t}{t^{max}} + c_1^{max} \tag{5.8}$$

where c_1^{min} is the minimum value that can be taken by c_1, c_1^{max} is the maximum value that can be taken by c_1, t is the current number of generations and belongs to the interval $[0, t^{max}]$ and t^{max} is the maximum number of generations. When $t = 0$ the value of c_1 is equal with c_1^{max} and when $t = t^{max}$ the value of c_1 is equal with c_1^{min}.

5.3.5 Adaptation of CSO for diets recommendation

In this subsection is described how the classical CSO algorithm can be adapted for discrete optimization problems, in particular for recommendation of diets. Discrete optimization problems are a special subset of the optimization problems in which the search space is discrete, or in other words for each dimension there is a finite set of states. In the approach that is used in this chapter the search space is represented by dishes that have well known nutritional values.

As in the classical version of the CSO algorithm the cats are initialized to random values. Those values are in intervals described by the minimum and the maximum values of the nutritional properties of the dishes from the search space. Each cat corresponds to a diet that is composed from a number of dishes and in order to determine the dishes from the search space that are the closest to the numerical values that describe the positions of the cats, the squared Euclidean distance is applied.

If the position vector of a cat is $\{(x_{1,1}, ..., x_{1,M}), ..., (x_{K,1}, ..., x_{K,M})\}$ where K is the number of dishes and M the number of nutritional values then the dish from the search space that corresponds to the k-th dish of the cat, where $k = 1, ..., K$, is the one for which the value of the function presented below has the minimum value:

$$F(cat, k, dish) = \sum_{i=1}^{M} (x_{k,i} - dish(nutrient_i))^2 \tag{5.9}$$

where $dish(nutrient_i)$ is the value of the i-th nutrient of the dish. The cat from the function $F(cat, k, dish)$ is from the swarm of cats and the dish is from the search space of dishes.

5.4 Application of CSO algorithm for recommendation of diets

5.4.1 Problem description

The generation of diets is a complex discrete optimization problem due to the fact that the search space is very large and for each person the recommended nutritional values for a day are different. These recommended nutritional values are computed by taking into consideration several properties of the person such as height, weight, age, food preferences and so on. Just to create an image of how many different types of foods exist, in [19] is listed information about approximately 250,000 foods. Additionally the world population is projected to reach 9.8 billion people by 2050 and due to the last technological advancements the people are expected to live longer than ever before. The variety of foods is also expected to be greater in the next years due to technological advancements and one of the major consequences is represented by the fact that the people are able to select diets that fit the best for their particular needs such as health improvement, weight loss or muscle gain.

5.4.2   How can the CSO algorithm be used for this problem?

In this subsection we present how the CSO algorithm can be used for recommendation of diets. The recommendation of diets using CSO was treated before by us in [20] but compared to the approach presented in that article, in this chapter we perform the following major changes: (1) we use another dataset of dishes which contains dishes for different types of diets such as diabetic and Mediterranean, (2) we consider only the nutritional values of the dishes when computing the fitness values of the diets and (3) we also consider how the tuning of the CSO parameters might influence the homogeneity of the diets. In this chapter we consider that a diet consists of four principal meals in a day namely, breakfast, lunch, dinner and supper. In addition each meal has three dishes: a main dish, a secondary dish and a dessert. In Figure 5.1 is presented the template of a diet. As can be seen in the figure a cat corresponds to a diet for one day and consists of one breakfast, one lunch, one dinner and one supper.

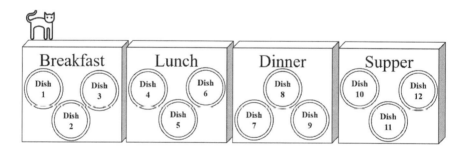

FIGURE 5.1
Illustrative template of a diet for the recommendation of diets optimization problem.

A brute force approach that considers all combinations of dishes is clearly too expensive in terms of computational resources such as time and memory allocation. The application of CSO for recommendation of diets is a great fit because it generates a near-optimal solution very fast, it produces different results at each run of the algorithm, thus ensuring the diversity of the diets, and it also has two major types of searching in the space of solutions namely, the seeking mode and the tracing mode, which overlap very well with the situations when the algorithm is applied for searching homogeneous diets and when the algorithm is applied for searching heterogeneous diets. By tuning the Mixture Ratio (MR) parameter, a number that indicates how many cats are in tracing mode and how many cats are in seeking mode, CDC which is the number of dimensions that will be mutated in the seeking mode and SPC, a flag used in seeking mode which specifies if the current position of the cat will be considered or not, it is possible to vary the degree of homogeneity of the generated recommendations of diets.

The mapping between the concepts used by CSO and the concepts used in the optimization problem is presented in Table 5.1.

TABLE 5.1

Mapping Between CSO Concepts and Optimization Problem Concepts

CSO Concept	Optimization Problem Concept
cat	a diet that consists of breakfast, lunch, dinner and supper
position	a vector that consists of $12 \times 4 = 48$ values
velocity	a vector that consists of $12 \times 4 = 48$ values
topology	ring topology
objective function	formula 5.10

The concepts are described in more detail next. A cat is mapped to a diet that has four main components namely, a breakfast, a lunch, a dinner and a supper, the position of the cat is represented by a vector that has 48 values (12 dishes × 4 nutritional values per dish), the velocity of the cat is represented similarly by a vector that has 48 values and the topology that is used in the local version of the algorithm is the ring topology which considers the left and the right neighbor cats. We consider that each dish has values for the following properties: (1) proteins (g), (2) lipids (g), (3) carbohydrates (g) and (4) energy (cal). The equation of the objective function is:

$$OF(cat) = OF(cat_{breakfast}) + OF(cat_{lunch})$$
$$+ OF(cat_{dinner}) + OF(cat_{supper}) \quad (5.10)$$

where:

$$OF(cat_{breakfast}) = \sum_{i=1}^{3} OF\left(dish_{breakfast,i}\right) \quad (5.11)$$

$$OF(cat_{lunch}) = \sum_{i=1}^{3} OF\left(dish_{lunch,i}\right) \quad (5.12)$$

$$OF(cat_{dinner}) = \sum_{i=1}^{3} OF\left(dish_{dinner,i}\right) \quad (5.13)$$

$$OF(cat_{supper}) = \sum_{i=1}^{3} OF\left(dish_{supper,i}\right) \quad (5.14)$$

and the objective function for a dish is:

$$OF(dish) = \sum_{i=1}^{4} (dish(nutrient_i) - dish(optimal_{value}(nutrient_i)))^2 \quad (5.15)$$

The proposed method presented in this chapter is based on the local version of the CSO algorithm and it consists of the following steps.

Step 1 The initial population of cats X which consists of N cats X_i with $i = 1, ..., N$ is created randomly. Each cat is a D dimensional vector where $D = 12 \times 4 = 48$. For each i-th cat X_i the vectors $Xbest_i$, $Lbest_i$ and V_i are created and then the vector $Gbest$ is created for the whole swarm of cats. Vector $Xbest_i$ indicates the best position of the cat X_i in the search space among the positions that were obtained so far. Vector $Lbest_i$ determines the best position reached by another cat among the NN nearest neighbors of cat X_i found until now. Vector V_i represents the velocity of cat X_i. $Gbest$ determines the global best solution that was found by the CSO algorithm. In the beginning the vector X_i is assigned to $Xbest_i$ vector for the i-th cat, and the vectors $Lbest_i$, V_i and $Gbest$ are initialized to zero.

Step 2 Each cat X_i is evaluated using the objective function $OF(.)$ described by the formula 1.10 and the objective of the CSO algorithm is to minimize the value of the objective function $OF(.)$. When the value of the computed $OF(.)$ function for a cat X_i is lower than the value of $Xbest_i$, the values from cat X_i are assigned to vector $Xbest_i$.

Step 3 For each cat X_i the best local cat $Lbest_i$ is selected from the NN nearest neighbors of that cat X_i. The similarity between pairs of cats is determined using the values returned by $OF(.)$ and two cats are more similar when their $OF(.)$ values are closer.

Step 4 In this step the positions of the cats are updated using the equations that correspond either to the seeking mode or to the tracing mode. For the seeking mode SMP copies are created, the position of each copy is updated using the formula:

$$X_{cn} = X_c \times (1 \pm SRD \times R) \quad (5.16)$$

and the position to move to is selected randomly from the set of SMP copies. For the tracing mode the velocities of the cats are updated using the formula:

$$v_{i,d} = v_{i,d} + R \times c_1 \times (Lbest_{i,d} - X_{i,d}) \quad (5.17)$$

and the positions of the cats are updated using the formula:

$$X_{i,d,new} = X_{i,d,old} + v_{i,d} \quad (5.18)$$

In the formulas presented above the parameters have the following significance: SMP is the seeking memory pool, X_{cn} is the new position of the copy, X_c is the position of the copy, SRD is the seeking range of the selected dimensions, R is a random number from the interval $[0, 1]$, $v_{i,d}$ is the value of the velocity of the i-th cat for the d-th dimension, c_1 is a constant, $Lbest_{i,d}$ is the value of the local best position of the i-th cat for the d-th dimension, $X_{i,d}$ the value of the position of the i-th cat for the d-th dimension, $X_{i,d,new}$ is the new position of the cat i for the d-th dimension, $X_{i,d,old}$ is the old position of the cat i for the d-th dimension and $v_{i,d}$ is the value of velocity of the i-th cat for the d-th dimension.

Step 5 The worst cat from the swarm is replaced by a cat that is created randomly. The worst cat is the one for which the value that is returned by $OF(.)$ has the highest value. The scope of this operation is to increase the convergence of the algorithm.

Step 6 If a better cat is found considering the value returned by $OF(.)$ then the value of the *Gbest* cat is updated with that cat. If the number of the current generation is less than the number of generations given as input then the algorithm jumps to the first step, otherwise the loop stops and the algorithm continues with the next step.

Step 7 The algorithm returns the diet that corresponds to the *Gbest* cat. The diet that corresponds to the *Gbest* cat is calculated using the procedure presented in more detail in the subsection that describes how CSO is adapted for diets recommendation.

5.4.3 Description of experiments

In the approach presented in this chapter is considered a dataset that contains dishes for two types of diets namely, the diabetic diet and Mediterranean diet. Some examples of dishes are Zucchini Saute and Italian Stewed Tomatoes. The main challenge is represented by the initial configuration of the parameters of the CSO algorithm in order to be applied successfully on the discrete optimization problem described in this chapter. Since some parameters depend very much on the dataset that is used as support, we present justifications for each choice of parameters.

We consider that the number of dimensions of the search space is equal to $12 \times 4 = 48$ (12 dishes and 4 nutritional values for each dish). The number of cats N is equal to 10 and the value of MR is equal to 0.3, thus at each generation of the algorithm 7 cats are in seeking mode and 3 cats are in tracing mode. The number of generations is chosen as 100 after a series of experiments, but the optimal value of this number can be determined experimentally if we consider as stopping condition the state when the generation of the algorithm for which the absolute value of the difference between the global best at that generation and the global best at the previous p-th generation is less than a threshold ϵ.

The optimal values for p and ϵ should also be determined experimentally. The value of p is from the set $\{1, 2, 3, ...\}$ and the value of ϵ should be a very small numerical value. A possible approach to determine the value of ϵ is to consider the values from the set $\{0.1, 0.01, 0.001, 0.0001, 0.00001, 0.000001, ...\}$.

The value of SMP is 5 and that means that 5 copies are generated for each cat in the seeking mode, the value of SRD is equal to 0.2, the value of CDC is $2 \times 11 = 22$ and the value of c_1 is 0.5.

The range of variability $\left[P_{i,j}^{min}, P_{i,j}^{max}\right]$ for the i-th cat for each j-th dimension corresponds to the minimum and the maximum values of the nutritional properties that correspond to that dimension. In addition we consider that the range of variability of the velocities is $\left[-2 \times P_{i,j}^{max}, 2 \times P_{i,j}^{max}\right]$.

5.4.4   Results obtained

In this subsection are presented experimental results for two types of diets namely, a diabetic diet and a Mediterranean diet. We chose these particular diets because they consider the health condition and this dimension will have a great importance in the future when selecting diets as the world population will increase significantly in the next years. For example in the case of the diabetic diet the prediction of diet recommendations is a hot research topic that was treated in literature using approaches such as type 2 fuzzy logic [21] and ontology and semantic matching [22]. In the case of the Mediterranean diet, some ingredients such as extra-virgin olive oil lower the incidence of Alzheimer's disease [23]. The optimal nutritional values for each type of diet are computed for a single day.

5.4.4.1   Diabetic diet experimental results

We consider the profile of a woman named Anna Smith that has diabetes. She is 60 years old, has a height of 179 centimeters and a weight of 75 kilograms. The recommended nutritional values for one day are computed considering the age, the height and the weight, and their values are: 136 proteins (g), 65 lipids (g), 203 carbohydrates (g) and 1936 energy (calories).

In Table 5.2 are listed the minimum and the maximum values of the nutrients of the dishes that compose the diabetic diet and in addition the optimal values of these nutrients for a day for Anna Smith.

The dishes that correspond to the global best cat are summarized in Table 5.3.

TABLE 5.2

Minimum and Maximum Values of the Nutrients of the Diabetic Diet Dishes and the Optimal Values of the Nutrients for a Day for Anna Smith.

Nutrient	Minimum Value / Dish	Maximum Value / Dish	Optimal Value / Day
proteins (g)	0.3	30	136
lipids (g)	0.1	21.3	65
carbohydrates (g)	6.6	36.7	203
energy (cal)	27	363	1936

TABLE 5.3

Diabetic Diet Dishes Recommendation for Anna Smith.

Dish	Proteins (g)	Lipids (g)	Carbs (g)	Energy (cal)
Breakfast				
Shiraz Salad	2.2	0.4	14.6	61.0
Grilled Chicken Citrus Salad	30.0	2.2	26.1	238.0
White Beans and Peppers	8.5	1.8	26.6	150.0
Lunch				
Banana Oat Energy Bars	3.6	4.0	20.0	124.0
Easy Roasted Peppers	1.8	0.5	10.8	55.0
Bean Soup With Kale	11.0	2.5	31.0	182.0
Dinner				
Curried Israeli Couscous	4.9	3.7	29.7	174.0
Curried Cottage Fries	3.8	4.1	28.5	162.0
Meyer Lemon Avocado Toast	3.6	1.2	11.8	72.0
Supper				
Black Bean Spread	6.8	0.7	16.3	96.0
Black Beans and Rice	6.3	0.9	27.1	140.0
Chickpea and Couscous Delight	5.4	0.9	29.3	147.0
Total	87.89	22.9	271.8	1601.0

5.4.4.2   Mediterranean diet experimental results

We consider the profile of a man named John Smith who is 80 years old, has a height of 176 centimeters and a weight of 85 kilograms. The corresponding nutritional values for one day are: 154 proteins (g), 71 lipids (g), 220 carbohydrates (g) and 2138 energy (cal).

In Table 5.4 are listed the minimum and the maximum values of the nutrients of the dishes that compose the Mediterranean diet and in addition the optimal values of these nutrients for a day for John Smith.

TABLE 5.4

Minimum and Maximum Values of the Nutrients of the Mediterranean Diet Dishes and the Optimal Values of the Nutrients for a Day for John Smith.

Nutrient	Minimum Value / Dish	Maximum Value / Dish	Optimal Value / Day
proteins (g)	1.6	41.7	154
lipids (g)	0.5	77.8	71
carbohydrates (g)	0.6	95	220
energy (cal)	49	853	2138

TABLE 5.5

Mediterranean Diet Dishes Recommendation for John Smith.

Dish	Proteins (g)	Lipids (g)	Carbs (g)	Energy (cal)
Breakfast				
Chef John's Tzatziki Sauce	2.2	3.4	2.5	49.0
Real Hummus	1.6	2.5	6.8	54.0
Kefir Yogurt (Sana)	5.9	2.9	8.4	84.0
Lunch				
Mediterranean Roast Chicken	39.8	26.6	81.6	724.0
Mediterranean Kale	4.6	3.2	14.5	91.0
Easy Greek Yogurt	7.9	2.3	10.7	95.0
Dinner				
Whole Wheat Pita Bread	7.4	1.1	17.1	101.0
Slow Cooker Mediterranean Stew	3.4	0.5	30.5	122.0
Cauliflower Tabbouleh	4.9	7.1	15.1	128.0
Supper				
Air Fryer Potato Wedges	2.3	5.3	19.0	129.0
Salmon Tartare	14.3	7.7	0.6	133.0
Asparagus Sformato	9.6	9.8	4.0	139.0
Total	103.9	72.4	210.79	1849.0

The dishes that correspond to the global best cat are summarized in Table 5.5.

5.5 Conclusions

In this paper the CSO algorithm was presented in detail, both in local version and in global version. Some modifications of the CSO algorithm such as velocity clamping, inertia weight, mutation operators, acceleration coefficient c_1 and adaptation for diets recommendation were demonstrated and discussed. Finally the CSO algorithm was applied to a real world problem namely, the recommendation of diets. Using the CSO algorithm we obtain near optimal results fast and we can also control the homogeneity of the recommended diets by varying several parameters such as MR which is the mixture ratio or CDC which is the count of dimensions to change. In this chapter at the beginning of the algorithm the cats are initialized randomly but as future research work we would like: (1) to consider cats that are initialized to specific diets that are recommended by nutritionists and to search similar diets in the neighborhood of those cats, (2) to parallelize the algorithm using big data frameworks such as Apache Spark [24] and (3) to include more parameters when recommending diets such as the food preferences, the quality of the foods and other characteristics of the foods like smell, taste or price.

References

1. S.-C. Chu, P.-W. Tsai, J.-S. Pan. "Cat Swarm Optimization" in *Yang Q., Webb G. (eds) PRICAI 2006: Trends in Artificial Intelligence. PRICAI 2006. Lecture Notes in Computer Science*, vol. 4099, 2006, pp. 854-858.

2. J. Kennedy, R. Eberhart. "Particle Swarm Optimization" in *Proceedings of ICNN'95 - International Conference on Neural Networks*, vol. 4, 1995, pp. 1942-1948.

3. X.-S. Yang, S. Deb. "Cuckoo Search via Levy flights" in *2009 World Congress on Nature & Biologically Inspired Computing (NaBIC)*, 2009, pp. 210-214.

4. X.-S. Yang. "A New Metaheuristic Bat-Inspired Algorithm" in *Gonzalez J.R., Pelta D.A., Cruz C., Terrazas G., Krasnogor N. (eds) Nature Inspired Cooperative Strategies for Optimization (NICSO 2010). Studies in Computational Intelligence*, 2010, pp. 65-74.

5. M. Dorigo, M. Birattari, T. Stutzle. "Ant colony optimization" in *IEEE Computational Intelligence Magazine*, vol. 1, 2006, pp. 28-39.

6. G.-G. Wang, S. Deb, L. dos S. Coelho. "Elephant Herding Optimization" in *2015 3rd International Symposium on Computational and Business Intelligence (ISCBI)*, 2015, pp. 1-5.

7. M. Yazdani, F. Jolai. "Lion Optimization Algorithm (LOA): A nature-inspired metaheuristic algorithm" in *Journal of Computational Design and Engineering*, vol. 3, 2016, pp. 24-36.

8. A. Sarangi, S. K. Sarangi, M. Mukherjee, S. P. Panigrahi. "System identification by Crazy-cat swarm optimization" in *2015 International Conference on Microwave, Optical and Communication Engineering (ICMOCE)*, 2015, pp. 439-442.

9. K. C. Lin, K. Y. Zhang, J. C. Hung. "Feature Selection of Support Vector Machine Based on Harmonious Cat Swarm Optimization" in *2014 7th International Conference on Ubi-Media Computing and Workshops*, 2014, pp. 205-208.

10. B. Crawford, R. Soto, N. Berrios, F. Johnson, F. Paredes. "Binary cat swarm optimization for the set covering problem" in *2015 10th Iberian Conference on Information Systems and Technologies (CISTI)*, 2015, pp. 1-4.

11. P.-W. Tsai, J.-S. Pan, S.-M. Chen, B.-Y. Liao, S.-P. Hao. "Parallel Cat Swarm Optimization" in *2008 International Conference on Machine Learning and Cybernetics*, 2008, pp. 3328-3333.

12. R. A. Jamous, A. A. Tharwat, E. El-Seidy, D. I. Dayoum. "Modifications of Particle Swarm Optimization Techniques and Its Application on Stock Market: A Survey" in *(IJACSA) International Journal of Advanced Computer Science and Applications*, vol. 6, no. 3, 2015, pp. 99-108.

13. J. C. Bansal, P. K. Singh, M. Saraswat, A. Verma, S. S. Jadon, A. Abraham. "Inertia Weight strategies in Particle Swarm Optimization" in *Proc. of the 2011 Third World Congress on Nature and Biologically Inspired Computing*, 2011, pp. 633-640.

14. Y. Shi, R. Eberhart. "A modified particle swarm optimizer" in *1998 IEEE International Conference on Evolutionary Computation Proceedings. IEEE World Congress on Computational Intelligence (Cat. No.98TH8360)*, 1998, pp. 69-73.

15. R. C. Eberhart, Y. Shi. "Tracking and optimizing dynamic systems with particle swarms" in *Proceedings of the 2001 Congress on Evolutionary Computation (IEEE Cat. No.01TH8546)*, 2001, pp. 94-100.

16. A. Nikabadi, M. Ebadzadeh. "Particle swarm optimization algorithms with adaptive Inertia Weight : A survey of the state of the art and a Novel method" in *IEEE Journal of Evolutionary Computation*, 2008.

17. M. Imran, R. Hashim, N. E. A. Khalid. "An overview of particle swarm optimization variants" in *Procedia Engineering*, vol. 53, 2013, pp. 491-496.

18. H. Wang, C. Li, Y. Liu, S. Zeng. "A hybrid particle swarm algorithm with Cauchy mutation" in *2007 IEEE Swarm Intelligence Symposium*, 2007, pp. 356-360.

19. United States Department of Agriculture, Agricultural Research Service, USDA Food Composition Databases, https://ndb.nal.usda.gov

20. D. Moldovan, P. Stefan, C. Vuscan, V. R. Chifu, I. Anghel, T. Cioara, I. Salomie. "Diet generator for elders using cat swarm optimization and wolf search" in *International Conference on Advancements of Medicine and Health Care through Technology*, 2016, pp. 238-243.

21. H. A. Mohammed, H. Hagras. "Towards developing Type 2 fuzzy logic diet recommendation system for diabetes" in *2018 10th Computer Science and Electronic Engineering (CEEC)*, 2018, pp. 56-59.

22. A. Arwan, M. Sidiq, B. Priyambadha, H. Kristianto, R. Sarno. "Ontology and semantic matching for diabetic food recommendations" in *2013 International Conference on Information Technology and Electrical Engineering (ICITEE)*, 2013, pp.

23. Y. Batarseh, H. Qosa, K. Elsayed, J. N. Keller, A. Kaddoumi. "Extra-virgin olive oil and oleocanthal reduce amyloid ß load in Alzheimer's disease mouse model" in *2016 32nd Southern Biomedical Engineering Conference (SBEC)*, 2016, pp. 92-92. 170-175.

24. B. Akil, Y. Zhou, U. Rohm. "On the usability of Hadoop MapReduce, Apache Spark & Apache flink for data science" in *2017 IEEE International Conference on Big Data (Big Data)*, 2017, pp. 303-310.

6

Chicken Swarm Optimization - Modifications and Application

Dorin Moldovan

Department of Computer Science
Technical University of Cluj-Napoca, Romania

Adam Slowik

Department of Electronics and Computer Science
Koszalin University of Technology, Koszalin, Poland

CONTENTS

6.1   Introduction

Chicken Swarm Optimization (CSO) was introduced in 2014 in [1] and it is a part of the family of algorithms that are generically called nature inspired algorithms. Some illustrative algorithms from this family of algorithms are Particle Swarm Optimization (PSO) [2], Ant Colony Optimization (ACO) [3], Cuckoo Search (CS) [4], Lion Optimization Algorithm (LOA) [5], Kangaroo Mob Optimization (KMO) [6] and Crab Mating Optimization (CMO) [7]. Some of these algorithms are well known and they represent the source of inspiration for other bio-inspired algorithms, while other algorithms are relatively new and they are adaptations of the original PSO algorithm for different types of animal behaviors. The CSO algorithm is inspired by the behavior of the chickens when they search for food and it is a bio-inspired algorithm that can be applied for solving various types of engineering problems that are characterized by a search space with many dimensions. Each solution is represented by a chicken that has a position and there are three types of chickens namely, roosters, hens and chicks. The algorithm has as a main objective the identification of the best chicken according to an objective function which depends on the optimization problem that is solved. Some hybrid algorithms from literature that are based on the CSO algorithm are: Bat-Chicken Swarm Optimization (B-CSO) [8], Cuckoo Search-Chicken Swarm Optimization (CS-CSO) [9] and Chicken Swarm Optimization-Teaching Learning Based Optimization (CSO-TLBO) [10]. The chapter is organized as follows: Section 6.2 presents a short description of the global version of the CSO algorithm, Section 6.3 illustrates modifications of the CSO algorithm, Section 6.4 presents the application of the CSO algorithm for falls detection in daily living activities [11], [12] and Section 6.5 presents the main conclusions.

6.2   Original CSO algorithm in brief

The original CSO algorithm in global version can be represented using the pseudo-code from Algorithm 7.

Algorithm 7 Pseudo-code of the original CSO.

1: create the initial population of chickens C_i ($i = 1, 2, ..., N$) randomly such that each chicken is represented by a D-dimensional vector
2: evaluate the fitness values of all N chickens
3: create the D-dimensional vector $Gbest$
4: assign the best chicken C_i to $Gbest$
5: $t = 0$
6: **while** $t < I_{max}$ **do**

7:    **if** t modulo $G == 0$ **then**

8:        rank the fitness values of the chickens and establish the hierarchical order of the swarm

9:        divide the swarm of chickens into several groups and determine the relations between the chicks and the associated mother hens in each group

10:    **end if**

11:    **for** $i = 1 : N$ **do**

12:        **if** $C_i ==$ rooster **then**

13:            $C_{i,j}^{t+1} = C_{i,j}^t \times \left(1 + \mathcal{N}(0, \sigma^2)\right)$

14:        **end if**

15:        **if** $C_i ==$ hen **then**

16:            $C_{i,j}^{t+1} = C_{i,j}^t + S_1 \times R_1 \times \left(C_{r_1,j}^t - C_{i,j}^t\right) + S_2 \times R_2 \times \left(C_{r_2,j}^t - C_{i,j}^t\right)$

17:        **end if**

18:        **if** $C_i ==$ chick **then**

19:            $C_{i,j}^{t+1} = C_{i,j}^t + FL \times (C_{m,j}^t - C_{i,j}^t)$

20:        **end if**

21:        update the value $C_{i,j}^{t+1}$ if it is not in the interval $[C_{min}, C_{max}]$

22:        evaluate the fitness value of the new solution

23:        if the new solution is better than *Gbest* then update *Gbest*

24:    **end for**

25:        $t = t + 1$

26: **end while**

27: return the *Gbest* as a result

6.2.1   Description of the original CSO algorithm

The pseudo code of the original CSO algorithm is presented in Algorithm 7 and in this subsection the algorithm is described in more detail. The inputs of the algorithm are represented by: D - the number of dimensions, N - the number of chickens, RN - the number of roosters, HN - the number of hens, CN - the number of chicks, MN - the number of mother hens, FL - a random number from the interval $[0.5, 0.9]$ which is used when the positions of the chicks are updated, G - a number that specifies how often the hierarchy of the chicken swarm is updated, I_{max} - the maximum number of iterations and $[C_{min}, C_{max}]$ - an interval that indicates the minimum and the maximum possible values of the positions of the chickens. The output of the algorithm is represented by the best position achieved by a chicken. In the global version of the algorithm this position is represented by the *Gbest* vector.

In *step 1* of the algorithm, the initial population of chickens C_i where $i = 1, 2, ..., N$ is created randomly such that each chicken is described by a D-dimensional vector. In *step 2* of the algorithm, each chicken is evaluated using an objective function $OF(.)$. In *step 3* the D-dimensional vector *Gbest* is created and the best chicken C_i is assigned to *Gbest* in the global version of the algorithm in *step 4*. In *step 5* of the algorithm, the current iteration t is initialized to 0 and the next steps are repeated for a number of iterations that is equal to I_{max}. If the current iteration is divisible without rest by G then the chickens are ranked considering their fitness values, the novel hierarchical order of the swarm is established, the swarm of chickens is divided into several groups, each group having a dominant rooster, and the relations between the chicks and the corresponding mother hens are determined for each group of chickens. The position of each chicken is updated considering the type of the chicken. The roosters update their positions using the formula:

$$C_{i,j}^{t+1} = C_{i,j}^t \times \left(1 + \mathcal{N}(0, \sigma^2)\right) \tag{6.1}$$

where $t+1$ is the next iteration, t is the current iteration, i is the index of the chicken, j is the index of the dimension and has values from the set $\{1, ..., D\}$ and $\mathcal{N}(0, \sigma^2)$ is a Gaussian with mean 0 and standard deviation σ^2. The formula for σ^2 is:

$$\sigma^2 = \begin{cases} 1 & \text{if } OF(C_i) \leq OF(C_k) \\ e^{\frac{OF(C_k) - OF(C_i)}{|OF(C_i)| + \epsilon}} & \text{otherwise} \end{cases} \tag{6.2}$$

where ϵ is a small positive constant which is used in order to avoid division by 0 and k is the index of a rooster that is selected randomly from the group of roosters. In the global version of the algorithm the hens update their positions using the formula:

$$C_{i,j}^{t+1} = C_{i,j}^t + S_1 \times R_1 \times \left(C_{r_1,j}^t - C_{i,j}^t\right) + S_2 \times R_2 \times \left(C_{r_2,j}^t - C_{i,j}^t\right) \tag{6.3}$$

where R_1 and R_2 are random numbers from $[0, 1]$, r_1 is the index of the hen's rooster mate according to the hierarchical order that is reestablished every G iterations, r_2 is the index of another rooster or of a hen selected randomly from the swarm such that $r_1 \neq r_2$ and S_1 and S_2 have the formulas:

$$S_1 = e^{\frac{OF(C_i) - OF(C_{r_1})}{|OF(C_i)| + \epsilon}} \tag{6.4}$$

$$S_2 = e^{(OF(C_{r_2}) - OF(C_i))} \tag{6.5}$$

In the above formulas ϵ is a small positive constant that is used in order to avoid division by 0, like in the case of the formula that is used for updating the positions of the roosters. In the local version of the algorithm the formula that is used in order to update the positions of the hens is:

$$C_{i,j}^{t+1} = C_{i,j}^t + S_1 \times R_1 \times \left(C_{r,j}^t - C_{i,j}^t\right) + S_2 \times R_2 \times \left(C_{h,j}^t - C_{i,j}^t\right) \tag{6.6}$$

where $r = r_1$ and h is the index of a hen that is selected randomly from the group of hens in which C_i belongs. If there is no such hen then the formula becomes:

$$C_{i,j}^{t+1} = C_{i,j}^t + S_1 \times R_1 \times \left(C_{r,j}^t - C_{i,j}^t\right) \tag{6.7}$$

The chicks update their positions using the formula:

$$C_{i,j}^{t+1} = C_{i,j}^t + FL \times \left(C_{m,j}^t - C_{i,j}^t\right) \tag{6.8}$$

where FL is a random number from the interval $[0.5, 0.9]$ and m is the mother hen of the chicken with the index i. The rest of the parameters have the same signification as in the case of the formulas (6.1) and (6.3). If the value of $C_{i,j}^{t+1}$ is not in the interval $[C_{min}, C_{max}]$ (step 21) then that value is updated. In *step 22* the fitness value of the new solution is computed using the objective function $OF(.)$ and in *step 23* if the new solution is better than the previous solution then the value of *Gbest* is updated. Finally in the last step of the algorithm the best solution is returned. In the case of the global version of CSO algorithm the best solution is *Gbest*.

6.3 Modifications of the CSO algorithm

In this section are presented illustrative modifications of CSO [13].

6.3.1 Improved Chicken Swarm Optimization (ICSO)

In this version [14] the chicks learn both from their mother and from the rooster of the group. The situation in which the chicks get trapped in local optima is avoided and the new formula that is used in order to update the positions of the chicks is:

$$C_{i,j}^{t+1} = w \times C_{i,j}^{t} + FL \times (C_{m,j}^{t} - C_{i,j}^{t}) + c \times (C_{r,j}^{t} - C_{i,j}^{t}) \qquad (6.9)$$

where w is the self-learning coefficient of the chick C_i, c is the learning factor and r is the index of the rooster of the group to which the chick C_i belongs. The value of w decreases from 0.9 to 0.4 in each iteration according to the formula [15]:

$$w = w_{min} \times \left(\frac{w_{max}}{w_{min}} \right)^{\frac{1}{1 + 10 \times \frac{t}{I_{max}}}} \qquad (6.10)$$

In the formula (6.10) w_{min} is the minimum value of w, w_{max} is the maximum value of w, t is the current iteration and I_{max} is the maximum number of iterations. FL is a random number from the interval $[0.5, 0.9]$ and the value of c is 0.4.

6.3.2 Mutation Chicken Swarm Optimization (MCSO)

The version presented in [16] introduces the following formula in order to update the positions of the chicks:

$$C_{i,j}^{t+1} = C_{i,j}^{t+1} \times \left(1 + \frac{1}{2} \times \eta \right) \qquad (6.11)$$

where η is a random value from the Gaussian distribution $\mathcal{N}(0,1)$ that has the mean 0 and the standard deviation 1. The objective of the introduction of the mutation operator is to avoid the situations in which the chicks get trapped in local optima.

6.3.3 Quantum Chicken Swarm Optimization (QCSO)

In this version of the algorithm [17] the formula that is used for updating the positions of the chicks is:

$$C_{i,j}^{t+1} = C_{i,j}^t + FL \times (C_{m,j}^t - C_{i,j}^t) \times \log\left(\frac{1}{rand}\right) \tag{6.12}$$

such that:

$$FL = FL_{max} - (FL_{max} - FL_{min}) \times \frac{t}{I_{max}} \tag{6.13}$$

where FL_{max} is the maximum possible value of FL, FL_{min} is the minimum possible value of FL, t is the current iteration, I_{max} is the maximum number of iterations and $rand$ is a random number from $(0,1)$.

6.3.4 Binary Chicken Swarm Optimization (BCSO)

This version of the algorithm [18] is used for solving discrete optimization problems in which the search space is discrete and the solution can be represented using a binary vector. The formula that is used for transforming the positions of the chickens in arrays of zeros and ones is:

$$B_{i,j} = \begin{cases} 1 & \text{if } S(C_{i,j}) \geq 0.5 \\ 0 & \text{if } S(C_{i,j}) < 0.5 \end{cases} \tag{6.14}$$

where S is a sigmoid function that has the formula:

$$S(x) = \frac{1}{1 + e^{-x}} \tag{6.15}$$

and $C_{i,j}$ is the position of the chicken C_i for the j-th dimension while $B_{i,j}$ is the binary version of the position of the chicken C_i for the j-th dimension.

6.3.5 Chaotic Chicken Swarm Optimization (CCSO)

The CCSO algorithm is introduced in [19] where the authors apply knowledge from chaos theory [20]. The application of chaos theory has as the main objective the improvement of the performance of the CSO algorithm. The objective of the chaotic maps is to solve the problem of generation of values for the random variables which are used when the positions of the chickens are updated and two illustrative maps are the tent map and the logistic map.

The equation for the tent map is [21]:

$$x^{t+1} = \begin{cases} 2 \times \lambda \times x^t & \text{if } 0 \leq x^t \leq \frac{1}{2} \\ 2 \times \lambda \times (1 - x^t) & \text{if } \frac{1}{2} \leq x^t \leq 1 \end{cases} \qquad (6.16)$$

where λ is a number from the interval $[0, 1]$, x^{t+1} is the new value and x^t is the current value. The equation for the logistic map is:

$$y^{t+1} = \mu \times y^t \times (1 - y^t) \qquad (6.17)$$

where μ is a value from the interval $[0, 4]$. The values of x^t and y^t are in the interval $[0, 1]$. When $\lambda = 1$ and $\mu = 4$, the maps are in the chaos regions. Each map leads to a different version of the CCSO algorithm as follows: the tent maps are used in the CCSO Tent Map version and the logistic maps are used in the CCSO Logistic Map version. These maps help in a better exploration of the search space and they also solve the problem of getting trapped in local optima.

6.3.6 Improved Chicken Swarm Optimization - Rooster Hen Chick (ICSO-RHC)

The abbreviation ICSO-RHC [22] is given by the fact that the improved CSO algorithm includes position update modes for roosters, hens and chicks, and in addition it also includes a strategy for population update. In the case of the roosters the equations that are used for updating the positions consider information about the positions of the hens. The equations which are used for updating the positions of the hens are improved considering the guidance of the hens towards elite individuals from the chicken swarm. The chicks update their positions considering not only the position of the mother hen but also the position of the rooster which is the head of the swarm they belong to. Finally, in order to ensure the diversity of the chicken swarm, the following update strategy is applied:

$$C_i^{t+1} = \begin{cases} C_i^t & \text{if } R_i \geq P_e \\ C_{min} + rand \times (C_{max} - C_{min}) & \text{if } R_i < P_e \end{cases} \qquad (6.18)$$

where $rand$ and R_i are random numbers from $[0, 1]$ selected uniformly and P_e is the elimination probability which is usually equal to 0.1.

6.4 Application of CSO for detection of falls in daily living activities

6.4.1 Problem description

In this subsection is described briefly the problem that is approached in this chapter using a modified version of the CSO algorithm. The problem is presented in more detail in [11] and in the approach presented in this article the problem is simplified. The input of the problem is represented by data generated from monitoring sensors placed on different parts of the body and the output can be of two types, 1 if the monitored subject performs a fall or 0 if the monitored subject performs a regular daily living activity. The problem is a classification problem and the classifications are performed using the classical Random Forest (RF) algorithm [23] from Apache Spark [24]. The dataset that is used as experimental support is taken from [11] and it contains information from different subjects that perform 16 types of daily living activities and 20 types of falls. Data is collected using 6 sensors and for each fall or activity 5 repetitions are performed.

For simplicity in this chapter in the experiments are considered only two subjects, a male subject and a female subject, and for each subject only one repetition out of five repetitions is considered for each fall or activity. This simplicity was introduced because the repetitions for each fall or activity are similar when they are performed by the same monitored subject and because the algorithm uses the same configuration of parameters for each subject. In a version of the algorithm that uses different parameter values for each subject it would be recommended to run experiments for various subjects, but in this chapter due to the fact that the algorithm uses the same configuration of parameters for both subjects, the application of data generated from the monitoring of two representative subjects is justified. The first subject is a male that weighs 81 kilograms, is 174 centimeters and is 21 years old. The second subject is a female that weighs 60 kilograms, is 165 centimeters in height and is 20 years old.

In Table 6.1 are presented the main characteristics of the data used in experiments after normalization for the male subject and for the female Psubject.

TABLE 6.1
Characteristics of the Data Used in Experiments.

Characteristic	Male	Female
number of features	126	126
number of labels	2	2
number of falls samples	5964	6739
number of daily living activities samples	7525	8249

6.4.2   How can the CSO algorithm be used for this problem?

The two major modifications that are performed in order to apply the CSO algorithm for falls detection using data generated from the monitoring of daily living activities using sensors placed on different parts of the body are:

- the modification of the global version of the CSO algorithm such that the positions of the chickens are transformed into arrays of zeros and ones;

- the evaluation of the fitness values of the chickens using a RF classifier in which the features of the data that is used in training are indicated by the positions of the chickens;

In this approach the fitness value of each chicken C_i is evaluated using the formula:

$$OF(C_i) = RF(train, test, B_i) \tag{6.19}$$

where C_i is the i-th chicken from the set of N chickens, RF is a Random Forest classifier, $train$ is the training data, $test$ is the testing data and B_i is the binary version of the position of C_i (see the formula 6.14). The validation data is data which is not used when the classification model is created and it is used for evaluating the performance of the classification model.

In the BCSO algorithm for classification of falls in daily living activities the line 22 from Algorithm 7 is replaced by the lines:

1: $B_{i,j}^{t+1} = 0$
2: **if** $S(C_{i,j}^{t+1}) \geq 0.5$ **then**
3:    $B_{i,j}^{t+1} = 1$
4: **end if**
5: $OF(C_i) = RF(train, test, B_i)$

The pseudo-code of the RF algorithm that is used for computing the fitness values of the chickens is presented in Algorithm 8. The input data is represented by $train$ - the training data, $test$ - the testing data, B - the binary version of the position of the chicken and T - the size of the forest. The output of the algorithm is represented by the accuracy of the classification model. Each step is described briefly in the pseudo-code of the algorithm. F represents the selected features according to B which is the binary position of the chicken and K is a number that should be much less than the number of features from the set F.

Algorithm 8 Pseudo-code of RF for Chicken Fitness Values Computation.

1: select randomly K features from the set of features F
2: compute the best node d using the best split which is calculated using the $train$ data and the set of features K
3: compute the daughter nodes using the best split
4: repeat steps 1, 2, 3 until a certain number of nodes are computed
5: repeat steps 1, 2, 3, 4 until the size of the forest is T
6: $accuracy = RFClassificationModel(test)$
7: return $accuracy$ as a result

6.4.3 Description of experiments

In the experiments are considered the following values for the configurable parameters: the population of chickens is equal to 10, the number of iterations is 30, the hierarchy of the swarm is updated every 5 iterations, FL_{min} is equal to 0.5, FL_{max} is equal to 0.9, the value of ϵ is 10^{-9}, the percentage of roosters is 20%, the percentage of hens is 60%, the percentage of chicks is 20% and $[C_{min}, C_{max}] = [-5.12, 5.12]$. The values -5.12 and 5.12 are the values which are used in the global version of the algorithm and using exactly these numerical values is not very important. Any interval $[-L, L]$ where L is a number greater than 0 can be considered as an alternative. In each iteration of the algorithm the fitness values of the chickens are computed considering the training data and the testing data and finally the chicken with the best fitness value is returned as the final result of the optimization problem. The objective of the optimization problem is to determine a subset of features from the entire set of features such that the performance of the classification algorithm is high. The original datasets have 126 features and that means that there are 2^{126} possible combinations of features. The checking of all those combinations is too expensive in terms of computational resources and thus the application of an algorithm that determines a near-optimal solution fast is justified. The fitness value of each chicken is described by the accuracy of a RF classification model that is trained using the training data and tested using the testing data. The value of the accuracy in terms of TP (True Positives), TN (True Negatives), FP (False Positives) and FN (False Negatives) is:

$$accuracy = \frac{TP + TN}{TP + TN + FP + FN} \tag{6.20}$$

The global best chicken is used further in order to train a RF classification model that is applied on the validation data. In addition to accuracy, the metrics that are used for the evaluation of the performance of the classification model are:

$$precision = \frac{TP}{TP + FP} \tag{6.21}$$

$$F1Score = \frac{2 \times TP}{2 \times TP + FP + FN} \tag{6.22}$$

$$recall = \frac{TP}{TP + FN} \tag{6.23}$$

The accuracy is a metric that describes the closeness of a value that is measured to a known value while the precision describes the closeness of two measurements to each other. The recall or the sensitivity is equal to the ratio of positive observations that are correctly predicted to all observations and the F1Score is the weighted average between the values of the precision and of the recall.

6.4.4   Results obtained

In Tables 6.2 and 6.3 are presented the values of the accuracy and of the number of selected features in the initial iteration of the algorithm and in the final iteration of the algorithm. In the case of the male subject the value of the accuracy increases with 0.004 after 30 iterations and in the case of the female subject the value of the accuracy increases with 0.008 after 30 iterations. In contrast the number of selected features increases by 3 in the case of the male subject while the number of selected features decreases by 3 in the case of the female subject. Even though in the case of the male subject the number of features increases, the number of features by which that number increases is a small number that represents approximately 2% of the total number of features.

TABLE 6.2
Accuracy and Number of Features in the Initial Iteration and in the Final Iteration of the Optimization Algorithm for the Male Subject.

Criteria	First Iteration	Last Iteration
accuracy	0.974	0.978
number of features	66	69

TABLE 6.3
Accuracy and Number of Features in the Initial Iteration and in the Final Iteration of the Optimization Algorithm for the Female Subject.

Criteria	First Iteration	Last Iteration
accuracy	0.975	0.983
number of features	59	56

Table 6.4 presents the values for accuracy, precision, F1Score and recall obtained in the validation data after the RF classification model is trained using the configurations indicated by the chickens that represent the values returned by the optimization algorithm for the male and for the female subjects. In the case of the male subject the value of the accuracy 0.980 is better than the values returned by the BCSO algorithm in each of the 30 iterations and which are in the interval [0.974, 0.978]. On the other hand in the case of the female subject the value of the accuracy 0.976 is in the interval [0.975, 0.983]. In both cases the value of the accuracy is better than the value of the accuracy from the first iteration of the BCSO algorithm.

TABLE 6.4
Evaluation Metrics Results That Correspond to the Best Chicken Returned by the Optimization Algorithm.

Metric	Male	Female
accuracy	0.980	0.976
precision	0.971	0.959
F1Score	0.985	0.973
recall	1.000	0.987

6.4.5   Comparison with other classification approaches

In Table 6.5 are presented the results obtained using other approaches in the case of the male subject and in Table 6.6 are presented the results obtained using other approaches in the case of the female subject. The tables also contain the values obtained after the application of the BCSO approach. The experiments were performed in Konstanz Information Miner (KNIME) [25] and like in the case of the BCSO approach, the training data is 60% and the testing data is 20%. The accuracy column contains the accuracy results of the classification models using the standard configurations from KNIME in the case of the other approaches and in the case of the BCSO approach the accuracy column contains the value of the accuracy from the last iteration of the BCSO algorithm.

The two features selection approaches that are compared are an approach in which the features are selected using Principal Component Analysis (PCA) with a value for the minimum information to preserve equal to 95% and an approach in which the features are selected using Correlation Filter (CF) with a threshold equal to 55%. The values for the configuration parameters of PCA and CF were selected after a series of experiments such that the number of selected features would be approximately equal to the number of features returned by the BCSO algorithm.

As can be seen in Tables 6.5 and 6.6 the approaches in which the applied classification algorithm is RF and the features are selected using PCA and CF are better than the approach in which the classification algorithm is RF and the features are selected using BCSO. Even though the methods that are based on Gradient Boosted Trees (GBT) are better than the approach described in this chapter, when the data is classified using Decision Trees (DT) the method based on RF and BCSO is better and this proves once again the fact that in some cases the appropriate use of BCSO in combination with a classical classification algorithm is better than the use of some combinations of classical classification and features selection algorithms.

TABLE 6.5

Comparison of the Results Obtained Using the BCSO Approach with Other Approaches for the Male Subject.

Classification Algorithm	Features Selection Approach	Accuracy	Number of Features
GBT	CF	0.995	55
GBT	PCA	0.988	51
DT	CF	0.976	55
DT	PCA	0.965	51
RF	BCSO	0.978	69
RF	CF	0.997	55
RF	PCA	0.993	51

TABLE 6.6
Comparison of the Results Obtained Using the BCSO Approach with Other Approaches for the Female Subject.

Classification Algorithm	Features Selection Approach	Accuracy	Number of Features
GBT	CF	0.995	60
GBT	PCA	0.989	50
DT	CF	0.980	60
DT	PCA	0.975	50
RF	BCSO	0.983	56
RF	CF	0.996	60
RF	PCA	0.996	50

Even though the results that are presented in this chapter cannot be compared perfectly with the results from other articles due to the fact that for each subject only one test data was used for each monitoring sensor and the feature that describes the temperature was eliminated due to the fact that it has constant values, compared to the results from [11] in which 10-fold cross validation is used, the values of the accuracy returned in the last iteration of the BCSO algorithm, namely 0.978 for the male subject and 0.983 for the female subject, are better or comparable with the values of the accuracy obtained in the case of Dynamic Time Warping (DTW) and of Artificial Neural Network (ANN). In the case of DTT the value of the accuracy is 0.978 and in the case of the ANN the value of the accuracy is 0.956.

However the comparison cannot be considered highly accurate because in the approach presented in this chapter the ratio between the number of samples from the training data and the number of samples from the testing data is $60 : 20 = 3 : 1$ and in the case of the 10-fold cross validation the ratio is $9 : 1$. Moreover in this chapter only one trial out of five trials is considered for each fall or activity for each monitored subject. These major differences might be enough to justify why the results obtained in the approach based on BCSO are not as good as the ones obtained in [11] using Support Vector Machines (SVM), K-Nearest Neighbors (K-NN), Bayesian Decision Making (BDM) and Least Squares Method (LSM) which return accuracy results better than 0.991.

6.5 Conclusions

In this chapter we described the original version of the CSO algorithm in brief, we presented illustrative modifications of the algorithm such as ICSO, MCSO, QCSO, BCSO, CCSO and ICSO-RHC, and we discussed how this algorithm can be applied for features selection in classification problems in which the data is characterized by a big number of features and has two labels. The CSO algorithm in its binary version was applied further for the problem of falls classification in daily living activities. The CSO algorithm results are good classification results and the performance can be improved by varying the configuration parameters or by applying a modified version of the algorithm in order to avoid the situations in which the chicks are trapped in local optima. As future research work we want to: (1) propose different modifications of the CSO algorithm that are not approached in literature yet but which were approached in the case of other bio-inspired algorithms, (2) apply the CSO algorithm in other engineering problems and (3) conduct a more complex research in order to determine for each modification of the CSO algorithm the most representative classes of engineering problems that can be solved using that modification.

References

1. X. Meng, Y. Liu, X. Gao, H. Zhang. "A New Bio-inspired Algorithm: Chicken Swarm Optimization" in *Proc. of International Conference in Swarm Intelligence ICSI 2014: Advances in Swarm Intelligence*, Lecture Notes in Computer Science, vol. 8794, Springer, 2014, pp 86-94

2. C. Xiang, X. Tan, Y. Yang. "Improved Particle Swarm Optimization algorithm in dynamic environment" in *Proc. of the 26th Chinese Control and Decision Conference (2014 CCDC)*, 2014, pp. 3098-3102.

3. G. Ping, X. Chunbo, L. Jing, L. Yanqing. "Adaptive ant colony optimization algorithm" in *Proc. of the 2014 International Conference on Mechatronics and Control (ICMC)*, 2014, pp. 95-98.

4. S. Dhabal, S. Tagore, D. Mukherjee. "An improved Cuckoo Search Algorithm for numerical optimization" in *Proc. of the 2016 International Conference on Computer, Electrical & Communication Engineering (ICCECE)*, 2016, pp. 1-7.

5. R. Babers, A. E. Hassanien, N. I. Ghali. "A nature-inspired metaheuristic Lion Optimization Algorithm for community detection" in *Proc. of the 2015 11th International Computer Engineering Conference (ICENCO)*, 2015, pp. 217-222.

6. D. Moldovan, I. Anghel, T. Cioara, I. Salomie, V. Chifu, C. Pop. "Kangaroo mob heuristic for optimizing features selection in learning the daily living activities of people with Alzheimer's" in *Proc. of the 2019 22nd International Conference on Control Systems and Computer Science (CSCS)*, 2019, pp. 236-243.

7. V. R. Chifu, I. Salomie, E. S. Chifu, A. Negrean, M. Antal. "Crab mating optimization algorithm" in *Proc. of the 2014 18th International Conference on System Theory, Control and Computing (ICSTCC)*, 2014, pp. 353-358.

8. S. Liang, T. Feng, G. Sun, J. Zhang, H. Zhang. "Transmission power optimization for reducing sidelobe via bat-chicken swarm optimization in distributed collaborative beamforming" in *Proc. of the 2016 2nd IEEE International Conference on Computer and Communications (ICCC)*, 2016, pp. 2164-2168.

9. S. Liang, T. Feng, G. Sun. "Sidelobe-level suppression for linear and circular antenna arrays via the cuckoo search–chicken swarm optimisation algorithm" in *IET Microwaves, Antennas & Propagation*, vol. 11, 2017, pp. 209-218.

10. S. Deb, K. Kalita, X.-Z. Gao, K. Tammi, P. Mahanta. "Optimal placement of charging stations using CSO-TLBO algorithm" in *Proc. of the 2017 Third International Conference on Research in Computational Intelligence and Communication Networks (ICR-CICN)*, 2017, pp. 84-89.

11. A. T. Ozdemir, B. Barshan. "Detecting falls with wearable sensors using machine learning techniques" in *Sensors*, vol. 14, no. 6, 2014, pp. 10691-10708.

12. A. T. Ozdemir. "An analysis on sensor locations of the human body for wearable fall detection devices: principles and practice" in *Sensors*, vol. 16, no. 1161, 2016, pp. 1-25.

13. S. Deb, X.-Z. Gao, K. Tammi, K. Kalita, P. Mahanta. "Recent studies on Chicken Swarm Optimization algorithm: a review (2014–2018)" in *Artificial Intelligence Review*, 2019, pp. 1-29.

14. D. Wu, F. Kong, W. Gao, Y. Shen, Z. Ji. "Improved Chicken Swarm Optimization" in *Proc. of 2015 IEEE International Conference on Cyber Technology in Automation, Control, and Intelligent Systems (CYBER)*, 2015, pp. 681-686.

15. G. Chen, J. Jia, Q. Han. "Study on the strategy of decreasing inertia weight in particle swarm optimization algorithm" in *Journal of Xi'an Jiao Tong University*, vol. 40, no. 1, 2006, pp. 53-61.

16. K. Wang, Z. Li, H. Cheng, K. Zhang. "Mutation chicken swarm optimization based on nonlinear inertia weight" in *Proc. of the 2017 3rd IEEE International Conference on Computer and Communications (ICCC)*, 2017, pp. 2206-2211.

17. X. B. Meng, H. X. Li. "Dempster-Shafer based probabilistic fuzzy logic system for wind speed prediction" in *Proc. of the 2017 International Conference on Fuzzy Theory and its Applications (iFUZZY)*, 2017, pp. 1-5.

18. M. Han, S. Liu. "An improved Binary Chicken Swarm Optimization algorithm for solving 0-1 knapsack problem" in *Proc. of the 2017 13th International Conference on Computational Intelligence and Security*, 2017, pp. 207-210.

19. K. Ahmed, A. E. Hassanien, S. Bhattacharyya. "A novel chaotic chicken swarm optimization algorithm for feature selection" in *2017 Third International Conference on Research in Computational Intelligence and Communication Networks (ICRCICN)*, 2017, pp. 259-264.

20. X. Wang, L. Liu, Y. Zhang. "A novel chaotic block image encryption algorithm based on dynamic random growth technique" in *Optics and Lasers in Engineering*, vol. 66, 2015, pp. 10-18.

21. B. Wang, W. Li, X. Chen, H. Chen. "Improved chicken swarm algorithms based on chaos theory and its application in wind power interval prediction" in *Mathematical Problems in Engineering*, no. 1240717, 2019, pp. 1-10.

22. J. Wang, Z. Cheng, O. K. Ersoy, M. Zhang, K. Sun, Y. Bi. "Improvement and application of chicken swarm optimization for constrained optimization" in *IEEE Access*, vol. 7, 2019, pp. 58053-58072.

23. Y. Xu. "Research and implementation of improved random forest algorithm based on Spark" in *Proc. of the 2017 IEEE 2nd International Conference on Big Data Analysis (ICBDA)*, 2017, pp. 499-503.

24. G. Gousios. "Big data software analytics with Apache Spark" in *Proc. of the 2018 IEEE/ACM 40th International Conference on Software Engineering: Companion (ICSE-Companion)*, 2018, pp. 542-543.

25. L. Feltrin. "KNIME an open source solution for predictive analytics in the geosciences [software and data sets]" in *IEEE Geoscience and Remote Sensing Magazine*, vol. 3, no. 4, 2015, pp. 28-38.

7

Cockroach Swarm Optimization – Modifications and Application

Joanna Kwiecien

AGH University of Science and Technology
Department of Automatics and Robotics, Krakow, Poland

CONTENTS

7.1 Introduction

During the last decade, a number of computational intelligence algorithms have been developed. The cockroach swarm optimization algorithm discussed here, is one of many interesting metaheuristic search algorithms inspired by nature. The CSO algorithm was described by Chen and Tang in 2010 [1]. It mimics the food foraging behavior of swarms of cockroaches and involves the procedures concerning the individuals' movement within the group. A coordination system of a group of cockroaches, being a decentralised system, is responsible for the organisation of tasks required for resolving a specific task.

Cockroaches, like many other animals, appear to employ some mechanisms by which a swarm may assemble. They make decisions based on their interaction with peers. Although cockroaches are known to have a variety of habits, most available relate to swarming, chasing, dispersing, and ruthless behavior. The CSO algorithm belongs to a group of population methods, because it applies a population of current solutions which can generate new solutions after proper selection and modification. Therefore, in its structure one can find three kinds of simple behaviors applied to search space: chase swarming that reflects that cockroaches can communicate with each other, dispersion that reflects scattering or escaping from light (it may improve the survival rate), and ruthlessness, when the strong cockroach eats the weak one randomly at intervals of time. Cockroaches change their positions with time. Hence, the CSO algorithm uses a population of solutions for every iteration, and the cockroach updates its solution based on either the best solution obtained (within its visual scope or the best solution obtained so far) or a random movement.

When applying the CSO algorithm, it is necessary to make decisions regarding the determination of its structure to be suitable for the solution of a given problem, as well as carefully select the quantitative and qualitative parameter values, being typical for the algorithm under consideration, followed by an algorithm performance evaluation. In order to cope with the application of the CSO algorithm to solve the traveling salesman problem, some adaptations of the algorithm are presented. It should be mentioned that many researchers described the traveling salesman problem in respect to its practical use, and the TSP is one of the most famous combinatorial problems. A lot of real problems can be formulated as the TSP problem [2]. As indicated in [3], path planning and wayfinding tasks as TSP-like tasks are quite common in human navigation.

The rest of this chapter is organized as follows: in Section 7.2 a brief introduction to the CSO algorithm is presented. The pseudo-code for the basic version of CSO algorithm is also discussed. Section 7.3 provides some modifications of the CSO algorithm including inertia weight, stochastic constriction factor, hunger component, and neighborhood scheme. In Section 7.4, the application of the CSO algorithm for the traveling salesman problem is described. Therefore, we present how to adapt the CSO algorithm to solve this problem, concerning the movement performance. Various modifications of movements, e.g. approaching the best local or global individual with the use of crossover operators in chase-swarming, or relocating based on 2-opt moves are used. Finally, in Section 7.5 some concluding remarks are provided.

7.2 Original CSO algorithm in brief

7.2.1 Pseudo-code of CSO algorithm

The main steps of the CSO algorithm are outlined in the pseudo-code form (Algorithm 9).

Algorithm 9 Pseudo-code of the original CSO.

1: determine the $D - th$ dimensional objective function $OF(.)$
2: initialize the CSO algorithm parameter values such as k – the number of cockroaches in the swarm, *visual* – the visual scope, *step* - the step of cockroaches, *Stop* – termination condition
3: generate a population of k individuals, the $i - th$ individual represents a vector $X_i = (x_{i1}, x_{i2}, ..., x_{iD})$
4: **for** each $i - th$ cockroach from swarm **do**
5:     evaluate quality of the cockroach X_i using $OF(.)$ function
6: **end for**
7: find the best solution P_g in initial population
8: **while** termination condition not met **do**
9:     **for** $i = 1$ to k **do**
10:         **for** $j = 1$ to k **do**
11:             **if** the objective function of the $j - th$ cockroach is better than the objective function of the $i - th$ cockroach, within its visual scope **then**
12:                 move cockroach i towards j
13:             **end if**
14:             **if** the position of cockroach i is local optimum **then**
15:                 move cockroach i towards P_g
16:             **end if**
17:         **end for** j
18:     **end for** i
19:     **if** new solution is better than P_g **then**
20:         update P_g (cockroach i is a current global solution)
21:     **end if**
22:     **for** $i = 1$ to k **do**
23:         move cockroach randomly using formula
24:         $X_i = X_i + rand(1, D)$
25:         **if** the new position is better than P_g **then**
26:             update P_g
27:         **end if**
28:     **end for**
29:     select cockroach h randomly
30:     replace cockroach h by the best global cockroach
31: **end while**
32: return the best one as a result

7.2.2 Description of the original CSO algorithm

The CSO algorithm starts with a fixed number of cockroaches and in each cycle of the algorithm, the cockroaches' positions are updated to search for a better solution. In the CSO algorithm, individuals have a randomly generated position at the beginning. The search space, in which the cockroaches move, has dimension D. Therefore, the position of the $i-th$ individual in the CSO algorithm can be represented by D-dimensional vector.

At the start, we should have defined an objective function $OF(.)$ (step 1) and the CSO algorithm parameters (step 2) such as number of cockroaches (k), visual range $(visual)$, $step$ - a fixed value, and the stopping criterion $(Stop)$. The visibility parameter denotes the visual distance of cockroaches. In the third step, we have to initialize the swarm with random solutions. The $i-th$ individual denotes a D dimensional vector $X_i - (x_{i1}, x_{i2}, ..., x_{iD})$ for $i = 1, 2$, ..., k. The location of each cockroach is a potential solution, and the objective function value corresponding to each solution is calculated (step 5). After such evaluation, the swarm's global best position is kept and considered as global optimum P_g (step 7). In step 8, the main loop of the CSO approach is started until the stopping criterion is fulfilled. The stopping criterion can include, for example, the maximum number of iterations, the number of iterations without improvement, the computational time, obtaining an acceptable error of a solution.

In the chase-swarming procedure (steps 9 to 18), the locally best individuals form small swarms and follow the best one. In its basic version, the individuals perform movement towards that one that enjoys the best target function value, relating to the range of their visibility. Therefore, the new position of strongest cockroaches P_i moving forward to the global optimum P_g can be computed as (step 12):

$$X_i = X_i + step \cdot rand \cdot (P_g - X_i) \qquad (7.1)$$

The locally best individuals create small swarms and follow the best one in the group. Note that within this procedure, the new direction of each individual X_i moving to its local optimum P_i in the range of its visibility (step 15) is obtained from:

$$X_i = X_i + step \cdot rand \cdot (P_i - X_i) \qquad (7.2)$$

There is a possibility that a cockroach moving in a small group will be strongest if it finds a better solution, because individuals follow in other ways than their local optimum. If it is better than P_g with respect to the objective function $OF(.)$, the global solution will be updated (step 20). A lonely cockroach within its own scope of visibility, is its local optimum and it moves forward to the best global solution. In addition, a dispersion procedure is incorporated into the running process (steps 22 to 28). The dispersion of individuals is performed as random movements (step 24) using the formula: $X_i = X_i + rand(1, D)$, where $rand(1, D)$ is a D-dimensional random vector (D is the space dimension) within a certain range. It is a simulation of the appearance of a light or a brisk movement. In step 26, the position of the best cockroach is updated. From time to time, the ruthless behavior can occur when the current best individual replaces a randomly chosen individual (step 30). This situation corresponds to the depletion of nutritional resources that may lead to cannibalism among cockroaches.

When the stopping criterion has been met, the algorithm will stop and the best position of cockroaches will be returned as the result obtained after all cycles of the CSO.

7.3 Modifications of the CSO algorithm

Since its introduction in 2010 [1], the CSO algorithm has had some improvements aimed at guaranteeing stability and improving diversity. Here, some modifications of the CSO algorithm are briefly described.

7.3.1 Inertia weight

Modified CSO [4] extends CSO and introduces an inertia weight w in chase-swarming behavior. Other procedures remain without changes. Therefore, in the chase-swarming procedure, the cockroach position which moves forward to the global optimum P_g is changed as:

$$X_i = w \cdot X_i + step \cdot rand \cdot (P_g - X_i) \tag{7.3}$$

and each individual X_i moves to its local optimum P_i in the range of its visibility as follows:

$$X_i = w \cdot X_i + step \cdot rand \cdot (P_i - X_i) \tag{7.4}$$

Here, w is used for affecting the convergence in CSO. While a large inertia weight w supports a global search, a small value favours a local search. According to [4], it is a constant, and the value of this factor usually is from the range $[0.5, 1]$.

7.3.2 Stochastic constriction coefficient

To maintain the stability of a swarm, a stochastic constriction factor (SCF) to control cockroach movements during chase-swarming procedure was proposed in [5]. SCF supports generation of different values as a constriction factor in cycle (each iteration). In order to insure the speed and convergence of the CSO algorithm, the chase-swarming procedure with the introduction of SCF can be expressed as follows:

$$X_i = \begin{cases} \xi(X_i + step \cdot rand \cdot (P_i - X_i)), & \text{if } X_i \neq P_i \\ \xi(X_i + step \cdot rand \cdot (P_g - X_i)), & \text{if } X_i = P_i \end{cases} \tag{7.5}$$

7.3.3 Hunger component

An improved cockroach swarm optimization (ICSO) including hunger behavior was presented in [6]. It prevents local optimum and enhances diversity of population. In the algorithm, hunger behavior occurs after the chase-swarming procedure.

A variable t_h is called a hunger threshold. It is defined, when the cockroach is hungry and this threshold is reached. It is a random number [0,1]. The formula for hunger behavior is given by:

$$x_i = x_i + (x_i - ct) + x_f \qquad (7.6)$$

where x_i is cockroach position, $(x_i\ ct)$ denotes cockroach migration from its current position, c denotes the controlling speed of the migration at time t, and x_f is food location.

7.3.4   Global and local neighborhoods

Because after some iterations all cockroaches are near around P_g, two kinds of neighborhood structures were presented in [7]. One, namely the local structure, where each cockroach current position at the $G - th$ population $(X_{i,G+1})$ is computed by the best position $(P_{i,G})$ found so far in a small neighborhood. In turn, the second structure can be the entire population at the $G - th$ current generation. Therefore, the new position of the $i - th$ cockroach $(F_{i,G})$ in chase-swarming behavior is formulated as follows:

$$F_{i,G} = \begin{cases} X_{i,G} + (P_{i,G} - X_{i,G}) + (X_{r1} - X_{r2}), & \text{if } rand(0,1) < 0.5 \\ X_{i,G} + (P_{g,G} - X_{i,G}) + (X_{r3} - X_{r4}), & \text{otherwise} \end{cases}$$
$$(7.7)$$

where the indices r_1, r_2, r_3, r_4 belong to [1, the number of cockroaches], $rand(0,1)$ is a uniformly distributed random number.

It should be mentioned that in $G+1$ generation, $X_{i,G+1}$ is chosen between $X_{i,G}$ and $F_{i,G}$ according to the following selection rule:

$$X_{i,G+1} = \begin{cases} F_{i,G}, & \text{if } OF(F_{i,G}) < OF(X_{i,G}) \text{ for minimization problems} \\ X_{i,G}, & \text{otherwise} \end{cases}$$
$$(7.8)$$

7.4   Application of CSO algorithm for traveling salesman problem

7.4.1   Problem description

The goal of the traveling salesman problem is to find the path of shorter length or minimum cost between all the requested points (cities), visiting each point exactly once and returning to the starting point. The salesman starts at a point, visiting all the points one by one and returns to the initial point. For a given set of cities from 1 to N and a cost matrix $C = [c_{i,j}]$, where each value $c_{i,j}$ represents the cost of traveling between cities i and j, each possible tour can be represented as a permutation $\pi = (\pi(1), ..., \pi(N))$, where $\pi(i) \in N$ represents the city visited in step i, $i = 1, ..., N$. Therefore, the goal is to find the minimal cost of the closed tour which minimizes the following objective function [8]:

$$f(\pi) = \sum_{i=1}^{N-1} c_{\pi(i),\pi(i+1)} + c_{\pi(N),\pi(1)} \qquad (7.9)$$

If distances $c_{i,j}$ and $c_{j,i}$ between two cities i and j are equal then we have a symmetric TSP, and it can be defined as an undirect graph. If the problem is an asymmetric TSP, it can be defined as a direct graph. If all cities are described by their coordinates (x,y) and distances between cities are Euclidean, we have a Euclidean TSP. As mentioned in Section 7.1, TSP has a great importance in many research and practical problems. It belongs to the class of $NP - complete$ problems and many approximate algorithms are used to solve its instances.

7.4.2 How can the CSO algorithm be used for this problem?

Below, a brief description of the CSO algorithm for solving TSP is provided according to [9]. A correct representation of the individual, distance or the method of the individuals' movement, with the determination of the quality of adjustment, constitute the main elements that require the process of algorithm adaptation to the problem being solved. For the adaptation, owing to the easy implementation, a representation of solutions with the permutation of the set of n cities is used, and the initial positions of individuals is generated randomly.

A cockroach, in the CSO algorithm, is a combination of the visited cities and a decreased fitness value is indicative of better individuals. The basic aspect of the cockroach swarm optimization algorithm adaptation consists in the definition of the scope of visibility and of the movement. When solving a traveling salesman problem, the position of a cockroach represents a solution to the TSP problem, coded as the permutation. It should be emphasized that the solution notation in the form of permutation requires a proper definition of movement. Therefore, the objective function for each individual, and the position of individuals are denoted by $f(\pi)$ and π, respectively. It should be noted that the solution quality is determined by the path length. Hence, upon generation of initial solutions, their quality is estimated. In this stage, a cockroach should evaluate its position itself first, before swarming with others. The smaller the value of $f(\pi)$ is, the better the quality solution is. The purpose of the subsequent steps is to improve solutions and find the path with shorter length. Therefore, k individuals are required to find a closed tour and encounter others in the searching process.

Next, the strategy of updating the current solution is generated within the chase-swarming procedure, where the new position of the cockroach is explored and evaluated. Movement in chase-swarming procedure can be determined by crossover operators known from genetic algorithms. It should be mentioned that traditional crossover methods may not provide a good tour; therefore we have to use such a crossover operator that ensures path modification without cycles or inconsistent components. In [9] partially-mapped crossover (PMX), order crossover (OX), and sequential constructive crossover (SCX) were investigated in the considered procedure. If the path of cockroach j is shorter than the tour of cockroach i, within its *visual*, then cockroach i moves towards j according to crossover operators. On the other case, if the path of cockroach i is local optimum (within its *visual*) then cockroach i goes to P_g (with crossover operator). When those cockroaches finish this procedure, they will compare the quality of the routes visited in the current position and the better one is selected as a candidate solution. It should be empasized that if two cockroaches have no common positions in their permutations, they cannot move with respect to each other.

In the basic version of CSO presented in the literature, the random step is involved in the dispersing procedure. It is helpful to escape from the local optimum. In the TSP, random movement can be represented by the swap operator that generates a new cockroach position by exchanging two randomly selected cities.

As was shown in [9], in order to increase the efficiency of the algorithm, an alternative approach could be used. Such modified dispersion based on 2-opt moves allows for the adaptation of the CSO algorithm to the TSP. The modification consisted in the reduction of the set of pairs of candidate edges, because only selected edges could be exchanged. The selection of edges for exchange was performed randomly, with the assumption that each edge in the graph was exchanged with at least one randomly selected edge. If a new solution obtained as a result of an edge without creating cycles or inconsistent components exchange was better than the previous one, it was saved as a current solution. This procedure can be outlined in the pseudo-code form (Algorithm 10).

Algorithm 10 Pseudo-code of the dispersion procedure.

1: **for** each individual **do**
2:    **for** each position t_a in solution **do**
3:       choose the random set of edges to remove (lk)
4:       **for** each edge $b = 1 : lk$ **do**
5:          remove edge $a = (t_a, t_{a+1})$ and b
6:          add new pair of edges according to *2-opt* moves
7:          calculate the path length
8:          **if** the new solution is better **then**
9:             update current solution
10:          **end if**

11:        **end for**
12:     **end for**
13: **end for**

The whole searching procedure of the best path ceases if the maximum number of iterations is met.

7.4.3 Description of experiments

The main purpose of the research on the CSO algorithm modification in the traveling salesman problem was to determine the evaluation of the usability of this method. For experiments, the selected benchmarks taken from the library TSPLIB available at http://www.iwr.uni-heidelberg.de/groups/comopt/software/TSPLIB95/ were used. The dimensions of the selected problems range from 51 to 225 cities.

In all cases the population size was considered to be the size of the test problem, *visual* was around 40 % of population size. In turn, the maximum number of iterations was set to 1000. It is important to remark that, in all experiments for each instance, fixed parameters were used during all iterations. During preliminary researches we were interested in testing three crossover operators and finding the best one for a sample instance containing 51 towns (*eil51*). An analysis of the influence of those operators on the quality of the results obtained was made (see [9]). For the random initial cockroach population and dispersion based on the 2-opt moves, it was concluded that the best result was obtained for the sequential crossover (SCX) in the chase-swarming procedure [9]. Therefore, all the results presented in this chapter are based on the SCX operator. For each case the CSO algorithm ran 20 times with the same control parameters. For each instance, global minimum values were computed for each run and compared with the reference values.

7.4.4 Results obtained

Owing to the application of the SCX operator and modification of the dispersion procedure, based on 2-opt movements it is possible to minimise the obtained errors of results. In the majority of the presented test cases, the average values of deviation from the reference values did not exceed 4%, which indicated potential possibilities of the CSO algorithm application in solving the traveling salesman problem.

Table 7.1 summarizes the best values of the path length on selected instances for 20 trials. The first column indicates the test instance, the second column (*Ref*) shows the reference valus, the third column (*Best*) presents the best solutions found by CSO, the fourth column (*Av*) gives the average solution of 20 independent runs. The last two columns *Dev_best* and *Dev_av* stand for the relative deviation of the found solution from the reference value (Ref), calculated as $Dev\_best = \frac{Best-Ref}{Ref} \cdot 100\%$ and $Dev\_av = \frac{Av-Ref}{Ref} \cdot 100\%$.

TABLE 7.1

Results of experiments obtained by CSO approach with SCX and directed dispersion.

Instance	Ref	Best	Av	Dev best [%]	Dev av [%]
eil51	426	428.83	438.16	0.66	2.85
berlin52	7542	7544.37	7834.12	0.03	3.87
st70	675	677.11	695.53	0.31	3.04
eil76	538	552.39	571.05	2.68	6.14
pr76	108159	109049.00	110962.90	0.82	2.59
kroA100	21282	21381.30	21774.12	0.47	2.31
rat195	2323	2436.80	2499.78	4.90	7.61
ts225	126643	128881.00	129964.15	1.77	2.62

Analyzing the performance of this approach in comparison with the reference solutions (relative deviations of the found solutions from Table 7.1), one can see that for five instances (*eil*51, *berlin*52, *st*70, *pr*76, *kroA*100) CSO with SCX and directed dispersion finds the best solutions near optimal, with values of *Dev_best* less than 1 %. It should be mentioned that the best found solutions are the same as in [9]. In all of the test instances, the presented approach gives *Dev_av* between 2.31 % and 7.61 %. Therefore, CSO can produce quite good results, but it requires good modifications of key components and the settings of parameters.

7.5   Conclusions

In this chapter the CSO algorithm, and some of its modifications were presented. As shown in Section 7.4, the CSO algorithm can be used to solve the traveling salesman problem. As described the basic problem of the CSO adaptation to discrete optimization problems consists in the proper definition of the solution and the consequential modification of the individuals' movement performance manner. One may choose to adapt some operators of genetic algorithms. Results of the SCX operator in the chase-swarming procedure, together with the modified dispersion procedure (using the 2-opt movements) are described. Certain adaptations may be designed in such a way as to trigger a specific strategy of movements. When analyzing the results obtained, one can observe that by introducing presented modifications into the framework of the CSO algorithm, this method can produce satisfactory solutions. In [10], the use of other modifications for solving the TSP problem is demonstrated.

References

1. Z.Chen, H. Tang. "Cockroach swarm optimization" in *Proc. of 2nd International Conference on Computer Engineering and Technology (ICCET)*, 2010, pp. 652-655.

2. R. Matai, S.P. Singh, M.L. Mittal. "Traveling Salesman Problem: An Overview of Applications, Formulations, and Solution Approaches" in *Traveling Salesman Problem, Theory and Applications*, pp.1-24, InTech 2010.

3. J.M. Wiener, T. Tenbrink. "Traveling salesman problem: The human case" in *KI: Themenheft KI und Kognition*, vol.22, pp. 18-22, 2008.

4. Z. Chen. "A modified cockroach swarm optimization" in *Energy Procedia*, vol. 11, pp. 4-9, 2011.

5. I.C. Obagbuwa, A.O. Adewumi, A.A. Adebiyi. "Stochastic constriction cockroach swarm optimization for multidimensional space function problems" in *Mathematical Problems in Engineering*, Article ID 430949, 2014.

6. I.C. Obagbuwa, A.O. Adewumi. "An improved cockroach swarm optimization" in *The Scientific World Journal*, Article ID 375358, 2014.

7. L. Cheng, Y. Song, M. Shi, Y. Zhai, Y. Bian."A study on improved cockroach optimization algorithm" in *Proc. of the 7th International Conference on Computer Engineering and Networks*, 2017.

8. D. Davendra, I. Zelinka,R. Senkerik, M. Bialic-Davendra. "Chaos Driven Evolutionary Algorithm for Salesman Problem" in *Traveling Salesman Problem, Theory and Applications*, pp. 55-70, InTech 2010.

9. J. Kwiecien. "Use of different movement mechanisms in cockroach swarm optimization algorithm for traveling salesman problem" in *Artificial Intelligence and Soft Computing*, vol. 9693, pp. 484-493, 2016.

10. L. Cheng, Z. Wang, S. Yanhong, A. Guo. "Cockroach swarm optimization algorithm for TSP" in *Adv. Eng. Forum*, vol. 1, pp. 226-229, 2011.

8

Crow Search Algorithm - Modifications and Application

Adam Slowik

Department of Electronics and Computer Science
Koszalin University of Technology, Koszalin, Poland

Dorin Moldovan

Department of Computer Science
Technical University of Cluj-Napoca, Romania

CONTENTS

8.1 Introduction

Crow Search Algorithm (CSA) [2] is a novel bio-inspired algorithm introduced in 2016. The main source of inspiration of CSA is the behavior of intelligent crows in nature when they search for food. CSA presents similarities with the following three algorithms: Genetic Algorithms (GA) [3], Particle Swarm Optimization (PSO) [4] and Harmony Search (HS) [5]. Like in the case of GA and PSO, CSA is a non-greedy algorithm and like in the case of PSO and HS, CSA keeps the good solutions in memory.

Other representative algorithms from the family of the bio-inspired algorithms [6] are Ant Colony Optimization (ACO) [7], Elephant Search Algorithm (ESA) [8], Chicken Swarm Optimization (CSO) [9], Artificial Algae Algorithm (AAA) [10], Kangaroo Mob Optimization (KMO) [11] and Moth Flame Optimization (MFO) [12]. In the research literature CSA was applied for solving various types of optimization problems or optimization tasks such as the optimal power flow problem [13], electromagnetic optimization [14] and continuous optimization tasks [15].

Even though the algorithm is a relatively new bio-inspired algorithm, in literature there are already modifications of CSA which were applied in solving various types of optimization problems. A part of these modifications namely, Chaotic Crow Search Algorithm (CCSA) [16], Modified Crow Search Algorithm (MCSA) [17] and Binary Crow Search Algorithm (BCSA) [18] are presented in more detail in the section of the chapter which illustrates modifications of CSA.

The chapter has the following structure: Section 8.2 presents the original CSA in brief, Section 8.3 presents illustrative modifications of CSA, Section 8.4 presents the application of CSA for the tuning of the number of nodes of each layer of a deep neural network applied for jobs status prediction using as experimental support the CIEMAT Euler log and Section 8.5 presents the conclusions and future research directions.

8.2   Original CSA in brief

In this section is presented briefly the original version of CSA [2].

Algorithm 11 Pseudo-code of CSA.

1: initialize the population of crows C_i $(i = 1, ..., N)$
2: for each crow C_i calculate the fitness value using objective function $OF(.)$
3: initialize the memory of the crows M_i $(i = 1, ..., N)$
4: define the awareness probability AP and the flight length fl
5: $t = 0$
6: **while** $t < Iter_{max}$ **do**
7:      **for** $i = 1 : N$ **do**
8:          select a random value k from $\{1, ..., N\}$
9:          generate a random number r in $[0, 1]$
10:          **if** $r \geq AP$ **then**
11:              generate the new position of the crow using formula 8.1
12:          **else**
13:              generate the new position of the crow randomly
14:          **end if**
15:      **end for**
16:      for each crow C_i check the feasibility of the solution represented by C_i

17:       compute the fitness value (OF) for each crow

18:       update the memory for each crow

19:       $t = t + 1$

20: **end while**

21: return M_i from memory for which $OF(M_i)$ is minimal/maximal

---

In *step 1* the population of N crows C_i $(i = 1, ..., N)$ is initialized randomly in the D - dimensional space such that the values for each dimension are in the interval $[Min_j, Max_j]$ $(j \in [1, D])$. For each crow C_i the fitness value is computed using an objective function in *step 2* and in *step 3* the memory of the crows M_i $(i = 1, ..., N)$ is initialized. In this step $M_i = C_i$ for all $i \in [1, N]$. In *step 4* the awareness probability AP which is a number from $[0, 1]$ and the flight length fl are defined. These parameters are considered in the equations that are used for the generation of the new positions of the crows.

The current iteration t is initialized to 0 in *step 5* and then for a number of iterations equal to $Iter_{max}$ the steps 7-19 are executed. In steps 7-15 the positions of the crows are updated as follows: in *step 8* the value of k is selected randomly from $\{1, ..., N\}$ and in *step 9* a random number r is generated from $[0, 1]$. If the value of r is greater than or equal to AP (step 10) then the new position of C_i is generated using the formula:

$$C_{i,j} = C_{i,j} + r \times fl \times (M_{k,j} - C_{i,j}) \qquad (8.1)$$

where $j \in [1, D]$. Otherwise the new position of C_i is generated randomly in the D - dimensional search space (step 13). The feasibility of each crow C_i is checked in *step 16*, and in this chapter a position is considered feasible if the values for each dimension are in the interval $[Min_j, Max_j]$ $(j \in [1, D])$. If this condition is not respected then the new values are updated accordingly to Min_j or Max_j $(j \in [1, D])$. The fitness value for each crow C_i is computed in *step 17* and in *step 18* the memory is updated to the value of the position if the fitness value of the new position is better. The current iteration t is incremented in *step 19*. Finally, in *step 21*, the algorithm returns that value M_i from memory for which the corresponding fitness value is minimal or maximal depending on the type of the optimization problem.

8.3   Modifications of CSA

This section presents a selection of representative modifications of CSA [2].

8.3.1   Chaotic Crow Search Algorithm (CCSA)

In the case of the CCSA modification of CSA [16], the searching mechanism of CSA introduces various chaotic maps [20] and the formulas for some of the most representative chaotic maps applied in that article are presented next. In all three cases the initial value of x_0 is a random number from $[0, 1]$.

(a) Chebyshev chaotic map

$$x_{t+1} = cos\left(t \times cos^{-1}(x_t)\right) \tag{8.2}$$

where $x_t \subsetneq [0, 1]$.

(b) Logistic chaotic map

$$x_{t+1} = c \times x_t \times (1 - x_t) \tag{8.3}$$

where c is a number from $[0, 4]$ and $x_t \in [0, 1]$.

(c) Piecewise chaotic map

$$x_{t+1} = \begin{cases} \frac{x_t}{p} & \text{if } 0 \leq x_t < p \\ \frac{x_t - p}{\frac{1}{2} - p} & \text{if } p \leq x_t < \frac{1}{2} \\ \frac{1 - p - x_t}{\frac{1}{2} - p} & \text{if } \frac{1}{2} \leq x_t < 1 - p \\ \frac{1 - x_t}{p} & \text{if } 1 - p \leq x_t < 1 \end{cases} \tag{8.4}$$

where $p = 0.2$ and $x_t \in [0, 1]$.

The chaotic maps are applied in the equation that is used for updating the position of the crow as follows:

$$C_{i,j} = \begin{cases} C_{i,j} + m_i \times fl \times (M_{k,j} - C_{i,j}) & \text{if } m_z \geq AP \\ r_j \times (Max_j - Min_j) + Min_j & \text{otherwise} \end{cases} \tag{8.5}$$

where r_i and r from the original equations of CSA are substituted with m_i and m_z, values that are obtained using the chaotic map.

8.3.2   Modified Crow Search Algorithm (MCSA)

MCSA [17] addresses the premature convergence and the stagnation of the classical CSA. The AP and fl parameters are tuned using an approach based on information about population diversity and Gaussian distribution as follows:

$$AP = 0.02 \times |R| \tag{8.6}$$

$$fl = 0.4 \times |R| \times (1 - div) \tag{8.7}$$

where R is generated by $\mathcal{N}(0, 1)$, the Gaussian distribution with mean 0 and standard deviation 1, and div represents the mean of the population diversity for the current iteration after it is truncated in the interval $[0, 1]$ according to the approach described in more detail in [21].

8.3.3 Binary Crow Search Algorithm (BCSA)

This version is applied in the case of optimization problems in which the search space is discrete and the solutions of BCSA [18] can be represented as arrays of zeros and ones. The formula applied in order to transform the positions of the crows in binary vectors is:

$$B_{i,j} = \begin{cases} 1 & \text{if } F(C_{i,j}) \geq 0.5 \\ 0 & \text{otherwise} \end{cases} \tag{8.8}$$

where F is defined by the formula:

$$F(C_{i,j}) = \left| \frac{C_{i,j}}{\sqrt{1 + C_{i,j}^2}} \right| \tag{8.9}$$

such that $C_{i,j}$ is the position of the crow C_i for the j-th dimension and $B_{i,j}$ is the binary variant.

A possible alternative of BCSA would be one in which another function is applied in order to convert the positions of the crows in arrays of zeros and ones like in the case of Binary Chicken Swarm Optimization (BCSO) [19] where a sigmoid function that has the formula presented below is applied:

$$F(C_{i,j}) = \frac{1}{1 + e^{-C_{i,j}}} \tag{8.10}$$

Formula 8.9 describes a V-shape transfer function while formula 8.10 describes an S-shape transfer function. According to the experiments performed in [22] on another bio-inspired algorithm called Ant Lion Optimization (ALO), the approach based on the V-shape transfer function improves significantly the performance of the optimization algorithm.

8.4 Application of CSA for jobs status prediction

8.4.1 Problem description

The problem that is approached in this chapter as an illustrative application of CSA for the solving of engineering problems in the tuning of the number of nodes of each layer of a deep neural network that is applied for the prediction of the jobs status. The problem is complex because the deep neural network has 10 hidden layers and each layer has a number of nodes from the set $\{10, ..., 100\}$. A brute force approach would consider $(100 - 10 + 1)^{10} = 91^{10}$ different combinations of nodes, where 91 is the size of the set $\{10, ..., 100\}$ and 10 is the number of hidden layers, and consequently a brute force approach is not feasible because the deep neural network would require 91^{10} training operations, one training for each possible combination of nodes.

The problem described in this chapter is a multiobjective optimization problem. The first objective is the minimization of the number of nodes of the deep neural network and the second objective is the maximization of the accuracy of the deep neural network. The multiobjective function used by CSA has the formula:

$$OF(C) = w_1 \times \frac{no_{nodes}(dnn(C)) - min_{nodes}}{max_{nodes} - min_{nodes}} + w_2 \times (1 - accuracy(dnn(C)))$$

$$(8.11)$$

The parameters used in formula 8.11 have the following meaning: min_{nodes} is the minimum possible number of nodes of the hidden layers of the deep neural network and has the value $min_{nodes} = 10 \times 10 = 100$, max_{nodes} is the maximum possible number of nodes of the hidden layers of the deep neural network and has the value $max_{nodes} = 10 \times 100 = 1000$, $dnn(C)$ is the deep neural network trained and tested using a configuration of nodes for the hidden layers indicated by the position of the crow C, $no_{nodes}(dnn(C))$ is the number of nodes of $dnn(C)$ and $accuracy(dnn(C))$ is the accuracy of $dnn(C)$. Moreover, the weights w_1 and w_2 respect the relation:

$$w_1 + w_2 = 1 \tag{8.12}$$

Due to the fact that in the second term of formula 8.11 the value of the accuracy is subtracted from 1, the optimization problem presented in this chapter minimizes both criteria and consequently the ideal solution has the fitness value equal to 0 and the worst solution has the fitness value equal to 1. The optimization problem uses as experimental support the CIEMAT Euler log [1]. The main characteristics of that dataset are summarized in Table 8.1.

TABLE 8.1
Main Characteristics of the CIEMAT Euler log [1].

Characteristic	Value
monitoring duration	from November 2008 to December 2017
number of jobs	9263012
number of features	18

The significance of each feature is presented briefly in Table 8.2 and more details about each feature can be found using the reference [1].

The meaning of the status of each job [1] is presented in Table 8.3.

In this chapter only the last 100000 records are used in experiments due to the time constraints associated with the training of the deep learning model. Those records are selected as the last records because they describe the newest monitored data and they are also consecutive records in order to respect the timeseries nature of the monitored data. The training data consists of 80000 records and the testing data consists of 20000 records which are selected randomly from the set of 100000 records. Moreover, from the initial set of 18

TABLE 8.2
CIEMAT Euler Log Features Description.

Feature No.	Description	Value
1	the number of the job	integer
2	the submit time	seconds
3	the wait time	seconds
4	the run time	seconds
5	the number of processors used by the job	integer
6	the average CPU time used	seconds
7	the used memory	kilobytes
8	the requested number of processors	integer
9	the requested time	seconds
10	the requested memory	kilobytes
11	the status of the job	integer
12	the ID of the user	positive integer
13	the ID of the group	positive integer
14	the application number	positive integer
15	the queue number	positive integer
16	the partition number	positive integer
17	the number of the preceding job	positive integer
18	the think time from the preceding job	seconds

TABLE 8.3
The Meaning of the Status of Each Job.

Status	Meaning
0	the job failed
1	the job was completed successfully
2	a partial execution that will be continued
3	the job was completed (the last partial execution)
4	the job failed (the last partial execution)
5	the job was canceled

features the features $\{10, 16, 17, 18\}$ are removed because they are constant for all records. The influence of the other features on the final classification results was validated using the Boruta [23] algorithm for features selection from R. In Figure 8.1 are presented the features selection results after the Boruta algorithm was applied for all 100000 records.

The Boruta algorithm compares iteratively the importances of the attributes with the importances of the shadow attributes (permuted copies). The attributes which are significantly better than the shadowMax are admitted as confirmed and the attributes left without any decision are considered tentative. The remaining attributes are rejected. As shown in Figure 8.1, from

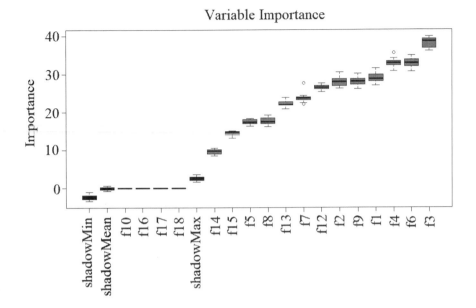

FIGURE 8.1

Features selection results after the application of the boruta algorithm.

the total number of 17 features, 13 features are confirmed and 4 features are rejected. The feature number 11 is the class label and consequently both for the training data and for the testing data the records have 13 features and 1 label. The labels of the records have only one status from the set $\{0, 1, 2, 3, 4, 5\}$ therefore the deep learning model has six possible outputs that correspond to each type of status.

8.4.2   How can CSA be used for this problem?

This subsection describes how CSA is used for the problem of tuning a deep neural network which is applied for job status prediction. The architecture of the neural network applied in this chapter is presented in Figure 8.2. The number of nodes of the first layer and the number of nodes of the last layer are static while the hidden layers have variable numbers of neurons given by the position of the crow associated with that deep neural network.

The deep learning model applied for the classification of the job status was developed in Python using Keras and TensorFlow and it is based on a sequential model with 1 epoch and a batch size equal to 64. An epoch describes the passing of the entire dataset forward and backward through the deep neural network only once and the batch size describes the total number of training samples that are present in a single batch or part of the training dataset. The values of the features are normalized prior to the application of the deep

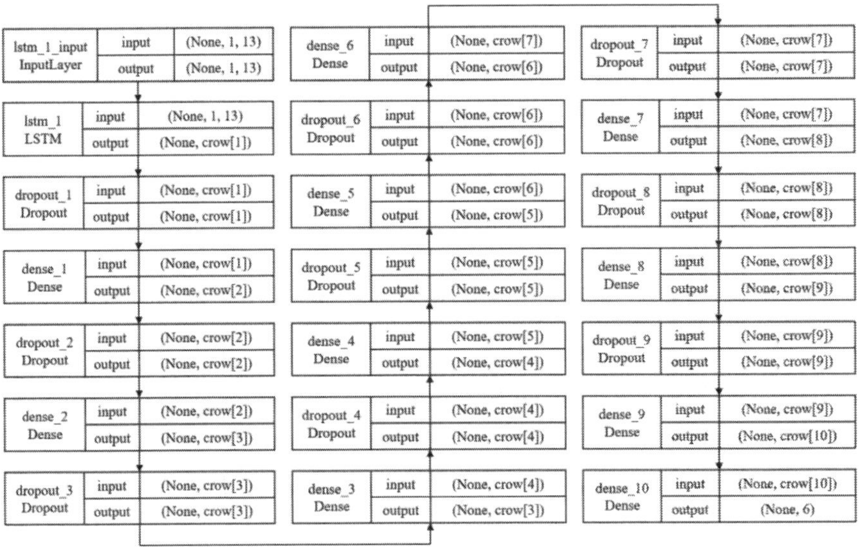

Layer		Shape		Layer		Shape		Layer		Shape
lstm_1_input InputLayer	input	(None, 1, 13)		dense_6 Dense	input	(None, crow[7])		dropout_7 Dropout	input	(None, crow[7])
	output	(None, 1, 13)			output	(None, crow[6])			output	(None, crow[7])
lstm_1 LSTM	input	(None, 1, 13)		dropout_6 Dropout	input	(None, crow[6])		dense_7 Dense	input	(None, crow[7])
	output	(None, crow[1])			output	(None, crow[6])			output	(None, crow[8])
dropout_1 Dropout	input	(None, crow[1])		dense_5 Dense	input	(None, crow[6])		dropout_8 Dropout	input	(None, crow[8])
	output	(None, crow[1])			output	(None, crow[5])			output	(None, crow[8])
dense_1 Dense	input	(None, crow[1])		dropout_5 Dropout	input	(None, crow[5])		dense_8 Dense	input	(None, crow[8])
	output	(None, crow[2])			output	(None, crow[5])			output	(None, crow[9])
dropout_2 Dropout	input	(None, crow[2])		dense_4 Dense	input	(None, crow[5])		dropout_9 Dropout	input	(None, crow[9])
	output	(None, crow[2])			output	(None, crow[4])			output	(None, crow[9])
dense_2 Dense	input	(None, crow[2])		dropout_4 Dropout	input	(None, crow[4])		dense_9 Dense	input	(None, crow[9])
	output	(None, crow[3])			output	(None, crow[4])			output	(None, crow[10])
dropout_3 Dropout	input	(None, crow[3])		dense_3 Dense	input	(None, crow[4])		dense_10 Dense	input	(None, crow[10])
	output	(None, crow[3])			output	(None, crow[3])			output	(None, 6)

FIGURE 8.2
Architecture of the deep neural network tuned using CSA applied for the prediction of the jobs status.

neural network in order to obtain better accuracy results. The first hidden layer is a LSTM (Long-Short Term Memory) layer and the other 9 layers are dense layers with *relu* activation. Between each 2 consecutive hidden layers a dropout equal to 0.5 is considered. The last layer has *softmax* activation and 6 outputs. The model is compiled using the *categorical_ crossentropy* loss and the *adam* optimizer.

The application of CSA for this problem leads to an adaptation of the classical version of the algorithm. The position of each crow is a vector with 10 values such that each value describes the number of nodes of each hidden layer and is a value from $[10, 100]$. These values are chosen in order to create a complex search space for which a brute force approach is not feasible. Moreover, these values are also chosen in order to prevent overfitting. The positions of the crows are discretized converting the real values from the interval $[10, 100]$ to integer values immediately after (step 16) of the original Algorithm 11 where the feasibility of the solutions is checked. This transformation is mandatory because each layer of the deep neural network must be characterized by an integer number of nodes. Therefore, for a crow C_i from the search space the discretized position is given by the vector:

$$([C_{i,1}], ..., [C_{i,D}]) \tag{8.13}$$

where $[]$ is the mathematical formula for the integer part and $(C_{i,1}, ..., C_{i,D})$ is the position of the crow C_i before the discretization phase.

8.4.3   Experiments description

Three types of experiments were considered depending on the criteria used in the seeking of the solutions in the search space. These criteria are summarized in Table 8.4.

TABLE 8.4

Description of the Criteria Applied in the CSA Experiments.

Criterion	Description
$(w_1, w_2) = (1.0, 0.0)$	the solutions are sought using only the first criterion
$(w_1, w_2) = (0.5, 0.5)$	the solutions are sought with regards to both criteria simultaneously
$(w_1, w_2) = (0.0, 1.0)$	the solutions are sought using only the second criterion

For each experiment 10-fold repetitions were performed and the values of the average (r_{avg}) and of the standard deviation (SD) were computed. The formulas of the two evaluation metrics are:

$$r_{avg} = \frac{\sum_{i=1}^{10} r_i}{10} \tag{8.14}$$

and

$$SD = \sqrt{\frac{\sum_{i=1}^{10} (r_i - r_{avg})^2}{9}} \tag{8.15}$$

where r_i with $i \in \{1, ..., 10\}$ is the result obtained in the i-th fold of the experiment.

For each criterion the search space is explored as follows:

- criterion 1 $(w_1 = 1.0, w_2 = 0.0)$: the number of nodes in each layer is optimized without considering the value of the accuracy;

- criterion 2 $(w_1 = 0.5, w_2 = 0.5)$: the number of nodes in each layer is optimized in order to obtain a higher accuracy;

- criterion 3 $(w_1 = 0.0, w_2 = 1.0)$: the accuracy is optimized without considering the number of nodes in each layer;

The performance of CSA is compared with the performance of PSO and of CSO and consequently both for PSO and for CSO three experiments which consist of 10-fold repetitions are performed. In Tables 8.5, 8.6 and 8.7 are presented the values of the configuration parameters used in experiments for CSA, PSO and CSO, respectively.

TABLE 8.5
Values of the CSA Configuration Parameters.

Configuration Parameter	Value
number of crows (N)	10
number of dimensions (D)	10
maximum number of iterations ($Iter_{max}$)	50
minimum possible value for crow positions (Min)	10
maximum possible value for crow positions (Max)	100
flight length (fl)	5
awareness probability (AP)	0.5

TABLE 8.6
Values of the PSO Configuration Parameters.

Configuration Parameter	Value
number of particles (N)	10
number of dimensions (D)	10
maximum number of iterations ($Iter_{max}$)	50
minimum possible value for particles positions (Min)	10
maximum possible value for particles positions (Max)	100
minimum possible value for particles velocity (V_{min})	−1
maximum possible value for particles velocity (V_{max})	1
cognitive coefficient value (c_1)	2
social coefficient value (c_2)	2

TABLE 8.7
Values of the CSO Configuration Parameters.

Configuration Parameter	Value
number of chickens (N)	10
number of dimensions (D)	10
maximum number of iterations ($Iter_{max}$)	50
hierarchical order updating period (G)	5
minimum value of the flight length (FL_{min})	0.5
maximum value of the flight length (FL_{max})	0.9
a number used for avoiding division by zero (e)	0.000000001
roosters percent (RP)	10%
hens percent (HP)	30%
minimum possible value for chickens positions (Min)	10
maximum possible value for chickens positions (Max)	100

8.4.4   Results

In Table 8.8 are presented the values of the average (r_{avg}), of the standard deviation (SD) and of the best result $(best)$ returned after the execution of the 10-folds of the experiments in the case for each combination of optimization algorithm and criterion.

TABLE 8.8

Summary of the Experimental Results for Each Optimization Algorithm and for Each Criterion.

Metric	Algorithm	Criterion 1	Criterion 2	Criterion 3
r_{avg}	CSA	0.02977	0.07541	0.05319
	PSO	0.27910	0.19207	0.05311
	CSO	0.09799	0.11799	0.05456
SD	CSA	0.04794	0.03521	0.00016
	PSO	0.05807	0.03067	0.00007
	CSO	0.08110	0.04796	0.00235
$best$	CSA	0.00000	0.02910	0.05295
	PSO	0.19444	0.13007	0.05299
	CSO	0.02333	0.06533	0.05349

In Tables 8.9, 8.10 and 8.11 are presented the configurations of the deep neural network which correspond to the best results returned by CSA, PSO and CSO, respectively.

TABLE 8.9

Summary of the Best Results Returned by CSA.

Criterion	Result	Crow Position
$(w_1, w_2) = (1.0, 0.0)$	0.00000	$(10, 10, 10, 10, 10, 10, 10, 10, 10, 10)$
$(w_1, w_2) = (0.5, 0.5)$	0.02910	$(10, 10, 10, 10, 10, 10, 10, 10, 10, 10)$
$(w_1, w_2) = (0.0, 1.0)$	0.05295	$(48, 100, 10, 98, 83, 90, 100, 90, 45, 77)$

TABLE 8.10

Summary of the Best Results Returned by PSO.

Criterion	Result	Particle Position
$(w_1, w_2) = (1.0, 0.0)$	0.19444	$(25, 50, 10, 18, 10, 31, 66, 13, 13, 39)$
$(w_1, w_2) = (0.5, 0.5)$	0.13007	$(21, 21, 10, 10, 21, 63, 10, 63, 29, 10)$
$(w_1, w_2) = (0.0, 1.0)$	0.05299	$(65, 66, 55, 66, 56, 35, 69, 53, 43, 68)$

In Figure 8.3 is presented the evolution of the OF(.) value for each criterion for CSA, PSO and CSO for the folds of the experiments that led to the best results which are summarized in Tables 8.9, 8.10 and 8.11.

TABLE 8.11
Summary of the Best Results Returned by CSO.

Criterion	Result	Chicken Position
$(w_1, w_2) = (1.0, 0.0)$	0.02333	$(11, 10, 10, 12, 10, 10, 28, 10, 10, 10)$
$(w_1, w_2) = (0.5, 0.5)$	0.06533	$(18, 10, 16, 10, 10, 22, 10, 10, 26, 10)$
$(w_1, w_2) = (0.0, 1.0)$	0.05349	$(83, 98, 57, 39, 44, 92, 12, 96, 46, 63)$

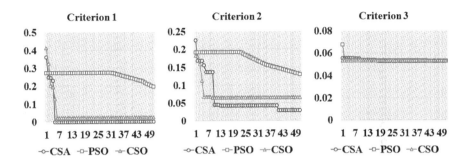

FIGURE 8.3
Evolution of the OF(.) value for CSA, PSO and CSO for each criterion for the folds of the experiments that returned the best results.

The best results were obtained by CSA for all three criteria and these results might be justified by the big value of the flight length (fl) which is equal to 5 and by the value of the awareness probability (AP) which is equal to 0.5. These values of the two configuration parameters lead to a premature convergence of the CSA algorithm. A similar behavior is presented by CSO which returns the second best results for the first criterion and for the second criterion. In the case of PSO the search space is exploited more due to the fact that the velocity of the particles is limited to values from the interval $[-1, 1]$, but that limitation also leads to a slow exploration rate of the particles. However, even with this limitation PSO returns a value almost identical to the value returned by CSA in the case of the third criterion.

8.5 Conclusions

This chapter described briefly the original version of CSA, presented modifications of CSA such as CCSA, MOCSA and BCSA, and finally showed how CSA can be applied for the tuning of the number of nodes of each layer of a deep neural network used for the prediction of jobs status. As future research work the following research directions are proposed: (1) the development of novel modifications of CSA having as a main source of inspiration the modifications of other classical bio-inspired algorithms, (2) the research of the hybridizations of CSA with other bio-inspired algorithms and (3) the application of CSA in the solving of other engineering problems.

References

1. The CIEMAT Euler log, `https://www.cse.huji.ac.il/labs/parallel/workload/l_ciemat_euler/index.html`

2. A. Askarzadeh. "A novel metaheuristic method for solving constrained engineering optimization problems: Crow search algorithm" in *Computers & Structures*, vol. 169, 2016, pp. 1-12.

3. J. Stender. "Introduction to genetic algorithms" in *IEE Colloquium on Applications of Genetic Algorithms*, 1994, pp. 1/1-1/4.

4. Eberhart, Y. Shi. "Particle swarm optimization: developments, applications and resources" in *Proc. of the 2001 Congress on Evolutionary Computation (IEEE Cat. No.01TH8546)*, 2001, pp. 81-86.

5. J. Zhang, P. Zhang. "A study on harmony search algorithm and applications" in *Proc. of the 2018 Chinese Control And Decision Conference (CCDC)*, 2018, pp. 736-739.

6. A. Darwish. "Bio-inspired computing: Algorithms review, deep analysis, and the scope of applications" in *Future Computing and Informatics Journal*, 2018, pp. 231-246.

7. Y. Pei, W. Wang, S. Zhang. "Basic Ant Colony Optimization" in *Proc. of the 2012 International Conference on Computer Science and Electronics Engineering*, 2012, pp. 665-667.

8. S. Deb, S. Fong, Z. Tian. "Elephant Search Algorithm for optimization problems" in *Proc. of the 2015 Tenth International Conference on Digital Information Management (ICDIM)*, 2015, pp. 249-255.

9. D. Moldovan, V. Chifu, C. Pop, T. Cioara, I. Anghel, I. Salomie. "Chicken Swarm Optimization and deep learning for manufacturing processes" in *Proc. of the 2018 17th RoEduNet Conference: Networking in Education and Research (RoEduNet)*, 2018, pp. 1-6.

10. M. Kumar, J. S. Dhillon. "Hybrid Artificial Algae Algorithm for global optimization" in *Proc. of the 2017 3rd International Conference on Advances in Computing, Communication & Automation (ICACCA) (Fall)*, 2017, pp. 1-6.

11. D. Moldovan, I. Anghel, T. Cioara, I. Salomie, V. Chifu, C. Pop. "Kangaroo Mob heuristic for optimizing features selection in learn-

ing the daily living activities of people with Alzheimer's" in *Proc. of the 2019 22nd International Conference on Control Systems and Computer Science (CSCS)*, 2019, pp. 236-243.

12. N. Jangir, M. H. Pandya, I. N. Trivedi, R. H. Bhesdadiya, P. Jangir, A. Kumar. "Moth-Flame optimization algorithm for solving real challenging constrained engineering optimization problems" in *Proc. of the 2016 IEEE Students' Conference on Electrical, Electronics and Computer Science (SCEECS)*, 2016, pp. 1-5.

13. A. Saha, A. Bhattacharya, P. Das, A. K. Chakraborty. "Crow search algorithm for solving optimal power flow problem" in *Proc. of the 2017 Second International Conference on Electrical, Computer and Communication Technologies (ICECCT)*, 2017, pp. 1-8.

14. L. dos Santos Coelho, C. Richter, V. C. Mariani, A. Askarzadeh. "Modified crow search approach applied to electromagnetic optimization" in *Proc. of the 2016 IEEE Conference on Electromagnetic Field Computation (CEFC)*, 2016, pp. 1-1.

15. P. A. Kowalski, K. Franus, S. Lukasik. "Crow Search Algorithm for continuous optimization tasks" in *Proc. of the 2019 6th International Conference on Control, Decision and Information Technologies (CoDIT)*, 2019, pp. 7-12.

16. G. I. Sayed, A. Darwish, A. E. Hassanien. "Chaotic crow search algorithm for engineering and constrained problems" in *Proc. of the 2017 12th International Conference on Computer Engineering and Systems (ICCES)*, 2017, pp. 676-681.

17. L. dos Santos Coelho, C. E. Klein, V. C. Mariani, C. A. R. do Nascimento, A. Askarzadeh. "Electromagnetic optimization based on gaussian crow search approach" in *Proc. of the 2018 International Symposium on Power Electronics, Electrical Drives, Automation and Motion (SPEEDAM)*, 2018, pp. 1107-1112.

18. R. C. T. De Souza, L. d. S. Coelho, C. A. De Macedo, J. Pierezan. "A V-shaped binary crow search algorithm for feature selection" in *Proc. of the 2018 IEEE Congress on Evolutionary Computation (CEC)*, 2018, pp. 1-8.

19. M. Han, S. Liu. "An improved Binary Chicken Swarm Optimization algorithm for solving 0-1 knapsack problem" in *Proc. of the 2017 13th International Conference on Computational Intelligence and Security*, 2017, pp. 207-210.

20. A. H. Abdullah, R. Enayatifar, M. Lee. "A hybrid genetic algorithm and chaotic function model for image encryption" in *AEU - International Journal of Electronics and Communications*, vol. 66, 2012, pp. 806-816.

21. L. S. Coelho, T. C. Bora, V. C. Mariani. "Differential evolution based on truncated Levy-type flights and population diversity measure to solve economic load dispatch problems" in *International Journal of Electrical Power & Energy Systems*, vol. 57, 2014, pp. 178-188.

22. M. Mafarja, D. Eleyan, S. Abdullah, S. Mirjalili. "S-shaped vs. V-shaped transfer functions for ant lion optimization algorithm in feature selection problem" in *CoRR*, vol. abs/1712.03223, 2017, pp. 1-7.

23. Package 'Boruta', https://cran.r-project.org/web/packages/Boruta/Boruta.pdf

9

Cuckoo Search Optimisation – Modifications and Application

Dhanraj Chitara

Department of Electrical Engineering
Swami Keshvanand Institute of Technology (SKIT), Jaipur, India

Nand K. Meena

School of Engineering and Applied Science
Aston University, Birmingham, B4 7ET, United Kingdom

Jin Yang

James Watt School of Engineering
University of Glasgow, Glasgow, G12 8LT, United Kingdom

CONTENTS

9.1   Introduction

The cuckoo search optimization (CSO) algorithm is one of the recently developed bio-inspired meta-heuristic algorithms which is capable and very effective in solving complex combinatorial optimization problems. It is inspired from the cuckoo's aging behavior based on the parasitized breeding mechanism of the cuckoos egg and Levy flight search principle. The algorithm was developed by Yang and Deb [1], inspired by the obligate brood parasitism of some the cuckoo species. The cuckoo is a very pretty bird that not only has an attractive and sweet voice but is also intelligent enough to have a forceful reproduction strategy by which adult cuckoos lay their eggs in the nests of other host birds or species. The implementation of Levy flight behavior in CSO, makes this algorithm superior over other meta-heuristic optimization techniques such as genetic algorithm and particle swarm intelligence. In this chapter, a brief of the CSO algorithm is discussed and then some of the improved variants of this algorithm are also discussed.

The organization of this chapter is as follows. In Section 9.2, a brief of the standard CSO algorithm is presented including the explanation of cuckoo breeding behavior and Levy flights. All steps of the CSO algorithm are also presented. Some of the modified variants of CSO algorithm suggested in literature, to overcome some of the limitations observed in its standard variant, are also discussed in Section 9.3. The sections include modified cuckoo search by varying step size only, and improved cuckoo search by varying both step size and probability index. In Section 9.4, the application of the modified CSO algorithm is demonstrated to optimally design the power system stabilizer (PSS) parameters of a multi-machine power system (MMPS).

9.2   Original CSO algorithm in brief

9.2.1   Breeding behavior of cuckoo

Cuckoos are fascinating birds with a sweet voice and aggressive reproduction strategy. Some cuckoo species like the Ani and Guira lay their eggs in common nests, though they may remove other eggs to increase the hatching probability of their own eggs. There are large numbers of cuckoo species that engage in the obligate brood parasitism by laying their eggs in the nests of other host birds, often other species.

In cuckoo search optimization, three main types of brood parasitism are adopted: intra-specific brood parasitism, cooperative breeding, and nest takeover. A little host bird can engage in direct conflict with the interfering cuckoos. If a host bird discovers that some or all eggs are not their own, they will either throw out these unknown eggs or simply leave their nest or build a new nest somewhere else. Some cuckoo species such as the new world brood-parasitic Tapera have evolved in such a way that female parasitic cuckoos are often very specialized to mimic egg colors and patterns of a few chosen host species [1, 2]. This reduces the probability of their eggs being abandoned and thereby increases their reproduction.

9.2.2 Levy flights

Levy flight is a random walk where step size has a Levy tailed probability distribution. The term Levy flight was coined by Benoit Mandelbrot who used a specific definition for distribution of step sizes. Ultimately, the term Levy flight has been used to refer to a discrete grid rather than continuous space. It gives a scale invariant property that is used to model the data for exhibiting clusters. In nature, many animals and insects follow the Levy flight properties. Recent studies by Reynolds and Frye demonstrated the behavior of fruit flies (Drosophila Melanogaster) which explore their landscape by using numerous series of straight flight paths/routes, followed by a sudden right angle turn which is a Levy-flight-style intermittent scale free [3, 4].

9.2.3 Cuckoo search optimization algorithm

In beginning of the algorithm, each egg of the nest represents a solution whereas cuckoo egg represents a new solution. The algorithm has been described by the breeding strategy of some cuckoo species in conjunction with Levy flight behavior of a few birds. In this study also, if a host bird searches the eggs that are not its own then they either throw these foreign eggs away or just abandon the nest and construct a new nest at other places. The standard version of CSO algorithm follows three rules, listed here.

- Each cuckoo places one egg at a time and drops it in an arbitrarily selected nest.

- The best nests with excellence fitness or eggs will carry forward to the next generation.

- The number of host nests is fixed and the egg laid by a cuckoo is detected by the host bird with a probability index $p_a \in (0, 1)$.

The new solution (cuckoo) $x_i^{(t+1)}$ is generated by application of Levy flight as

$$x_i^{(t+1)} = x_i^{(t)} + a \oplus Levy(\lambda) \tag{9.1}$$

where, $\epsilon > 0$ is a step size that should be related to the scale problem of interest. Mostly, the value of step size $a = 1$ is chosen. The product $\oplus$ means an entry wise walk during multiplication. A Levy flight is an arbitrary walk in which the steps are defined in terms of step length, which have a definite probability distribution with the direction of steps being isotropic and random. It can be defined as

$$Levy \sim u = t^{-\lambda}, (1 < \lambda \le 3) \tag{9.2}$$

Here, the steps essentially form a random walk process that follows a power law step-length distribution with a heavy tail. To speed up the local search, some of the new solutions should be generated by a Levy walk around the best solution obtained so far. Before starting the iteration process, CSO identifies the best fitness x_{best}. Equation (9.2) has infinite mean with infinite variance. The detection step Φ is expressed as

$$\Phi = \left[\frac{\Gamma(1 + \beta) \cdot \sin\left(\pi \frac{\beta}{2}\right)}{\Gamma\left(\left(\frac{1+\beta}{2}\right) \cdot \beta \cdot 2^{\frac{\beta-2}{2}}\right)} \right]^{\frac{1}{\beta}} \tag{9.3}$$

where, Γ denotes the gamma function. In this algorithm, the value of β is taken as 1.5 and the evolution phase of x_i begins by defining v as $v = x_i$ [2]. After this step, the required step is evaluated as

$$stepsize_i = 0.01 \left(\frac{u_i}{v_i} \right)^{\frac{1}{\beta}} \cdot (v - x_{best}) \tag{9.4}$$

However, a substantial fraction of new solutions generated by far field randomization are located far enough from the current best solution. It ensures that the algorithm will not be trapped in a local optima. The control parameters of the algorithm are scale factor (β) and probability index (p_a). The pseudo-code of standard CSO is presented in Algorithm 12.

Algorithm 12 Pseudo-code of standard CSO.

1: Define objective function $F(x), x = x^1, x^2, x^3, \ldots, x^d;$ ▷ d=dimension
2: set CSO algorithm parameters such as total number of host nests n, step size a, scale factor β, the probability index p_a, and maximum generation $T^{\max}$;
3: generate initial but feasible population of n host nests and compute fitness function $f_i = F(x_i)$ for each individual solution $i \in n$;
4: **while** $gen < T^{\max}$ **do**   ▷ generation starts here...
5:     randomly select a cuckoo i and change its solution by using Levy flights;
6:     compute its fitness value or function $f_i = F(x_i)$;
7:     select a nest j among available n nests;
8:     **if** $f_j > f_i$ **then**   ▷ for maximization
9:         replace the solution x_i with x_j and f_i with f_j;
10:     **end if**
11:     a fraction (p_a) of worse nests are rejected and new nests are created;
12:     retain the most best/excellent nests with better fitness value;
13:     rank the nests and determine the current best solution;
14: **end while**

9.3 Modified CSO algorithms

In this section, some of the popularly known improved variants of CSO algorithms are discussed.

9.3.1 Gradient free cuckoo search

The standard variant of CSO algorithm is able to find an optimum solution for most of the complex optimization problems however a fast convergence cannot be guaranteed because its search entirely depends on random walks. In order to overcome the limitation, a modified cuckoo search (MCS) was proposed in [5]. Two modifications have been suggested to improve the convergence rate that makes the method more useful for wide-range applications.

In the first modification, the Levy flight step size a is adjusted. In basic CSO [1], the step size a is considered as a constant with value $a = 1$. In this modification, the value of a decreases as the number of generations increases to enforce more localized search so that individuals or the eggs get closer to the optimal solution. Initially, the value of Levy flight step size $a = 1$ is chosen and, at every generation, a new Levy flight step is evaluated by using $\alpha = a/\sqrt{G}$, where G is the current generation. This exploratory search is only executed on the fraction of nests to be rejected. In the second modification, the information exchange between the eggs is also introduced to speed up the convergence. This information exchange between eggs is missing in the standard CSO algorithm; essentially, the searches are completed independently [2].

In this modified CSO version, a part of the eggs with the greatest fitness will be placed into a set of top eggs. For each of the top eggs, a second egg in this set is randomly chosen and a new egg is then created on the line connecting these two top eggs. An inverse of the golden ratio $\phi = (1 + \sqrt{5})/2$ is used to evaluate the distance along this line at which a new egg is placed. In a case when both eggs have the same fitness, a new egg is created at the midpoint. Here, a random fraction is used in place of the golden ratio.

There is a possibility that the same egg is picked twice in this step [5]. In this case, a local Levy flight search is implemented from the randomly picked nest with step size $\alpha = a/G^2$. There are two parameters, a fraction of nests to be abandoned and a fraction of nests supposed to create the top nests. For most of the benchmark functions, the fraction of nests to be abandoned and the fraction of nests located in the top nests are set to 0.75 and 0.25 respectively [5].

9.3.2 Improved cuckoo search for reliability optimization problems

In the original CSO algorithm, the parameters p_a, λ and a are presented to find global and local improved solutions [6]. For varying the convergence rate of the CSO algorithm, the parameters p_a and a play an important role in fine tuning of solutions. The values of both p_a and a are assumed to be constant in the standard version of CSO. The performance of this algorithm is usually found poor for a high value of a and low value of p_a. It also requires a considerably high number of generations. Alternatively, the convergence speed increases for a lesser value of a and high values of p_a however increases the chances of premature convergence. The main advantage of improved cuckoo search (ICS) over basic CSO is its control on the values of p_a and a. In early iterations, the values of p_a and a should be sufficiently high to increase the diversity. However, the values of these parameters should decrease in final generations for fine tuning of solution vectors [6].

In ICS, the values of p_a and a are dynamically changing with generations, as expressed in (9.5)-(9.7).

$$P_a(gen) = \frac{(p_a^{\max} - p_a^{\min})}{NI} \times gen \tag{9.5}$$

$$\alpha(gen) = \alpha^{\max} exp(c \cdot gen) \tag{9.6}$$

$$c = \frac{Ln\left(\frac{\alpha_{\min}}{\alpha_{\max}}\right)}{NI} \tag{9.7}$$

where, NI and gen are denoting the total number of generations and current generation respectively.

9.4 Application of CSO algorithm for designing power system stabilizer

9.4.1 Problem description

In recent years, low frequency electromechanical oscillations is one of the most frequently encountered problems in small-signal stability analysis of interconnected power systems [7]. A small disturbance can significantly affect the characteristic of electromechanical oscillations of generators, irrespective of its origin. These oscillations produce oscillatory instability and cause system separation if system damping is insufficient. The electromechanical oscillations are generally controlled by a power system stabilizer (PSS)[5]. The PSS parameters are optimally designed to improve the system damping.

9.4.2 Objective function and problem formulation

In this section, an eigenvalue-based multiobjective function is devised for simultaneous control of damping factor and damping ratio of a two-area four machine (TAFM) power system [7]. PSS parameters are designed in such a way that unstable and/or poorly damped open-loop (without PSS) eigenvalues are shifted to a specified D-shape zone in the left-half of the s-plane for a wide range of operating conditions tested under different scenarios of severe disturbances.

In order to damp out the low frequency oscillations, the parameters of PSS are designed in such a way that the eigenvalue-based multiobjective function expressed in (9.8) should be minimum. This will place the unstable and/or poorly damped eigenvalues of all operating conditions to a D-shape zone characterized by $\sigma_{i,j} \geq \sigma_0$ and $\zeta_{i,j} \leq \zeta_0$ in the left-half of the s-plane, as shown in Fig. 9.1.

$$J = \sum_{j=1}^{np} \sum_{\sigma_{i,j} \geq \sigma_0} \left(\sigma_0 - \sigma_{i,j} \right)^2 + \sum_{j=1}^{np} \sum_{\zeta_{i,j} \leq \zeta_0} \left(\zeta_0 - \zeta_{i,j} \right)^2 \qquad (9.8)$$

subjected to:

$$K_i^{\min} \leq K_i \leq K_i^{\max} \qquad (9.9)$$

$$T_{1i}^{\min} \leq T_{1i} \leq T_{1i}^{\max} \qquad (9.10)$$

$$T_{3i}^{\min} \leq T_{3i} \leq T_{3i}^{\max} \qquad (9.11)$$

where, np is the number of operating points considered in the design problem. $\sigma_{i,j}$ and $\zeta_{i,j}$ denote the damping factor (real part) and damping ratio of the ith eigenvalue of the jth operating point. The value of the desired damping factor σ_0 and damping ratio ζ_0 are selected according to the problem requirement. The aim of CSO is to determine the optimal set of PSS parameters so that the minimum settling time and overshoots of the system are achieved.

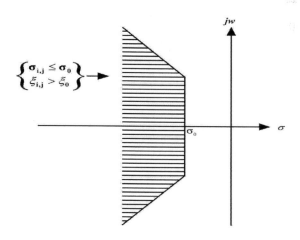

FIGURE 9.1
A D-shape zone in the left half of the s-plane where $\sigma_{i,j} \geq \sigma_0$ and $\zeta_{i,j} \leq \zeta_0$.

9.4.3   Case study on two-area four machine power system

This system comprises two generating areas connected by a 220-Km, 230-KV double circuit tie-line. The two areas 1 and 2 consist of two generators G_1, G_2 and G_3, G_4 respectively [7]. All generators' mechanical and electrical parameters are the same, except their inertia constants. These generators are represented by a fourth order non-linear model with a fast static excitation system. The single-line diagram and other details of the system are given in Fig. 9.2. Three test cases with different loading conditions have been considered to tune the PSS parameters for small-signal stability analysis, i.e., Case-1 as normal loading, Case-2 as light loading and Case-3 as high loading [8], as summarized in Table 9.1.

9.4.4   Eigenvalue analysis of TAFM power system without and with PSSs

The PSAT [9] is used for eigenvalue analysis of the system. Usually, a PSS is not required for the swing generator; therefore the PSS parameters of only

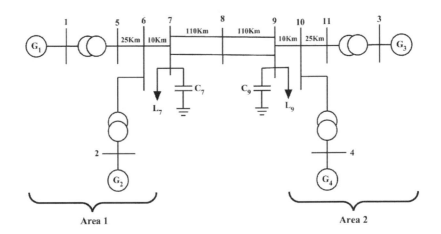

FIGURE 9.2
A single-line diagram of two-area four machine power system.

TABLE 9.1
Three operating conditions of TAFM power system.

Case-1	Case-2	Case-3
Nominal active power	Total active power decreasing by 20%	Total active power increasing by 20%
Nominal reactive power	Total reactive power decreasing by 15%	Total reactive power increasing by 15%

three generators G_1, G_2, G_4 are to be optimized as G_3 is the swing generator. The open-loop eigenvalues and damping ratios are calculated for only unstable and/or poorly damped modes of the system. Now, nine parameters of three generators $3 \times (K, T_1, T_2)$ are optimized by minimizing the objective function J by using CSO. The optimal values of these design parameters are presented in Table 9.2. The typical convergence characteristic of CSO is shown in Fig. 9.3. The figure shows that the algorithm is able to find the desired solution for which fitness function J is zero.

TABLE 9.2
Optimal values of PSS design parameters for three generators.

Generators	K	T1	T3
G1	99.421	0.014	0.116
G2	41.590	0.059	0.010
G4	36.781	0.049	0.035

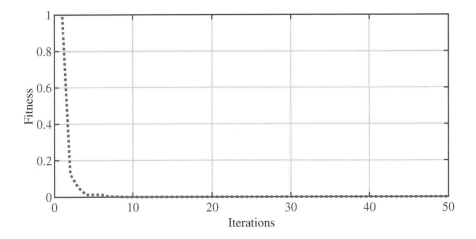

FIGURE 9.3
Convergence characteristic of CSO.

Furthermore, the comparison of eigenvalues and their damping ratios with and without PSSs for loading cases 1-3 is shown in Table 9.3. The table shows that all eigenvalues are shifted to a specified D-shape zone in the left half of the s-plane with improved damping factor and ratio.

9.4.5 Time-domain simulation of TAFM power system

In order to examine the performance of previously designed PSS controllers in terms of speed deviations under different scenarios of severe disturbances on

TABLE 9.3

Eigenvalues and damping ratios with and without PSSs for all operating cases.

Cases	Without PSS	With CSO designed PSS
Case-1	0.026±j3.803, -0.070	-1.157 ± j4.275, 0.26
	-0.541±j7.027, 0.076	-6.742±j4.788, 0.81
	-0.543±j6.810, 0.079	-7.733+j2.332, 0.95
Case-2	-0.068±j3.279, 0.021	-1.090±j3.595, 0.29
	-1.010±j6.380, 0.156	-7.472±j2.563, 0.94
	-0.535±j6.786, 0.078	-6.237±j4.167, 0.83
Case-3	0.160±j3.751, -0.042	-1.119±j5.322, 0.20
	0.042±j7.129, -0.005	-4.343±j2.917, 0.83
	-0.545±j6.803, 0.079	-5.049±j0.353, 0.99

TABLE 9.4

Scenarios of disturbances for testing the performance of designed PSSs on TAFM power system.

Scenarios	Detail
Scenario-1	A 9-cycle, 3-phase fault occurs at $t = 1$ sec on bus 7 without tripping the line 7-8 for Case-2
Scenario-2	A 12-cycle, 3-phase fault occurs at $t = 1$ sec on bus 9 without tripping the line 8-9 for Case-3

the TAFM power system are considered as shown in Table 9.4. For the sake of robustness, the number of cycles of operation increases with three-phase faults until designed controllers for the system fail.

Due to space limitation, specimen results for the comparison of speed deviations Δw_1, Δw_3 for Scenario-1 of Case-2 and the Δw_2, Δw_4 for Scenario-2 of Case-3 with designed PSS controllers are only shown in Fig. 9.4 (a)-(b) and 9.4(c)-(d) respectively. From this figure, it is clear that the system performance with designed PSSs is significantly improved for all severe loading cases and oscillations are quickly damped out. This illustrates that the CSO technique is capable of damping out the low frequency oscillations rapidly for a wide range of loading cases under severe scenarios of disturbances.

9.4.6   Performance indices results and discussion of TAFM power system

In addition to time-domain simulations, the effectiveness of designed PSS controllers is also analysed by determining two indices, known as integral

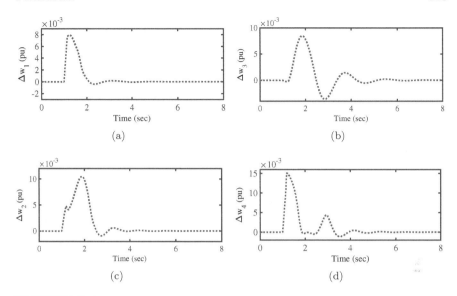

FIGURE 9.4
Comparison of speed deviations: (a) Δw_1 in scenario-1 of case-2; (b) Δw_3 in scenario-1 of case-2 ; (c) Δw_2 in scenario-2 of case-3 ; and, (d) Δw_4 in scenario-2 of case-3.

TABLE 9.5
Values of IAE and ITAE of the system for all scenarios and cases.

Cases	Scenario-1		Scenario-2	
	IAE	ITAE	IAE	ITAE
Case-1	15.8×10^{-3}	27.6×10^{-3}	20.7×10^{-3}	36.5×10^{-3}
Case-2	25.3×10^{-3}	50.2×10^{-3}	12.7×10^{-3}	21.4×10^{-3}
Case-3	12.3×10^{-3}	19.3×10^{-3}	44.4×10^{-3}	84.6×10^{-3}

of absolute error (IAE) and integral of time multiplied absolute value of error (ITAE), for two observed scenarios of different disturbances. These indices with designed PSS controllers are evaluated for each scenario of disturbances for loading cases 1-3 and presented in Table 9.5. From the table, it may be concluded that the designed PSS controllers provide enhanced damping to damp out low frequency local and inter-area modes of oscillations with less overshoot and settling time.

9.5 Conclusion

In this chapter, the CSO algorithm is presented to solve various optimization problems. In order to overcome some of the limitations observed in its standard variant, two modified variants have also been discussed. Finally, the application of the CSO algorithm to solve a real-life complex combinatorial problem is demonstrated. A modified CSO algorithm is used to optimally design PSS parameters for a wide range of operating conditions under severe scenarios of disturbances for damped out low frequency oscillations of a TAFM power system. The eigenvalue analysis, eigenvalue maps, time-domain simulations results and performance indices demonstrate that designed PSS controllers are capable of guaranteed robust performance of TAFM for a wide range of loading conditions under different scenarios of severe disturbances.

Acknowledgment

This project has received funding from the European Union's Horizon 2020 research and innovation programme under the Marie Sklodowska-Curie grant agreement No 713694.

References

1. X. S. Yang and S. Deb, "Cuckoo search via Levy flights. In: Proceedings of World Congress on Nature & Biologically Inspired Computing", NABIC, pp. 210-214, 2009.

2. X. S. Yang, *Engineering Optimization: An Introduction with Meta-heuristic Applications*, John Wiley and Sons; 2010.

3. L. F. Tammero and M. H. Dickinson, "The influence of visual landscape on the free flight behaviour of fruit fly Drosophila melanogaster", *The Journal of Experimental Biology*, vol. 205, pp. 327-343, 2002.

4. A. M. Reynolds and M. A. Frye, "Free-flight odor tracking in Drosophila is consistent with an optimal intermittent scale-free search", *PloS one*, vol. 2, no. 4, e354, 2007.

5. S. Walton, O. Hassan, K. Morgan, M.R. Brown, "Modified Cuckoo Search: A new Gradient Free Optimization Algorithm", *Chaos, Solutions & Fractals Nonlinear Science, and Non-equilibrium and Complex Phenomena*, Elsevier, vol. 44, pp. 710-718, 2011.

6. E. Valian, S. Tavakoli, S. Mohanna, A. Haghi, "Improved Cuckoo Search for Reliability Optimization Problems", *Computers & Industrial Engineering*, Elsevier, vol. 64, pp. 459-468, 2013.

7. P. Kundur, *Power System Stability and Control*, New York: Mc-Graw-Hill, 1994.

8. Amin Khodabakhshian and Reza Hemmati, "Multi-machine power system stabilizers design by using cultural algorithms", International Journal of Electrical Power & Energy Systems, vol. 44, pp. 571-580, September 2013.

9. F. Milano, Power System Analysis Toolbox. Version 2.1.6, 2010.

10

Improved Dynamic Virtual Bats Algorithm for Identifying a Suspension System Parameters

Ali Osman Topal

Department of Computer Engineering
Epoka University, Tirana, Albania

CONTENTS

10.1 Introduction

Dynamic Virtual Bats Algorithm (DVBA), by Topal and Altun [3], is another meta-heuristic algorithm inspired by bats' ability to manipulate frequency and wavelength of sound waves emitted during their hunt. In DVBA, a role-based search is developed to improve the diversification and intensification capability of the Bat Algorithm. There are only two bats: explorer and exploiter bat. While the explorer bat explores the search space, the exploiter bat makes an intensive search of the locale with the highest probability of locating the desired target. During the search bats exchange roles according to their positions.

Experimental results show that DVBA is suitable for solving most of the low dimensional problems. However, DVBA, similar to other evolutionary algorithms, has some challenging problems. The convergence speed of DVBA is slower than other population-based algorithms like PSO, GA, and BA.

Additionally, in high dimensional multimodal problems, escaping from the local optima traps becomes a difficult task for DVBA. Therefore, accelerating convergence speed and avoiding the local optima have become two important issues in DVBA. To minimize the impact of this weakness, an improved version of DVBA is proposed which accelerates convergence speed and avoids the local optima trap. To achieve both goals, a new search mechanism is proposed for the explorer bat [4]. This new search mechanism improves the search performance and gives DVBA more powerful exploitation/exploration capabilities. The rest of the chapter is organized as follows: in Section 10.2 original DVBA is summarized, in Section 10.3 we present improved DVBA, and in Section 10.4 IDVBA is used for identifying parameters of a suspension system. Finally, the conclusion is presented in Section 10.5.

10.2   Original Dynamic Virtual Bats Algorithm (DVBA)

The dynamic virtual bats algorithm (DVBA), proposed in 2014 for global numerical optimization, is a recently introduced optimization algorithm which imitates the bats' echolocation behavior in nature. When bats search out prey, they burst sound pulses with lower frequency and longer wavelengths so the sound pulses can travel farther. In this long range mode it becomes hard to detect the exact position of the prey; however, it becomes easy to search a large area. When bats detect prey, the pulses will be emitted with higher frequency and shorter wavelengths so that bats are able to update the prey location more often [6, 7]. In DVBA, two bats are used to imitate this hunting behavior. Each bat has its own role in the algorithm and during the search they exchange these roles according to their positions. These bats are referred as explorer bat and exploiter bat. The bat that is in a better position becomes the exploiter; meanwhile the other becomes the explorer. While the exploiter bat increases the intensification of the search around the best solution, the explorer bat will continue to explore other solutions.

In Fig. 10.1 and Fig. 10.2, the hunting strategy of a bat is simulated. The black triangle is the current solution (bat location), the black circles are the visited solutions on the search waves in this iteration. During the search, the explorer bat's search scope gets in its widest shape; the distance between the search waves and the angle between the wave vectors (red dashed arrows) get larger(see Fig. 10.1). On the contrary, if the bat becomes the exploiter bat, its search scope gets its narrowest shape; the distance between the search waves and the angle between the wave vectors get smaller (see Fig. 10.2). The number of the visited solutions is the same for both bats, just the distance between the solutions changes dynamically by using wavelength and frequency.

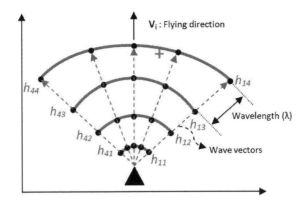

FIGURE 10.1
Exploration: Explorer bat is searching for prey with a wide search scope.

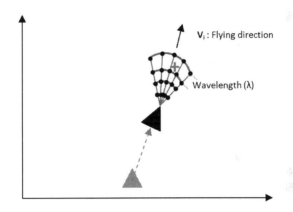

FIGURE 10.2
Exploitation: Exploiter bat is chasing prey with a narrow search space.

The wavelength and the distance between the solutions are proportional. The frequency and the angle between the wave vectors are inversely proportional.

The algorithm determines the best solution in the bat's search scope. If it is better than the current location (solution), the bat will fly to the better solution, decrease the wavelength and increase the frequency for the next iteration. These changes are targeted to increase the intensification of the search. Unless there is no better solution than the current solution in the search scope, the bat will stay on it, turn around randomly and keep scanning its nearby surrounding space. It will keep spinning in this position and expanding its search scope until it finds a better solution. Bats also exchange roles according to their position.

10.3   Improved Dynamic Virtual Bats Algorithm (IDVBA)

10.3.1   The weakness of DVBA

In DVBA, the explorer bat's search scope size is limited by the wavelength which might not be large enough to detect better solutions near its surrounding space. Thus, it is very likely that the explorer bat will be trapped in a local optima. In addition, the exploiter bat's search scope size becomes very small during the exploitation and it moves very slowly. Therefore it might need too much time to reach the global optima. These problems in DVBA have been eradicated by introducing probabilistic selection restart techniques in IDVBA.

10.3.2   Improved Dynamic Virtual Bats Algorithm (IDVBA)

To improve search performance and give DVBA more powerful search capabilities, two probabilistic selections are introduced: R - random flying probability and C - convergence probability in [4]. If the explorer bat is stuck in large local minima, it chooses to fly away from the trap randomly with a probability R related to number of unsuccessful attempts. In addition, it chooses to fly near to the exploiter bat with a probability C related to the number of escapes attempted from the traps. R and C are calculated as follows:

$$R_i^{t+1} = R_i^t[1 - 1/(\gamma_r \times trial_i)], \tag{10.1}$$

$$C_i^{t+1} = C_i^t[1 - 1/(\gamma_c \times rfly_i)], \tag{10.2}$$

where γ_c and γ_r are constant. $trial_i$ and $rfly_i$ denote the number of unsuccessful attempts and the number of random restarts, respectively. Obviously, the higher the $trial_i$ is, the greater the probability that the explorer bat might fly away from the trap to a random solution in the search space. As the unsuccessful attempts increase, the random flying probability R_i decreases and the possibility of $rand() < R_i$ being true (line 21 in Algorithm 13) increases. This can help the explorer bat to escape from the local optima trap rapidly. However, the explorer bat should not leave the trap without exploring the nearby surrounding space. Thus, γ_r should be chosen carefully.

The convergence probability C gives the possibility to the explorer bat to converge with the exploiter bat, instead of exploring a random position. Thus, the exploitation speed will be increased rapidly around the best position. Time after time the explorer bat visits the exploiter bat to speed up the exploitation process and then flies away randomly to keep up the exploration process. This also increases the exploitation capability of IDVBA. As shown in Eq. 10.2, the

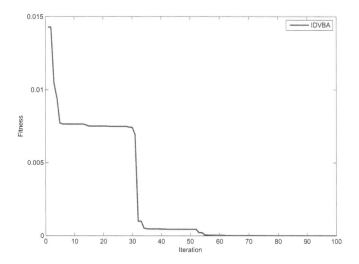

FIGURE 10.3
The convergence characteristic of IDVBA on 2D Rastrigin function.

convergence probability C is inversely proportional to the number of random restarts $rfly_i$ done by the explorer bat. Thus, increasing random restarts will increase the probability of $rand() > C$ that allows the explorer bat to visit the exploiter bat more often (line 30-31 in Algorithm 13).

In Fig. 10.3 the convergence characteristic of IDVBA is shown. Here, IDVBA is trying to optimize the multi-model Rastrigin function in 100 iterations. It can be easily seen, when the explorer bat visited the exploiter bat. As is expected, both bats collaborate to scrutinize the best found solution together, which increases the convergence speed rapidly during the visit.

Algorithm 13 IDVBA pseudo code. f_{gbest} is the global best solution and d is the number of dimensions. The code we discuss in the text is in boldface.

1: Objective function $f(x)$, $x = (x_1, ..., x_d)^T$
2: Initialize the bat population $x_i(i = 1, 2)$ and v_i
3: Initialize wavelength λ_i and frequency f_i
4: Define the parameters: β, the scope width variable b
5: Initialize the number of the waves(j) and search points(k)
6: Find f_{gbest} based on the bats' starting positions
7:   $C_i = R_i = 1.0$
8: **while** $(t <$ Max number of iterations$)$ **do**
9:     **for each** bat **do**
10:       Create a sound waves scope
11:       Evaluate the solutions on the waves
12:       Choose the best solution on the waves, h_*

13:     **if** $f(h_*)$ is better than $f(x_i)$ **then**
14:         Move to h_*, update x_i
15:         Change v_i towards the better position
16:         Decrease λ_i and increase $freq_i$
17:         $trial_i = 0$
18:     **end if**
19:     **if** $f(x_i)$ is not better than the best solution f_{gbest} **then**
20:         Calculate the random flying probability R_i by Eq.10.1
21:         If $rand() < R_i$ then
22:             Change the direction randomly
23:             $trial_i = trial_i + 1$
24:         **else**
25:             Restart the search from a random position
26:             Reset R_i and $trial_i$
27:             $rfly_i = rfly_i + 1$
28:             Calculate the C_i by Eq.10.2
29:         **end if**
30:         **if** $rand() > C_i$ **then**
31:             Produce a new solution around the exploiter bat.
32:             Reset C_i and $rfly_i$
33:         **end if**
34:         Increase λ_i and decrease $freq_i$
35:     **end if**
36:     **if** $f(x_i)$ is the best found solution **then**
37:         Minimize λ_i and maximize $freq_i$
38:         Change the direction randomly
39:         Reset $rfly_i$ and $trial_i$
40:     **end if**
41:     **end for**
42:     Rank the bats and find the current best x_{gbest}
43: **end while**

In our experiments, we set $\gamma_r = 12$ and $\gamma_c = 10$. By using these values, in a 100 iterations search, the explorer bat is given the chance to fly randomly 6 to 8 times and also, it is given the chance to visit the exploiter bat 3 to 6 times. IDVBA is given as in Algorithm 13. The differences from DVBA are shown in underlined font.

10.4   Application of IDVBA for identifying a suspension system

In this section IDVBA is applied to identify parameters of a suspension system. A quarter-car model was presented in response to the vertical force for the suspension system [2] as shown in Figure 10.4.

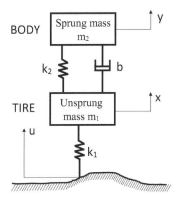

FIGURE 10.4
A quarter-car model of suspension system.

In this figure, m_1 is the mass of the tire, k_1 is the tire stiffness, m_2 is the mass of the vehicle, k_2 is the suspension stiffness, and b is the damping coefficient. u is the road displacement which is our input and y is the vertical displacement of sprung mass m_1 and that is our output.

The dynamic equations of motion for the displacements x and y are shown as follows:

$$m_1\frac{d^2x}{dt^2} + b(\frac{dx}{dt} - \frac{dy}{dt}) + k_1(x - u) + k_2(x - y) = 0 \qquad (10.3)$$

$$m_2\frac{d^2y}{dt^2} + b(\frac{dy}{dt} - \frac{dx}{dt}) + k_2(y - x) = 0 \qquad (10.4)$$

Our aim is to identify/estimate the parameters of a quarter-car suspension system. The basic idea of parameter estimation is to compare the real system output with the estimated model output for the same input [5]. We check how well the estimated model response fits the actual system response. To test the estimated model system we will use excitation signals that correspond to a realistic excitation of the system in real life. Consequently, in order to estimate the system parameters, an impulse signal is given as the excitation signal to a real and estimated model as shown in Figure 10.5 [2]. Then, the outputs are given as inputs to generate the objective function f, where the fitness will be calculated. We used the sum of squares error between the real and estimated model responses as objective function:

$$f = \sum_{n=0}^{N}(y[n] - y'[n])^2 \qquad (10.5)$$

where, $y[n]$ and $y'[n]$ are real and estimated values in each sample, respectively, and N is number of samples. IDVBA will try to minimize the sum of squares

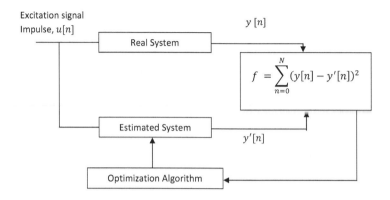

FIGURE 10.5
The estimation process.

error f by changing the parameters(m_1, m_2, k_1, k_2, and b) of the estimated model.

The quarter-car's nominal parameters are summarized as follows [1]:

$m_1 = 26kg$,
$m_2 = 253kg$,
$k_1 = 90000N/m$,
$k_2 = 12000N/m$,
$b = 1500N.sec/m$

The estimated model's parameters ranges are set as follows [1]:

$20 \leq m_1 \leq 30$,
$200 \leq m_2 \leq 300$,
$8500 \leq k_1 \leq 90000$,
$10000 \leq k_2 \leq 15000$,
$1200 \leq b \leq 1700$.

In order to test the efficiency of IDVBA, it is compared with the original DVBA. We used Equation 10.5 as objective function. In Equation 10.5, previously mentioned nominal parameters are used to find the displacement (y) of the real system for an impulse input signal. For the comparison, we set N=500 samples and iterations 100. The algorithms parameters are set as follows:

number of wave vectors: j=5,
number of search points: k=6,

number of search points on the search scope: $jxk = 30$ for each bat
β: 100,
scope wideness variable: b=20
And also, for IDVBA we set $\gamma_r = 12$ and $\gamma_c = 10$.

In optimization algorithms, the step size plays a very important role in terms of convergence speed and accuracy. If it is very small, accuracy can increase, however the convergence speed might slow down. In DVBA, the step size ρ is used to change the distance between the search points on the bat's search scope as shown in Figure 10.1. To decide the step size, we should pay attention to the size of the search space. If the parameters of the optimization problem have different ranges, like the quarter-car's parameters, different step sizes should be used for each parameter. For example, if we decide the step size based on k_1, it would be too big for m_1. While k_1 moves slowly, m_1 will just jump from its maximum to minimum values. So, ρ has different values for each parameter.

The test results are shown in Table 10.1 in terms of the best fitness values (BFV), the worst fitness values (WFV), the mean, and the standard deviation (STDEV) of the results found over the 20 independent runs by each algorithm. The best values are typed in bold.

Furthermore, Figure 10.6 illustrates the convergence speed of these algorithms for 50 iterations. The horizontal axis of the graph is the number of iterations, and the vertical axis is the mean of function values.

In Table 10.2, the parameters that are estimated by IDVBA, DVBA, and the actual parameters are shown.

TABLE 10.1
Comparison of DVBA and IDVBA identification of a quarter-car suspension system by using 100 iterations. BFV, WFT, Mean, and STDEV" indicate the best fitness value, the worst fitness value, the mean, and standard deviation of the results found over the 20 independent runs by each algorithm.

	BFV	**WFV**	**Mean**	**STDEV**
DVBA	3.2593e-06	5.2664e-04	1.1746e-04	2.2907e-04
IDVBA	**2.3480e-09**	**1.9909e-05**	**6.8526e-06**	**9.1657e-06**

TABLE 10.2
Comparison of DVBA and IDVBA on identification of quarter-car suspension system parameters with the real parameters by using 100 iterations.

	DVBA	IDVBA	Actual Values
m_1	27.73	24.72	26.00
m_2	232.74	257.29	253.00
k_1	86187.95	89931.03	90000.00
k_2	11005.63	12220.11	12000.00
b	1367.52	1531.63	1500.00

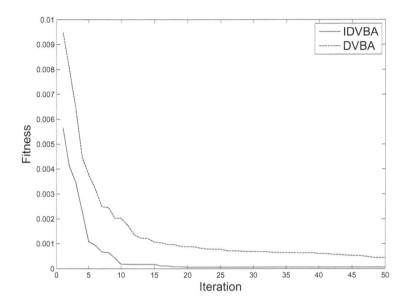

FIGURE 10.6
The comparision of convergence speed of DVBA and IDVBA.

As reported in Table 10.1, IDVBA performs better for identification of suspension parameters. From Figure 10.6, it is explicitly observed that IDVBA found effective results with the fastest convergence speed, which means that IDVBA using the improved search mechanism is more competent in tackling complex problems.

Since, we used only one impulse signal as excitation signal, we can say that the estimation is pretty close to the actual system parameters. To increase the accuracy of the estimated parameters, a series of impulses with random amplitudes, like Gaussian band-limited white noise can be used as an excitation signal to the real and the modeled system. It will take more time to compute, but give more chances for the algorithm to increase accuracy.

10.5   Conclusions

In this chapter, the original DVBA and the improved DVBA are presented. We used a quarter-car model system to show the performance of the algorithms in system identification. We can say that slightly changing the explorer bat's behaviour in DVBA, increased the convergence speed and the accuracy of the algorithm. The results show that IDVBA is superior to the original DVBA and it is a promising optimization algorithm for system identification. As future work, there might be some improvements on the exploiter bat to increase the detailed search as well.

References

1. H.Peng, R.Strathearn, A.G.Ulsoy, "A novel active suspension design technique-simulation and experimental results" in *Proc. of American Control Conference, IEEE*, vol. 1, 1997, pp. 709-713.

2. A.Alfi, and M.M.Fateh, "Parameter identification based on a modified PSO applied to suspension system" in *Journal of Software Engineering & Applications, JSEA*, 2010, vol.3, pp.221-229.

3. A.O.Topal , O.Altun, "Dynamic virtual bats algorithm (DVBA) for global numerical optimization" in *Proc. of the IEEE International Conference on Congress on Intelligent Networking and Collaborative Systems (INCoS)*, 2014, pp. 320-327 .

4. A.O.Topal, Y.E.Yildiz, M.Ozkul, "Improved dynamic virtual bats algorithm for global numerical optimization" *Proc. of the World Congress on Engineering and Computer Science.* Vol. 1, pp.462-467, 2017.

5. S.Lajqi, J.Gugler, N.Lajqi, A.Shala, R.Likaj, "Possible experimental method to determine the suspension parameters in a simplified model of a passenger car" *International Journal of Automotive Technology*, vol. 13(4), pp. 615-621, 2012.

6. C.Chandrasekar, "An optimized approach of modified bat algorithm to record deduplication." *International Journal of Computer Applications* 62.1, 2013.

7. M.Airas, "Echolocation in bats." *Proc. of Spatial Sound Perception and Reproduction.* The postgrad seminar course of HUT Acoustics Laboratory, pp. 1-25, 2003.

11

Dispersive Flies Optimisation: Modifications and Application

Mohammad Majid al-Rifaie

School of Computing and Mathematical Sciences, University of Greenwich
Old Royal Naval College, London, United Kingdom

Hooman Oroojeni M. J.

Department of Computing
Goldsmiths, University of London, London, United Kingdom

Mihalis Nicolaou

Computation-based Science and Technology Research Center
The Cyprus Institute, Nicosia, Cyprus

CONTENTS

11.1 Introduction

Information exchange and communication between individuals in swarm intelligence manifest themselves in a variety of forms, including: the use of different

145

update equations and strategies; deploying extra vectors/components in addition to the individuals' current positions; and dealing with tunable parameters. Ultimately the goal of the optimisers is to achieve a balance between global exploration of the search space and local exploitation of potentially suitable areas [1, 2] in order to guide the optimisation process.

One of the main motivations for studying dispersive flies optimisation (DFO) [3] is the algorithm's minimalist update equation which only uses the flies' position vectors for the purpose of updating the population. This is in contrast to several other population-based algorithms and their variants which besides using position vectors, use a subset of the following set of components (i.e. vectors): velocities and memories (personal best and global best) in particle swarm optimisation (PSO) [4], mutant and trial vectors in differential evolution (DE) [5], pheromone, heuristic vectors in Ant Colony Optimisation (ACO) [6], and so forth [7]. In addition to using only the position vectors at any given time, the only tunable parameter in DFO, other than the population size, is the *disturbance threshold*, Δ, which controls the component-wise restart in each dimension. This is again contrary to many well-known algorithms dealing with several (theoretically or empirically driven) *tunable parameters*, such as: learning factors, inertia weight in PSO, crossover or mutation rates, tournament and elite sizes, constricting factor in DE and/or Genetic Algorithms (GA) [8], heuristic strength, greediness, pheromone decay rate in ACO, impact of distance on attractiveness, scaling factor and speed of convergence in Firefly algorithm (FF) [9], and so on.

It is worthwhile noting that DFO is not the only minimalist algorithm and there have been several attempts to present 'simpler', more compact algorithms to better understand the dynamic of the population's behaviour as well as the significance of various communication strategies. Perhaps one of the most notable minimalist swarm algorithms is bare bones particle swarms (BB-PSO) [10] whose collapse has been studied in [11] along with the introduction of a dimensional jump or restart to the aforementioned algorithm, thus proposing the bare bones with jumps algorithm (BBJ). Another bare bones algorithm is barebones differential evolution (BBDE) [12] which is a hybrid of the barebones particle swarm optimiser and differential evolution, aiming to reduce the number of parameters; however, still with more components and parameters, and often at the expense of reduced performance level[1]. It is well understood that swarm intelligence techniques, including but not limited to PSO, are dependant on the tuning of their parameters (e.g. PSO relies on the choice of inertia weight [16, 17])[2]; as a result, adjusting a growing number of parameters becomes increasingly complex.

DFO has been applied to various problems, including but not limited to medical imaging [19], optimising machine learning algorithms [20, 21], train-

---

[1]Some of these shortcoming have been studied and at points overcome but at the expense of introducing additional parameters (see [13, 14, 15]).

[2]In a related work, Clerc [18] defines acceleration coefficient in order for PSO to be independent of computing the inertia weight.

ing and optimising deep neural network [22], computer vision and quantifying symmetrical complexities [23], building non-identical organic structures for gaming [24], identification of animation key points from 2D-medialness maps [25] and analysis of autopoiesis in computational creativity [26].

In this chapter, initially, DFO is briefly introduced in Section 11.2; then some of the modifications to the algorithms are discussed in Section 11.3; subsequently, the application of DFO in training neural networks for detecting a false alarm in ICU is presented in Section 11.4.

11.2   Dispersive flies optimisation

The swarming behaviour of the individuals in DFO consists of two tightly connected mechanisms, one is the formation of the swarms and the other is its breaking or weakening. The position vector of a fly in DFO is defined as:

$$\vec{x}_i^t = \left[x_{i0}^t, x_{i1}^t, ..., x_{i,D-1}^t \right], \qquad i \in \{0, 1, 2, ..., \textit{N-1}\} \tag{11.1}$$

where i represents the ith individual, t is the current time step, D is the problem dimensionality, and N is the population size. For continuous problems, $x_{id} \in \mathbb{R}$ (or a subset of $\mathbb{R}$).

In the first iteration, when $t = 0$, the dth component of the ith fly is initialised as:

$$x_{id}^0 = \text{U}(x_{\min,d}, x_{\max,d}) \tag{11.2}$$

This, effectively generates a random value between the lower $(x_{\min,d})$ and upper $(x_{\max,d})$ bounds of the respective dimension, d.

On each iteration, dimensions of the position vectors are independently updated, taking into account:

- current fly's position

- current fly's best neighbouring individual (consider ring topology, where each fly has a left and a right neighbour)

- and the best fly in the swarm.

Therefore, the update equation is

$$x_{id}^{t+1} = x_{i_n d}^t + u(x_{sd}^t - x_{id}^t) \tag{11.3}$$

where,

- x_{id}^t: position of the ith fly in d^{th} dimension at time step t

- $x_{i_n d}^t$: position of $\vec{x}_i^t$'s best *neighbouring* individual (in ring topology) in d^{th} dimension at time step t

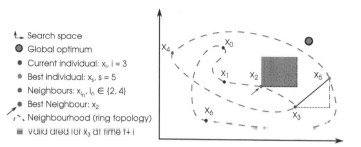

FIGURE 11.1
Sample update of x_i, where $i = 3$ in a 2D space.

- x_{sd}^t: position of the *swarm*'s best individual in the dth dimension at time step t and $s \in \{0, 1, 2, ..., N\text{-}1\}$

- $u \sim U(0, 1)$: generated afresh for each dimension update.

The update equation is illustrated in an example in Fig. 11.1 where $\vec{x}_3$ is to be updated. The algorithm is characterised by two main components: a dynamic rule for updating the population's position (assisted by a social neighbouring network that informs this update), and communication of the results of the best found individual to others.

As stated earlier, the position of members of the swarm in DFO can be restarted, with one impact of such restarts being the displacement of the individuals which may lead to discovering better positions. To consider this eventuality, an element of stochasticity is introduced to the update process. Based on this, individual dimensions of the population's position vectors are reset if a random number generated from a uniform distribution on the unit interval $U(0, 1)$ is less than the *restart threshold*, Δ. This guarantees a restart to the otherwise permanent stagnation over likely local minima.

Algorithm 14 summarises the DFO algorithm. In this algorithm, each member of the population is assumed to have two neighbours (i.e. ring topology).

Algorithm 14 Dispersive flies optimisation.

1: **procedure** DFO $(N, D, \vec{x}_{\min}, \vec{x}_{\max}, f)^*$
2:   **for** $i = 0 \rightarrow N - 1$ **do**   ▷ **Initialisation**: Go through each fly
3:     **for** $d = 0 \rightarrow D - 1$ **do**   ▷ Initialisation: Go through each dimension
4:       $x_{id}^0 \leftarrow U(x_{\min,d}, x_{\max,d})$ ▷ Initialise d^{th} dimension of fly i
5:     **end for**
6:   **end for**
7:   **while** ! termination criteria **do**   ▷ Main DFO loop
8:     **for** $i = 0 \rightarrow N - 1$ **do**   ▷ **Evaluation**: Go through each fly
9:       $\vec{x}_i.\text{fitness} \leftarrow f(\vec{x}_i)$

10:     **end for**
11:     $\vec{x}_s = \arg\min[f(\vec{x}_i)], \quad i \in \{0, 1, 2, \ldots, N-1\}$ ▷ Find best fly
12:     **for** $i = 0 \rightarrow N-1$ and $i \neq s$ **do**   ▷ **Update** each fly except the best
13:         $\vec{x}_{i_n} = \arg\min[f(\vec{x}_{(i-1)\%N}), f(\vec{x}_{(i+1)\%N})]$ ▷ Find best neighbour
14:         **for** $d = 0 \rightarrow D-1$ **do**   ▷ Update each dimension
15:             **if** $U(0,1) < \Delta$ **then**   ▷ **Restart mechanism**
16:                 $x_{id}^{t+1} \leftarrow U(x_{\min,d}, x_{\max,d})$ ▷ Restart within bounds
17:             **else**
18:                 $u \leftarrow U(0,1)$
19:                 $x_{id}^{t+1} \leftarrow x_{i_n d}^{t} + u(x_{sd}^{t} - x_{id}^{t})$ ▷ **Update** the dimension value
20:                 **if** $x_{id}^{t+1} < x_{\min,d}$ or $x_{id}^{t+1} > x_{\max,d}$ **then**   ▷ Out of bounds
21:                     $x_{id}^{t+1} \leftarrow U(x_{\min,d}, x_{\max,d})$ ▷ Restart within bounds
22:                 **end if**
23:             **end if**
24:         **end for**
25:     **end for**
26:     **end while**
27:     **return** $\vec{x}_s$
28: **end procedure**

---

* INPUT: swarm size, dimensions, lower/upper bounds, fitness function.

11.3   Modifications in DFO

In this section some of the immediate modifications to the algorithm are discussed. These modifications and variations were proposed after studying the behaviour of the population when applied to a number of problems.

11.3.1   Update equation

Diversity is the degree of convergence and divergence, which is defined as a measure to study the population's behaviour with regard to exploration and exploitation. There are various approaches to measure diversity. The average distance around the population centre is shown to be a robust measure in the presence of outliers and is defined as [2]:

$$\text{DIVERSITY}^t = \frac{1}{N}\sum_{i=1}^{N}\sqrt{\sum_{d=1}^{D}(x_{id}^{t} - \bar{x}_d^{t})^2} \tag{11.4}$$

$$\bar{x}_d^t = \frac{1}{N}\sum_{i=1}^{N}x_{id}^t. \tag{11.5}$$

FIGURE 11.2

Diversity values in different update equations. Illustrating diversity values in one dimension when DFO, with $\Delta = 0$, is applied to the Sphere function: $f(\vec{x}) = \sum_{d=1}^{D} x_d^2$.

It is well understood that the disturbance threshold, Δ, impacts the diversity of the swarm directly. Assuming a predetermined value, diversity is proportionally applied to the swarm throughout the optimisation process. By merely changing the update equation, diversity could be altered. In addition to the standard update equation, local-only and global-only update equations can be seen below:

$$\text{Standard:} \quad x_{id}^{t+1} = x_{i_n d}^t + u(x_{sd}^t - x_{id}^t) \tag{11.6}$$

$$\text{Local-only:} \quad x_{id}^{t+1} = x_{i_n d}^t + u(x_{i_n d}^t - x_{id}^t) \tag{11.7}$$

$$\text{Global-only:} \quad x_{id}^{t+1} = x_{sd}^t + u(x_{sd}^t - x_{id}^t) \tag{11.8}$$

Diversity values of the swarm in each iteration of the above-mentioned update equations are illustrated in Fig. 11.2. For instance, in order to have an initially more "pronounced" exploration phase, the local-only update equation could be used. On the other hand, if an immediate convergence of the population is desirable, global-only structure can be utilised[3]. Other configurations have been tried by researchers depending on the problem, for instance, in [22], Eq. 11.9, below, is used, which removes current positions of each fly from the update equation, relying entirely on the best neighbour $(\vec{x}_{i_n})$ and the best member of the population $(\vec{x}_s)$. In other words, this update equation, takes into account the position of the best neighbouring individual and the position of the best fly in the entire swarm to determine the updated position:

$$x_{id}^{t+1} = x_{i_n d}^t + u(x_{sd}^t - x_{i_n d}^t) \tag{11.9}$$

11.3.2   Disturbance threshold, Δ

Disturbance threshold, Δ, controls the diversity of the population and, in turn, influences the hill-climbing process. The standard DFO algorithm has

---

[3]Note: some of these features are shared amongst other swarm intelligence algorithms.

proposed a simple, single valued Δ for the entire dimensions. Therefore each dimension is exposed to the same probability of being restarted. However, depending on the nature of the problems and taking into account the dimensions' varying degree and need for exploration, a dimension-dependent Δ can be used. This would allow a tailored restart mechanism if such knowledge is available a-priori.

Furthermore, in order to adjust the level of exploration and exploitation, Δ can be tuned during the optimisation process based on the performance of the algorithm on a given problem. For instance, in [22], a mechanism is proposed for Δ to be re-adjusted depending on various performance factors and the swarm behaviour. This method is noted to have resulted in increased performance over the problem of detecting false alarms in Intensive Care Units (ICU).

The immediate impact of the disturbance threshold, as mentioned before, is the possibility of identifying better (and previously unexplored) solutions. This is achieved by randomly generating a value within the lower and upper bounds. Depending on the problems, the new position could be derived by other methods[4] where the updated value is not simply generated randomly between the bounds but rather tied to other associable values such as the current position of the fly, position of the best fly in the swarm, the best neighbouring fly or a combination of these. Following on this ethos, in [22], the position of the updated value is determined by the following:

$$x_{id}^{t+1} = \mathcal{N}(x_{i_n d}^t, \Delta^2) \qquad (11.10)$$

therefore, restarting the value of the dimension based on a sample from a Gaussian with the mean set to $x_{i_n d}^t$ and variance to Δ^2.

11.4   Application: Detecting false alarms in ICU

Deep Neural Networks (DNN) and Deep Learning (DL) have been increasingly popular in solving various problems which have been otherwise difficult to address. However, one of main challenges is to design and train suitable neural networks.

In this section, a neuroevolution-based approach for training neural networks is used and applied to the problem of detecting false alarms in Intensive Care Units (ICU) based on physiological data. A common tool for training or

---

[4]Note that the disturbance threshold is applied on each single dimension of each fly in each iteration independently, and there are instances where generating a value between the lower and upper bounds results in unsuitable solutions (exploration), which are then re-absorbed by the swarm to previously known solutions. This, therefore, may result in the mechanism becoming less effective.

optimising a neural network is backpropagation (BP) with stochastic gradient-based learning. However, recent research [27] has shown that gradient-free, population-based algorithms can outperform the classical gradient based optimisation methods. This section demonstrates the use of DFO in optimising a deep neural network which is applied to the detection of false alarms in ICU. This application is reported in more detail in [22]. The overall goal is to reduce the number of suppressed true alarms by deploying and adapting DFO.

11.4.1   Problem description

The focus of this experiment is detecting abnormalities in the heart function, called arrhythmias that can be encountered in both healthy and unhealthy subjects. The ICU is equipped with monitoring devices that are capable of detecting dangerous arrhythmias, namely asystole, extreme bradycardia, extreme tachycardia, ventricular tachycardia and ventricular flutter/fibrillation.

Depending on the Association for the Advancement of Medical Instrumentation (AAMI) guidelines, appropriate measures should be taken within 10 seconds of the commencement of the event as these arrhythmias might cause death [28]. Triggering the alarm when an arrhythmia occurs may improve the chance of saving lives. Mis-configuring, defective wiring, staff manipulation, and patient manipulation or movement may increase the false alarm ratio to 86%. Clinically, 6% to 40% of the ICU alarms proved to have lower priority and do not require immediate measures [29]. False alarms stimulate the patient's mental discomfort [30] and the clinical staff's desensitisation thereby causing a slower response to the triggered alarms [31]. True alarms that have high priority and need an urgent response are only 2 - 9% of all ICU alarms [32]. Therefore, false alarm detection and elimination is an essential area of research.

The Physionet 2015 challenge [33] provides a training data set containing 750 recordings that are available to the public, as well as 500 hidden records for scoring. This dataset consists of life-threatening arrhythmia alarm records that have been collected from four hospitals in the United States and Europe. The recordings are sourced from the devices made by three major manufacturing companies of intensive care monitor devices. Each record is five minutes or five minutes and 30 seconds long at 250Hz and contains only one alarm. A team of expert annotators labeled them 'true' or 'false'. The commencement of the event is within the last ten seconds of the recordings.

In this challenge, participants can submit their code which would be evaluated according to two type of events: event 1 (real time) and event 2 (retrospective). All recordings have a sample rate of 250Hz and contain two ECG leads and one or more pulsative waveform (RESP, ABP or PPG). The ECGs may contain noise, and pulsatile channels may contain movement artefacts and sensor disconnections. In this experiment, the focus is on event 1, where data is trimmed to have a length of five minutes and a subset of 572 records

TABLE 11.1

Subset of Physionet dataset (572 out of 750 recordings) that contains the ECG leads II and V and PLETH signal. This subset is used to ensure that the NN models are trained on identical leads and pulsatile waveform.

Disease Name	True Alarm	False Alarm
Asystole	17	77
Bradycardia	35	37
Tachycardia	90	4
Ventricular Flutter/Fibrillation	6	40
Ventricular Tachycardia	54	212

are chosen. These contain the ECG leads II and V and PLETH signal. This decision is made to ensure that the NN models are trained on identical leads and pulsatile waveform. In this subset, there are 233 True alarms and 339 False alarms. In each n-fold, the dataset is divided into training, testing and validation using 70% (400), 20% (114), and 10% (58) respectively. Table 11.1 describes the dataset.

11.4.2   Using dispersive flies optimisation

In this experiment the flies' positions, which are the weights associated with the neural network (NN), are termed $\vec{w}$. In the first iteration, $t = 0$, the ith vector's d^{th} component is initialised as $w_{id}^0 = \mathcal{N}(0,\ 1)$, where $\mathcal{N}$ denotes the Gaussian distribution. Therefore, the population is randomly initialised with a set of weights for each in the search space.

In each iteration of the original DFO equation, the components of the NN weights vectors are independently updated, taking into account the component's value, the corresponding value of the best neighbouring individual with the best Physionet score (consider ring topology), and the value of the best individual in the whole swarm. Therefore the updated equation is:

$$w_{id}^{t+1} \quad = \quad w_{i_n d}^t + u(w_{sd}^t - w_{id}^t) \tag{11.11}$$

where $w_{i_{nd}}^t$ is the weight (position) value of $\vec{w}_i^t$'s best *neighbouring* individual in the dth dimension at time step t, w_{sd}^t is the value of the *swarm*'s best individual in the dth dimension at time step t, and $u = U(0,1)$ is a random number generated from the uniform distribution between 0 and 1. In this experiment, few configuration changes are applied to DFO for deep network optimisation. The adapted update equation takes into account the corresponding value of the best neighbouring individual and the value of the best in the whole swarm:

$$w_{id}^{t+1} \quad = \quad w_{i_n d}^t + u(w_{sd}^t - w_{i_{nd}}^t) \tag{11.12}$$

In DFO, swarms disturbance is regulated by *the disturbance threshold*, Δ, to control the behaviour of the population (exploration or exploitation) in the search space. One of the impacts of these disturbances is the displacement of the individuals, which may lead to discovering a better Physionet score through finding better weights for the NN. Based on this, the individual components of the population's weights' vectors are reset if a random number, u is less than Δ. This approach guarantees a disturbance to the otherwise permanent stagnation over a possible local maximum. In the original DFO equation, the disturbance is done by updating the parameter with a random number in the acceptable range of lower and upper bounds. In this experiment, a parameter's disturbance is correlated with the current Δ and the best neighbour; that is, w_{id}^{t+1} is sampled from a Gaussian with the mean set to $w_{i_n d}$ and variance to Δ^2.

Algorithm 15 summarises the adapted DFO algorithm and Figure 11.5 demonstrates Δ's behaviour in 1000 iterations.

Algorithm 15 Adapted DFO for Training.

Input: population size N, model structure L, network weights $\vec{w}_i$, length of weights vector D, loss function $f(.)$.

1: $\Delta = 1$
2: **while** not converged **do**
3:     $\vec{w}_s = \arg\max\left[f(L(\vec{w}_i))\right], \quad i \in \{1, \ldots, N\}$
4:     **for** $i = 1 \to N$ and $i \neq s$ **do**
5:         $\vec{w}_{i_n} = \arg\max\left[f(L(\vec{w}_{i-1})), f(L(\vec{w}_{i+1}))\right]$
6:         **for** $d = 1 \to$ D **do**
7:             **if** $(U(0,1) < \Delta)$ **then**
8:                 $w_{id}^{t+1} \leftarrow \mathcal{N}(w_{i_n d}^t, \Delta^2))$
9:             **else**
10:                $w_{id}^{t+1} \leftarrow w_{i_n d}^t + u(w_{sd}^t - w_{i_n d}^t)$
11:             **end if**
12:         **end for**
13:         Dynamically update Δ (see Section 11.4.3.2)
14:     **end for**
15: **end while**

Output: Best fly's weight vector, $\vec{w}_s$.

11.4.3 Experiment setup

This part details the experiment setup comprising the model configuration and DFO configuration.

11.4.3.1 Model configuration

The experiment conducted here utilises two neural network architectures: a stack of dense layers, and a stack of convolutional and dense layers. The model structures are described in Tables 11.2 and 11.3 in detail. In NN, forward propagation includes a set of matrix operations where parameters include weights that connect NN layers to each other, and bias. The focus is on finding optimal weights of the network that provide the highest classification accuracy

TABLE 11.2

Dense Model Structure.

Layer Name	No of Neurons	Weights Shape	Total Weights	Bias
Dense 1 (Input)	64	(1, 64)	64	64
Dense 2	64	(64, 64)	4096	64
Dense 3	32	(2112, 32)	67584	32
Dense 4 (Output)	2	(32, 2)	64	2

TABLE 11.3

Convolution-Dense Model Structure.

Layer Name	No of Neurons	Weights Shape	Total Weights	Bias
Convolution 1 (Input)	32	(2, 1, 32)	64	32
Dense 2 (Input)	64	(32, 64)	2048	64
Dense 3	64	(64, 64)	4096	64
Dense 4	32	(1024, 32)	32768	32
Dense 5 (Output)	2	(32, 2)	64	2

for this task. Depending on the type of neurons and input shape in a NN, the shape of the output weights of that neuron varies. This study uses two models: model 1 consists of four dense layers, and model 2 one convolution and four dense layers. For instance, in model 1 of our study, the first layer is dense with 64 neurons. In the first layer, input shape is (-1, 33, 1) and the output weight is (1, 64) where 1 is the third axis of the layer's input shape, and 64 is the number of neurons in this layer: $[(-1, 33, 1) \times (1, 64)] + (64) = (-1, 33, 64)$

Model 2 includes a convolution as the first layer. This model has a similar input shape (-1, 33,1). The first layer has 32 neurons, and the shape of connecting weights to the next layer is (-1, 2, 1, 32) where 2 is the convolution window size, 1 is the third axis of the input shape, and 32 is the number of neurons in the convolution layer.

11.4.3.2 DFO configuration

In the experiment reported here, DFO is used to find the optimal weights in both NN models. The number of parameters (i.e. weights and bias) in models 1 and 2 are 71970 and 39234 respectively (see Tables 11.2 and 11.3). Each member of the population (fly) has a set of parameters representing the weights (including biases) of the NN model. Once all the parameters of each fly are initialised and loaded onto the NN model, the fitness (score) of each fly is calculated using 5-fold cross-validation, by using the mean Physionet score. After each iteration, each fly's best neighbour, and the best fly in the swarm are identified. The best fly holds the highest Physionet score amongst the population.

Before updating each weight, a value u is sampled from a uniform distribution $U(0, 1)$. If u is less than Δ, the weight is updated with the fly's best neighbour as focus, therefore, $w_{id}^{t+1} \leftarrow \mathcal{N}(w_{i_n d}^t, \Delta^2)$, otherwise, the fly's weight is updated with the focus on the best fly in the swarm $w_{id}^{t+1} \leftarrow w_{i_n d}^t + u(w_{sd}^t - w_{i_n d}^t)$.

An additional mechanism is also proposed to control the value of Δ. In the first phase or the parameter optimisation, Δ is set to 1 to bias the algorithm towards exploration; this process continues until there is no improvement in k iterations. Following this, Δ is set to zero, allowing the flies to converge to the best location. Once again, if no improvement is noticed in the Physionet score, Δ is increased by a random number between 0 and δ^5 ($\Delta \leftarrow \Delta + U(0, \delta)$). The algorithm is then set to run, if there is an improvement followed by a k iteration state of idleness, Δ is set to 0 again to exploit the recent finding. Alternatively if there is no improvement after the allowed idle time-frame, Δ is incremented further. In a situation where $\Delta > 1$, Δ is set to $U(0, 1)$, and the process continues as explained above until the termination points, which is $3,000$ iterations. Figures 11.3, 11.4 and 11.5 demonstrate the trend of improvement in accuracy, Physionet score and variations of Δ over the first 1000 iterations.

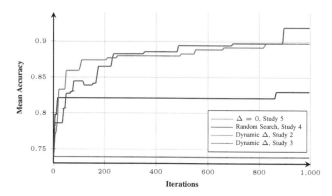

FIGURE 11.3
Model 1: 5-fold cross validation mean accuracy over 1000 iterations. Mean accuracy trend for random search, standard DFO, updated DFO, with dynamic and constant disturbance threshold (Δ).

11.4.4   Results

The result of this experiment, using DFO as a NN optimiser, is compared against the winning and 5th ranked entries in the Physionet challenge 2015 (Table 11.4), as well as the same NN architectures optimised with BP instead of the proposed gradient-free DFO training scheme.

As a benchmark, the accuracy and Physionet scores of [34] and [35] are used; they achieve $(87.24\%, 85.50\%)$ and $(87.78\%, 80.09\%)$ respectively. Various implementations of DFO, as well as random search are also investigated. For all experiments, 5-fold cross-validation is used. The DFO implementa-

---

[5]Note that throughout this in this paper, we use $k = 50$ and $\delta = 0.5$.

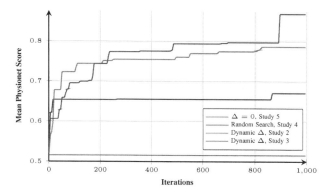

FIGURE 11.4

Model 1: 5-fold cross validation mean Physionet score over 1000 iterations. Mean Physionet score trend for random search, standard DFO, updated DFO, with dynamic and constant disturbance threshold (Δ).

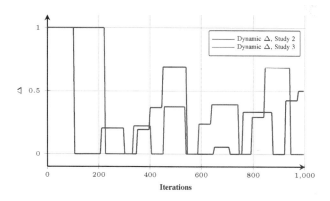

FIGURE 11.5

Model 1: 5-fold cross validation disturbance threshold (Δ) trend. Visualising Δ trend with considering dynamic and constant disturbance threshold (Δ) over 1000 iterations.

tion for Model 1 and 2 achieved the highest score among the other results (see Table 11.4). The resulting accuracy and Physionet scores are (91.91%, 86.77%) and (91.88%, 86.81%) respectively, therefore achieving better results than both back propagation-trained NN, as well as the challenge entries.

The behaviour of DFO, while having the constant Δ value of 0 and random search, is also investigated. Their accuracy and Physionet scores are (73.89%, 51.53%) and (84.16, 68.03) respectively. The models optimised via neuroevolution with DFO outperforming both (i) the networks trained by BP, as well as (ii) the winning entry of the Physionet challenge.

TABLE 11.4

Accuracy and Physionet score over 5-fold cross validation for first [34] and fifth rank [35] in Physionet challenge 2015, NN optimised with BP and adapted DFO algorithm with constant and dynamic disturbance threshold (Δ).

Author	Method	Mean Accuracy	Physionet 2015 Score
By [35]	SVM, BCTs, DACs	87.24% (+/- 2)	85.50% (+/- 3)
By [34]	Fuzzy Logic	87.78% (+/- 4)	80.09% (+/- 8)
Our Study 1	Dense Network, BP	87.70% (+/- 3)	75.35% (+/- 7)
Our Study 2	Dense NN, Standard DFO, Dynamic Δ, using Eq. 11.11	90.02% (+/- 3)	79.23% (+/- 3)
Our Study 3	**Dense NN, Adapted DFO, Dynamic Δ, using Eq. 11.12**	**91.91% (+/- 4)**	**86.77% (+/- 4)**
Our Study 4	Dense NN, Random Search	84.16% (+/- 3)	68.03% (+/- 5)
Our Study 5	Dense NN, Standard DFO, $\Delta = 0$ (i.e. no disturbance)	73.89% (+/- 2)	51.53% (+/- 4)
Our Study 6	Convolution-Dense NN, BP	88.21% (+/- 3)	76.34% (+/- 6)
Our Study 7	Convolution-Dense NN, Standard DFO, Dynamic Δ, using Eq. 11.11	89.15% (+/- 4)	77.21% (+/- 5)
Our Study 8	**Convolution-Dense NN, Adapted DFO, Dynamic Δ using Eq. 11.12**	**91.88% (+/- 2)**	**86.81% (+/- 4)**
Our Study 9	Convolution-Dense NN, Random Search	82.39% (+/- 5)	66.40% (+/- 7)
Our Study 10	Convolution-Dense NN, Standard DFO, $\Delta = 0$ (i.e. no disturbance)	73.90% (+/- 3)	52.53% (+/- 2)

11.5   Conclusions

This paper presents the standard DFO along with some of the existing and potential modifications to the algorithm: mainly the update equation, as well as its disturbance mechanism. These modifications are proposed in order to achieve a balance between exploration and exploitation at various stages of the optimisation process. The paper also presents an application of DFO in optimising a neural network which is applied real-world data from a Physionet challenge with promising results, outperforming both the classical backpropagation method as well as the winning entries of the corresponding Physionet challenge.

References

1. I.C. Trelea. "The particle swarm optimization algorithm: convergence analysis and parameter selection". Information Processing Letters, vol. 85(6), pp. 317-325, 2003.

2. O. Olorunda, A.P. Engelbrecht. "Measuring exploration/exploitation in particle swarms using swarm diversity" in *Proc. of IEEE Congress on Evolutionary Computation, CEC 2008*, pp. 1128-1134, 2008.

3. M.M. al-Rifaie. "Dispersive Flies Optimisation" in *Proc. of the 2014 Federated Conference on Computer Science and Information Systems*, vol. 2, pp. 529-538, 2014.

4. J. Kennedy. "The particle swarm: social adaptation of knowledge" in *Proc. of IEEE International Conference on Evolutionary Computation*, pp. 303-308, 1997.

5. R. Storn, K. Price. "Differential evolution–a simple and efficient heuristic for global optimization over continuous spaces". Journal of Global Optimization, vol. 11(4), pp. 341-359, 1997.

6. M. Dorigo, G.D. Caro, L.M. Gambardella. "Ant algorithms for discrete optimization". Artificial Life, vol. 5(2), pp. 137-172, 1999.

7. M.M. al-Rifaie. "Perceived Simplicity and Complexity in Nature" in AISB 2017: Computational Architectures for Animal Cognition, pp. 299-305, 2017.

8. T. Back, D.B. Fogel, Z. Michalewicz. *Handbook of Evolutionary Computation*. IOP Publishing Ltd. 1997.

9. X.-S. Yang. "Firefly algorithms for multimodal optimization" in *Proc. of International Symposium on Stochastic Algorithms*, pp. 169-178, 2009.

10. J. Kennedy. "Bare Bones Particle Swarms" in *Proc. of Swarm Intelligence Symposium, 2003 (SIS'03)*, pp. 80-87, 2003.

11. T. Blackwell. "A study of collapse in Bare Bones Particle Swarm Optimisation". IEEE Transactions on Evolutionary Computing, vol. 16(3), pp. 354-372, 2012.

12. M.G.H. Omran, A.P. Engelbrecht, A. Salman. "Bare bones differential evolution". European Journal of Operational Research, vol. 196(1), pp. 128-139, 2009.

13. M.M. al-Rifaie, T. Blackwell. "Bare Bones Particle Swarms with jumps". Lecture Notes in Computer Science, vol. 7461, pp. 49-60, 2012.

14. R.A. Krohling. "Gaussian particle swarm with jumps" in *Proc. of The 2005 IEEE Congress on Evolutionary Computation*, vol. 2, pp. 1226-1231, 2005.

15. M.M. al-Rifaie, T. Blackwell. "Cognitive Bare Bones Particle Swarm Optimisation with jumps". International Journal of Swarm Intelligence Research (IJSIR), vol. 7(1), pp. 1-31, 2016.

16. F. van den Bergh. "An Analysis of Particle Swarm Optimizers". PhD Thesis, University of Pretoria, South Africa, 2002.

17. F. van den Bergh, A.P. Engelbrecht. "A study of particle swarm optimization particle trajectories". Information Sciences, vol. 176(8), pp. 937-971, 2006.

18. M. Clerc, J. Kennedy. "The particle swarm-explosion, stability, and convergence in amultidimensional complex space". IEEE Transactions on Evolutionary Computation, vol. 6(1), pp. 58-73, 2002.

19. M.M. al-Rifaie, A. Aber. "Dispersive Flies Optimisation and Medical Imaging" in Recent Advances in Computational Optimization, Studies in Computational Intelligence book series (SCI, volume 610), pp. 183-203, Springer, 2016.

20. H. Alhakbani. "Handling Class Imbalance Using Swarm Intelligence Techniques, Hybrid Data and Algorithmic Level Solutions". PhD Thesis, Goldsmiths, University of London, London, United Kingdom, 2018.

21. H.A. Alhakbani, M.M. al-Rifaie. "Optimising SVM to classify imbalanced data using dispersive flies" in *Proc. of the 2017 Federated Conference on Computer Science and Information Systems, FedC-SIS 2017*, pp. 399-402, 2017.

22. H. Oroojeni, M.M. al-Rifaie, M.A. Nicolaou. "Deep neuroevolution: training deep neural networks for false alarm detection in intensive care units" in *Proc. of European Association for Signal Processing (EUSIPCO) 2018*, pp. 1157-1161, 2018.

23. M.M. al-Rifaie, A. Ursyn, R. Zimmer, M.A.J. Javid. "On symmetry, aesthetics and quantifying symmetrical complexity" in *Proc. of International Conference on Evolutionary and Biologically Inspired Music and Art*, pp. 17-32, 2017.

24. M. King, M.M. al-Rifaie. "Building simple non-identical organic structures with dispersive flies optimisation and A* path-finding" in *Proc. of AISB 2017: Games and AI*, pp. 336-340, 2017.

25. P. Aparajeya, F.F. Leymarie, M.M. al-Rifaie. "Swarm-Based Identification of Animation Key Points from 2D-medialness Maps" in Computational Intelligence in Music, Sound, Art and Design, pp. 69-83, Springer, 2019.

26. M.M. al-Rifaie F.F. Leymarie, W. Latham, M. Bishop. "Swarmic autopoiesis and computational creativity". Connection Science, pp. 1-19, 2017.

27. F.P. Such, V. Madhavan, E. Conti, J. Lehman, K.O. Stanley, J. Clune. "Deep Neuroevolution: Genetic Algorithms Are a Competitive Alternative for Training Deep Neural Networks for Reinforcement Learning". arXiv preprint arXiv:1712.06567, 2017.

28. Association for the Advancement of Medical Instrumentation and others. "Cardiac monitors, heart rate meters, and alarms". American National Standard (ANSI/AAMI EC13: 2002) Arlington, VA, pp. 1-87, 2002.

29. S.T. Lawless. "Crying wolf: false alarms in a pediatric intensive care unit". Critical Care Medicine, vol. 22(6), pp. 981-985, 1994.

30. S. Parthasarathy, M.J. Tobin. "Sleep in the intensive care unit". Intensive Care Medicine, vol 30(2), pp. 197-206, 2004.

31. M.-C. Chambrin. "Alarms in the intensive care unit: how can the number of false alarms be reduced?" Critical Care, vol. 5(4), pp. 184, 2001.

32. C.L. Tsien, J.C. Fackler. "Poor prognosis for existing monitors in the intensive care unit". Critical Care Medicine, vol. 25(4), pp. 614-619, 1997.

33. G.D. Clifford, I. Silva, B. Moody, Q. Li, D. Kella, A. Shahin, T. Kooistra, D. Perry, R.G. Mark. "The PhysioNet/Computing in Cardiology Challenge 2015: reducing false arrhythmia alarms in the ICU" in *Proc. of Computing in Cardiology Conference (CinC)*, pp. 273-276, 2015.

34. F. Plesinger, P. Klimes, J. Halamek, P. Jurak. "False alarms in intensive care unit monitors: detection of life-threatening arrhythmias using elementary algebra, descriptive statistics and fuzzy logic" in *Proc. of Computing in Cardiology Conference (CinC)*, pp. 281-284, 2015.

35. C.H. Antink, S. Leonhardt. "Reducing false arrhythmia alarms using robust interval estimation and machine learning" in *Proc. of Computing in Cardiology Conference (CinC)*, pp. 285-288, 2015.

12

Improved Elephant Herding Optimization and Application

Nand K. Meena

School of Engineering and Applied Science
Aston University, Birmingham, United Kingdom

Jin Yang

School of Engineering and Applied Science
Aston University, Birmingham, United Kingdom

CONTENTS

12.1   Introduction

Elephants show complex social and emotional bonding with their family groups as compared to many of the animals. These families are headed by an old and experienced female called a 'matriarch' or 'leader', whereas, the male elephants prefer to live a solitary life hanging around with their male friends, which could be from other herds or families. Interestingly, elephants have a powerful communication link with other herds or elephants by seismic waves travelling through or just over the ground. A herd can send the information about food, water, and sometimes warning of hungry predators or natural disasters as well. The other elephants can sense these signals through their feet or ear flapping and take action accordingly. The swarming or herding behaviour of elephants promotes the elephant herding optimization (EHO).

In 2015, Wang et al. [1, 2] proposed the EHO method inspired by the herding behaviour of elephants. It is a swarm intelligence based meta-heuristic optimization method. In this method, two behaviours of elephants are modelled into some set of mathematical equations. One is the clan updating operator, in which all the elephants of a clan update their positions according to their matriarch position and the matriarch also updates its own position by following the positions of that clan. The second operator is male elephant separation operator. In this operation, the male elephant is separated from the clan and a new position is obtained irrespective of the clan and leader elephant.

After its development, the EHO is applied to solve the diversified real-life optimization problems. A multi-level optimal image thresholding is determined in [3], as optimal threshold value determination has been found to be a hard optimization problem. In [4], the EHO method is applied to determine the optimal parameters tuning of the support vector machine. In order to establish best possible monitoring, of all required targets, with minimum number of static drones, the optimal drone deployment problem is solved in [5] by using the EHO method. Due to the large number of control points, a unmanned aerial vehicle path planning problem is solved in [6] by using the EHO method. In Correia et al. [7], the EHO algorithm is applied to solve an energy-supported source determination problem of wireless sensors networks. An EHO-based PID controller is designed in [8] for load frequency of power systems. The EHO, while suggesting some improvements, is also applied in [9] to solve the mixed-integer, non-linear and non-convex optimization problem of distributed generation (DG) allocation in distribution networks.

In this chapter, the modified EHO is applied to solve the optimal economic dispatch of microgrids composed of multiple distributed energy resources (DERs). The chapter is organized as follows. In Section 12.2, the brief of EHO algorithm is presented for basic understanding of the method. In Section 12.3, some modifications in standard EHO are discussed in order to improve the performance of this method. The application of modified EHO is applied to solve the optimal economic dispatch problems of microgrids to minimize the daily operating cost of the systems, followed by conclusions.

12.2 Original elephant herding optimization

In this section, the standard version of EHO is briefly discussed. To solve the global optimization problems, the swarm-intelligence behaviour of elephant herding is modelled into some set of mathematical equations based on three idealized rules. These rules are as follows, 1) The elephant herding is composed of pre-defined clans and each clan consists of a pre-defined fixed number of elephants; 2) In each generation, a fixed number of male elephants leave the clans to live independent life or with other males in the area; and 3) The elephants' population of each clan live or move under the leadership of their respective matriarch. As mentioned, the EHO method has two operators, discussed below.

12.2.1 Clan updating operator

This operator is specifically designed to update the position of elephants in each clan, which is influenced by the position of the leader elephant of that clan. The position of jth elephant of cth clan, except best and worst elephants, can be updated as

$$p_{jc}^{t+1} = p_{jc}^t + \alpha \times (p_{best} - p_{jc}^t) \times r \qquad (12.1)$$

where, p_{jc}^t represents the position of jth elephant of clan c in tth generation. The p_{best}, α and r denote the position of the matriarch, scaling factor varies between 0 to 1, and a random number following the uniform distribution respectively. The position of matriarch is not updated by using (12.1) therefore, is updated as

$$p_{jc}^{t+1} = \beta \times p_{center,c} \qquad (12.2)$$

$$p_{center,c} = \frac{\sum_{j=1}^{n_c} p_{jc}^t}{n_c} \qquad (12.3)$$

where, $\beta \epsilon [0,1]$ and n_c represent the scaling factor and number of elephants in cth clan respectively.

12.2.2 Separating operator

The male elephants are likely to live a solitary life when they grow up; therefore, a male separating operator is designed in EHO. In this operator, the elephant individual having worst fitness is removed from the population and a new elephant is generated as

$$p_{worst,jc}^{t+1} = p_{min,c} + rand \times (p_{max,c} - p_{min,c} + 1) \qquad (12.4)$$

where, $p_{min,c}$, $p_{max,c}$, and $rand$ are the minimum and maximum allowed position limits of clan c, and uniformly distributed random number respectively.

12.3 Improvements in elephant herding optimization

It has been observed that the standard version of EHO has some inherent limitations. In this section, some modifications are discussed to improve the performance of EHO algorithm.

12.3.1 Position of leader elephant

In standard EHO, the positions of leader elephants are updated according to the mean position, i.e. $p_{center,c}$, of the clan, as expressed in (12.2). In [9, 10], it has been investigated that by doing this, the clan is misguided as a result of that searching ability of the method. In order to overcome this limitation, the position of matriarch in (12.2), is updated as

$$r_{jc}^{t+1} = p_{best} + \beta \times p_{center,c} \tag{12.5}$$

12.3.2 Separation of male elephant

In EHO, when the male elephant leaves its family group, a new born baby is added to the clan in order to keep the number of elephants constant in a clan. The new born babies are placed at a randomly generated position, as suggested in (12.4). However, it has been observed from the herding behaviour of elephants that females always keep their babies near to fittest females groups, rather placing them, at random positions. In this improvement, the new born baby will be placed near the fittest female elephant of the clan. The position of the new elephant is updated as in [9]

$$p_{worst,jc}^{t+1} = \mu \times p_{fittest,jc} \tag{12.6}$$

where, μ and $p_{fittest,jc}$ represent the random number that varies from 0.9 to 1.1 and fittest female elephant j of clan c respectively.

12.3.3 Chaotic maps

The chaotic map is also one of the approaches to improve the performance of swarm intelligence based techniques. In this approach, the random numbers or values are replaced by chaotic maps. It is found that chaotic maps provide a random number without repetitions and ergodicity which thus improves the solution searching ability of swarm-based methods. In [11], two unlike one-dimensional maps are considered to improve the performance of the EHO method namely, chaotic circle and sinusoidal maps. The chaotic circle map is defined as

$$\theta_{k+1} = \theta_k + \Omega - \frac{K}{2\pi}\sin(2\pi\theta_k) \tag{12.7}$$

where, θ_{k+1} is computed in mod 1. To generate the chaotic sequence between zero and 1, the value of the circle map parameters are as follows: $K = 0.5$ and $\Omega = 0.2$. The chaotic sinusoidal map can be expressed as

$$\theta_{k+1} = A \cdot \sin(\pi\theta_k) \tag{12.8}$$

where, $A \epsilon (0, 1]$ and $x \epsilon (0, 1)$. In chaotic EHO, the maps explained in (12.7) and (12.8) are used to produce chaotic sequence numbers which replace the random numbers used in (12.1) and (12.4).

12.3.4 Pseudo-code of improved EHO algorithm

The pseudo-code of improved EHO (IEHO) is presented in Algorithm 16. In order to explain the implementation of the algorithm, all the steps of this algorithm are discussed in this section. At the start of IEHO, a function $OF(.)$ is prepared which produces the fitness value. The parameters of IEHO such as number of clans N, number of elephants in each clan, n_c, are initialized in step-2. The number of clans will be equal to the number of variables to be optimized, whereas, the number of elephants is the population size of the swarm. In step-3, the algorithm parameters are initialized such as α, β, maximum number of generations G_{max} etc. The upper and lower limits of all N variables are provided in step-4, i.e., $[p_{min,c}, p_{max,c}] \ \forall \ c = 1, 2, \ldots, N$. In order to initialize the algorithm, an array $(n_c \times N)$ of random but feasible population of elephants is generated in steps from 5 to 11, the fitness of each individual is calculated in step 10. For chaotic EHO, $rand$ in step-8 is replaced by chaotic sequence number, as discussed in Section 12.3.3.

In step 12, the best and worst elephants are determined based on calculated fitness. The generation of EHO starts from step-14 and the new position of each elephant is updated in one of the steps from {19, 22, 24}, followed by their fitness calculation in step 28. In improved EHO, the steps 19 and 24 are replaced by (12.5) and (12.6) respectively, discussed in Sections 12.3.1 and 12.3.2. For chaotic EHO, the random numbers '$rand$' will be replaced by the chaotic number sequence. In step 30, the new best and worst elephant are identified and then the matriarch elephant is updated if its fitness is better than the previous best elephant.

Algorithm 16 Pseudo-code of IEHO.

1: determine the objective function $OF(.)$
2: set the number of clans N, where $c\epsilon[1, N]$, and number of elephants in each clan n_c
3: determine the values of scaling factors α, β, and maximum number of generations, G_{max}
4: set the lower and upper bounds for each variable/clan c, $[p_{min,c}, p_{max,c}]$
5: randomly generate the positions for all elephants in each clan, as follows
6: **for** each j-th elephant **do**
7:     **for** each c-th clan **do**
8:         $pp_{jc} = p_{min,c} + (p_{max,c} - p_{min,c}) \cdot rand$
9:     **end for**
10:     $Fitness_j = OF(pp_j)$
11: **end for**
12: determine the best and worst elephants with their locations $bestloc$ and $worstloc$
13: set generation $t = 1$;
14: **while** $t \le G_{max}$ **do**
15:     update the position of elephants in all clans, as follows

16:     **for** each j-th elephant **do**
17:         **for** each c-th clan **do**
18:             **if** any$((j \neq bestloc)\&(j \neq worstloc))$ **then**
19:                 $pp\_new_{jc} = pp_{jc} + alpha \cdot (pbest_c - pp_{jc}) \cdot rand$
20:             **else if** $j = bestloc$ **then**
21:                 $pp_{center,c} = mean(pp_c)$
22:                 $pp\_new_{jc} = beta \cdot pp_{center,c}$
23:             **else if** $j - worstloc$ **then**
24:                 $pp\_new_{jc} = p_{min,c} + (p_{max,c} - p_{min,c}) \cdot rand$
25:             **end if**
26:         **end for**
27:         evaluate the fitness of new individual j as
28:         $Fitness_j = OF(pp\_new_j)$
29:     **end for**
30:     determine the new best and worst elephants
31:     **if** Is new best better than previous best **then**
32:         replace the best individual with new one
33:     **end if**
34:     reset old population $pp = pp\_new$
35:     reset iteration $t = t + 1$
36: **end while**
37: return the *pbest* as a result

The old population of elephants will be replaced by new population for the next generation, if any. The best solution will be printed at the end of the generation.

12.4   Application of IEHO for optimal economic dispatch of microgrids

12.4.1   Problem statement

A simple microgrid system is shown in Fig. 12.1, composed of a diesel generator (DG), battery energy storage system (BESS), and solar photovoltaic (PV) along with some interpretable and critical loads. The total load demand $P_d(h) = 2\text{MW}$, PV power generation $P_{pv}(h) = 748\text{kW}$, available state of charge in BESS $E_{bess}(h) = 2\text{MWh}$, scheduled power of utility $P_{sch}(h) = 1\text{MW}$ are the given data in hour h. The utility allows a maximum of 20% overdraw (OD) and under-draw (UD) of scheduled power in any hour. Determine the optimal dispatch of these energy resources for the minimum operating cost of hour h, expressed as

$$C(h) = C_{dg}(h) + C_{utility}(h) \qquad (12.9)$$

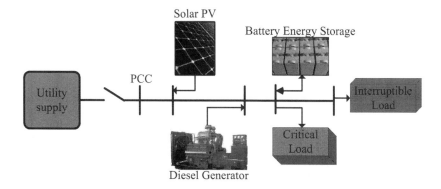

FIGURE 12.1
Microgrid system.

$$C_{dg}(h) = 0.3312 + 0.0156 P_{dg}(h) + 0.0003 P_{dg}^2(h) \; \$ \tag{12.10}$$

$$C_{utility}(h) = \big(P_{sch}(h) + \big|P_{sch}(h) - P_{utility}(h)\big|\big) \times E_{utility}(h) + \\ \big(\big|P_{sch}(h) - P_{utility}(h)\big|\big) \times E_{utility}^{OD/UD} \tag{12.11}$$

s. t.

Diesel generator limits

$$10 \text{ kW} \le P_{dg}(h) \le 100 \text{ kW} \quad \forall \; h \tag{12.12}$$

BESS power dispatch limits

$$-1000 \text{ kW} \le P_{bess}(h) \le 1000 \text{ kW} \quad \forall \; h \tag{12.13}$$

Power balance/equality constraint

$$P_{utility}(h) = P_d(h) - P_{pv}(h) - P_{bess}(h) - P_{dg}(h) \quad \forall \; h \tag{12.14}$$

Utility supply limits

$$0.8 P_{sch} \text{ kW} \le P_{utility}(h) \le 1.2 P_{sch} \text{ kW} \quad \forall \; h \tag{12.15}$$

Utility power constraint

$$P_{utility}(h) \ge 0 \quad \forall \; h \tag{12.16}$$

Energy balance constraints of BESS

$$E_{bess}(h+1) = E_{bess}(h) - \left[\frac{\sigma P_{bess}(h)}{\eta_{dis}} + \eta_{ch}(1-\sigma) \cdot P_{bess}(h) \right] \quad \forall \; h \tag{12.17}$$

where, $P_{dg}(h)$, $P_{bess}(h)$, $P_{utility}(h)$, $E_{utility}(h)$, $E_{utility}^{OD/UD}$, η_{ch}/η_{dis}, α, and σ represent the DG power generation, BESS power generation, utility power supply, power selling price of utility, OD/UD penalty price, charging/discharging efficiencies of BESS, and binary decision variable for charging of BESS respectively. The values of other parameters, used in the study, are $E_{utility} = 0.033$ $/kWh, $E_{utility}^{OD/UD} = 10$ $/kWh, and $\eta_{dis/ch} = 0.90$ etc.

12.4.2 Application of EHO to solve this problem

In order to solve this microgrid economic dispatch problem, we need to create an optimization model and then EHO can be applied to solve this problem. For doing this, we need to identify the optimization variables, i.e clans, for EHO. From the problem statement, it can be observed that there could be three optimization variables for a given hour h, such as DG power generation (P_{dg}), battery power dispatch (P_{bess}) and utility power supply ($P_{utility}$) etc. The other parameters cannot be considered as optimization variables because we do not have control over these. For example, solar PV generation P_{pv} which depends on environmental factors and load demand P_d depends on consumers, which are already given in the problem statement. It may also observed that PV and BESS are owned by microgrid and no fuel charges are given for these, as they might have fixed annual O&M costs, therefore they not considered in cost calculation of (12.9).

Furthermore, it may also be observed that $P_{utility}$ can also be determined from (12.14), if other variables are known. Therefore, two variables $[P_{dg}\ P_{bess}]$ are needed to optimize for the minimum operating cost of the microgrid system. The number of clans in EHO will be $N = 2$ with their lower $[p_{min,dg}\ p_{min,bess}]$ and upper $[p_{max,dg}\ p_{max,bess}]$ bounds being $[10\ -1000]$ and $[100\ 1000]$ respectively, see (12.12) and (12.13). A contour plot of the microgrid's unconstrained operating cost varying with BESS and DG dispatches is shown in Fig. 12.2. The figure shows that the dispatch of different distributed resources and utility grid depends on the load demand.

12.4.3 Application in Matlab and source-code

In this section, the EHO method discussed in Section 12.3.4 is applied to determine the minimum operating cost of the microgrid. The Matlab source-code of OF(.) for the same is presented in Listing 12.1. In this listing, steps 2 to 10 initialize the problem constants or parameters. Step 11 expresses the quadratic cost of DG, as defined in (12.9). Similarly, step 12 calculates the demand and supply mismatch of the microgrid as expressed in (12.14) and we did not consider the $P_{utility}$ as an optimization variable. Basically, it would be the excess load demand which is not supplied by local generation such as PV, DG and BESS or excess generation if demand is less than the generation.

However, it may be observed that microgrid schedules its hourly demand with utility, i.e. P_{sch}. Technically, the utility allows the microgrid to maintain scheduled demand with a maximum error margin of 20%. Therefore, the microgrid has to fulfill the promised constraints expressed in (12.15) and (12.16). In steps 14 to 19, these constraints are expressed. Here, we used a penalty based approach in which high microgrid cost is assigned in step 18, if these

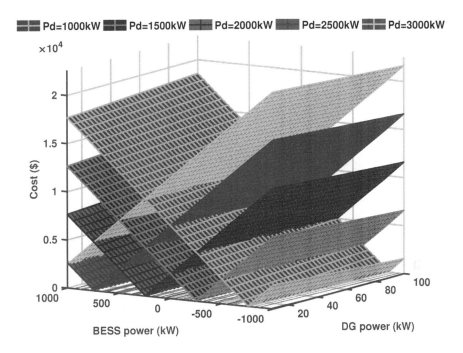

FIGURE 12.2
Operating cost of microgrid at different load demand (unconstrained).

constraints are violated. The penalty based approach smoothly rejects the bad
solution (not satisfying constraints) without or less affecting the initialization
of algorithm. These constraints can also be maintained by directly checking
and eliminating such solutions in the optimization process.

```
1  function [FIT] = OF_Microgrid(pop)
2  Ebess=2000;    % BESS sos level in this hour kWh
3  Psch=1000;     % utility scheduled demand of hour in kW
4  Pd = 2000;     % kW Load demand
5  Ppv = 748;     % kW solar power
6  Euti = 0.033;  % $/kWh utility energy price
7  Eod_ud=10;     % OD/UD penalty price $/kWh
8  neta=0.90;     % BESS charging/discharging efficiency
9  X = pop(1);    % Dispatch of diesel generator (kW)
10 Y = pop(2);    % Dispatch of BESS (kW)
11 C_dg = 0.3312 + 0.0156*X + 0.0003*X^2; %Running cost of DG
12 P_utility = Pd-Ppv-Y-X; % Power to/from utility grid
13 % constraints on utility power
14 if P_utility > 0 && P_utility >0.8*Psch && P_utility <1.2*Psch
15 P_utility = P_utility;
16 else
17 % High penalty for not satisfying equality and non-equality
        constraints of utility power
18 P_utility = randi([1e6 1e10],1,1);
19 end
20 %Cost calculation of power drawing from the grid...
21 C_utility = (Psch + abs(Psch-P_utility))*Euti + abs(Psch - P_utility)
        *Eod_ud;
```

```
22 %update the state of charge of BESS
23 if Y>=0
24 sigma=1;
25 else
26 sigma=0;
27 end
28 Ebes_new = Ebess -((sigma*Y)/neta + neta*(1-sigma)*Y);
29 %Total running cost of microgrid (To be minimized)
30 FIT = C_dg + C_utility;
```

Listing 12.1
Definition of microgrid cost function $F(.)$ in Matlab.

In steps 23 to 27, a binary decision variable σ is generated in order to update the state of charge of the battery for the next hour, i.e. $h+1$, followed by energy update in step 28. Finally, the total controlled operating cost is expressed in step 30. The function $F(.)$ is minimized by applying the IEHO method and the optimal dispatch of DG and BESS are determined. The minimum operating cost and optimal dispatch of DG and BESS obtained by IEHO are expressed as

$P_{dg} = 25.7692$ kW
$P_{bess} = 226.0874$ kW
minimum $C = 35.3705$ \$
The other calculated variables are as follows
SOC of BESS for next hour, $E_{bess}(h+1) = 1748.80$ kWh
Power supplied by utility, $P_{utility} = 1000.14$ kW.

From the results, it has been observed that all the constraints are satisfied by the proposed model and IEHO effectively determined the minimum operating cost of microgrid system.

12.5   Conclusions

In this chapter, the EHO algorithm is discussed along with some suggested improvements in order to improve the performance of the standard variant of the method. For its implantation, the pseudo-code is discussed point-wise and then a real-life microgrid optimization problem is formulated and solved to demonstrate its competitiveness to solve an engineering optimization problem. The Matlab source-code of the problem is also explained. The performance

TABLE 12.1
Some of the performance parameters of IEHO for solving the microgrid economic dispatch problem.

Best fitness	Worst fitness	Mean fitness	Standard deviation
35.3705	40.5471	35.9114	1.5981

parameters of the EHO method are presented in Table 12.1. The tables shows the best fitness, worst, and mean fitnesses along with standard deviation. It has been observed that the method performs well for the microgrid economic dispatch problem.

Acknowledgement

This work was supported by the Engineering and Physical Sciences Research Council (EPSRC) of United Kingdom (Reference Nos.: EP/R001456/1 and EP/S001778/1).

References

1. G. Wang, S. Deb and L. d. S. Coelho. "Elephant Herding Optimization," *3rd International Symposium on Computational and Business Intelligence (ISCBI)*, Bali, 2015, pp. 1-5. doi: 10.1109/ISCBI.2015.8

2. G.G. Wang, S. Deb, X.Z. Gao and L.D.S. Coelho. "A new meta-heuristic optimisation algorithm motivated by elephant herding behaviour". *International Journal of Bio-Inspired Computation*, vol. 8(6), pp.394-409.

3. E. Tuba, A. Alihodzic and M. Tuba. "Multilevel image thresholding using elephant herding optimization algorithm". *14th International Conference on Engineering of Modern Electric Systems (EMES)*, Oradea, 2017, pp. 240-243. doi: 10.1109/EMES.2017.7980424

4. E. Tuba and Z. Stanimirovic. "Elephant herding optimization algorithm for support vector machine parameters tuning". *9th International Conference on Electronics, Computers and Artificial Intelligence (ECAI)*, Targoviste, 2017, pp. 1-4. doi: 10.1109/ECAI.2017.8166464

5. I. Strumberger, N. Bacanin, S. Tomic, M. Beko and M. Tuba. "Static drone placement by elephant herding optimization algorithm". *25th Telecommunication Forum (TELFOR)*, Belgrade, 2017, pp. 1-4. doi: 10.1109/TELFOR.2017.8249469

6. A. Alihodzic, E. Tuba, R. Capor-Hrosik, E. Dolicanin and M. Tuba. "Unmanned aerial vehicle path planning problem by adjusted elephant herding optimization". *25th Telecommunication Forum (TELFOR)*, Belgrade, 2017, pp. 1-4. doi: 10.1109/TELFOR.2017.8249468

7. S.D. Correia, M. Beko, L.A. Da Silva Cruz and S. Tomic. "Elephant herding optimization for energy-based localization". *Sensors* vol. 18, 2018, pp. 2849.

8. D. K. Sambariya and R. Fagna, "A novel Elephant Herding Optimization based PID controller design for load frequency control in power system," 2017 International Conference on Computer, Communications and Electronics (Comptelix), Jaipur, 2017, pp. 595-600.

9. N. K. Meena, S. Parashar, A. Swarnkar, N. Gupta and K. R. Niazi. "Improved elephant herding optimization for multiobjective DER accommodation in distribution systems". *IEEE Transactions on Industrial Informatics*, vol. 14(3), 2018, pp.1029-1039.

10. S. Parashar, A. Swarnkar, K. R. Niazi and N. Gupta. "Modified elephant herding optimisation for economic generation coordination of DERs and BESS in grid connected micro-grid". *The Journal of Engineering*, vol. 13, 2017, pp. 1969-1973, 2017. doi: 10.1049/joe.2017.0673

11. E. Tuba, R. Capor-Hrosik, A. Alihodzic, R. Jovanovic and M. Tuba. "Chaotic elephant herding optimization algorithm". *IEEE 16th World Symposium on Applied Machine Intelligence and Informatics (SAMI)*, Kosice, 2018, pp. 000213-000216. doi: 10.1109/SAMI.2018.8324842

12. M. E. Baran and F. F. Wu. "Network reconfiguration in distribution systems for loss reduction and load balancing". *IEEE Transactions on Power Delivery*, vol. 4(2), 1989, pp. 1401-1407.

13

Firefly Algorithm: Variants and Applications

Xin-She Yang

School of Science and Technology
Middlesex University, London, United Kingdom

CONTENTS

13.1 Introduction

The original firefly algorithm (FA) was first developed by Xin-She Yang in late 2007 and early 2008 [1], based on the flashing characteristics of tropical fireflies. Since then, the FA has been applied to a wide range of applications [2], and it has also been extended to multiobjective optimization [3].

In addition, FA has been extended to its corresponding discrete variants so as to solve discrete and combinatorial optimization [4]. Modifications have been carried out to solve a navigation problem [5], and some randomly attracted neighborhood search has also been introduced [6]. Some qualitative and quantitative analysis about the FA may partly explain why FA works well in practice [7].

For more challenging problems such as protein structure prediction, FA was also attempted for lattice models [8]. Scheduling problems on grid computing has been solved using FA [9], and self-adaptive decision-making for robot swarms has been carried out [10]. Furthermore, FA-based feature selection has been used for network intrusion detection [11].

Though randomization has been used for most algorithms, chaos can have some advantages over standard randomization. For example, a fractional calculus-based FA has been developed to solve parameter estimation problems [12], and a chaos-based firefly algorithm can be used to solve large-sized steel dome design problems [13], and other optimization problems [14]. A set of compact FA variants has been developed with various modifications and enhancements [15].

Representations of solution vectors are usually in real numbers. An alternative is to use other representations such as quaternion [16], which shows some improved efficiency for certain types of problems.

Other interesting applications of FA include optimization of modular granular neural networks [17], resource-constrained project scheduling [18], feature selection [19], and clustering [20]. For a relatively comprehensive review on the firefly algorithm and its application, please refer to the edited book [21].

This chapter focuses on the variants of FA and their applications. Let us briefly review the standard FA first.

13.2   Firefly algorithm

A solution vector x to an optimization problem can be considered as the position of a firefly in a D-dimensional space with D independent variables. That is

$$x = [x_1, \ x_2, \ x_3, \ ..., \ x_D], \tag{13.1}$$

where we have used a row vector that can be changed into a column vector by a transpose (T).

13.2.1   Standard FA

The main equation for the firefly algorithm is to calculate the new position vector x_i^{t+1} at iteration t such that

$$x_i^{t+1} = x_i^t + \beta_0 e^{-\gamma r_{ij}^2}(x_j^t - x_i^t) + \alpha \epsilon_i^t, \tag{13.2}$$

which has three terms on the right-hand side. The first term is the current position x_i^t, while the second term is the attraction term where β_0 is the attractiveness at zero distance and γ is an absorption coefficient. The third term is a perturbation term with a scaling factor α and the perturbation is carried out by drawing a random vector ϵ_i^t from a normal distribution with a zero mean and unity variance.

The Cartesian distance r_{ij} is defined as

$$r_{ij} = ||\boldsymbol{x}_i - \boldsymbol{x}_j||_2 = \sqrt{\sum_{k=1}^{D} (x_{i,k} - x_{j,k})^2}, \quad (13.3)$$

where $x_{i,k}$ means the kth component or variable of the solution vector $\boldsymbol{x}_i$.

For the parameter values, we usually use $\beta_0 = 1$ in most applications. However, γ should be linked to the scale L of the problem under consideration. So we often set

$$\gamma = \frac{1}{L^2}. \quad (13.4)$$

For example, if a variable varies from -5 to $+5$, its scale is $L = 10$, so we can use $\gamma = 1/10^2 = 0.01$. If there is no prior knowledge of the scale of the problem, $\gamma = 0.01$ to 1 can be used as an initial guess.

On the other hand, the scaling factor α controls the step size of the movements of the fireflies. Ideally, α should be gradually reduced. In most applications, we can use

$$\alpha = \alpha_0 \theta^t, \quad (13.5)$$

where $0 < \theta < 1$ is a constant, and α_0 is the initial value of α. For example, we can use $\theta = 0.9$ to 0.99 and $\alpha_0 = 1$ [1,21].

13.2.2   Special cases of FA

Though there is only a single update equation (13.2) in FA, it still has much richer dynamics, and a few other algorithms can somehow be considered as special cases of FA.

- In the case of $\gamma = 0$, the attractiveness coefficient becomes a constant β_0. This means that all fireflies are visible and can be seen by other fireflies. Eq. (13.2) becomes

$$\boldsymbol{x}_i^{t+1} = \boldsymbol{x}_i^t + \beta_0(\boldsymbol{x}_j^t - \boldsymbol{x}_i^t) + \alpha\epsilon_t^t. \quad (13.6)$$

If we impose further that $\alpha = 0$ (no additional randomization), the above equation becomes a key equation in differential evolution (DE) if we now interpret β_0 as the learning or mutation factor F in DE. On the other hand, if we replace $\boldsymbol{x}_j^t$ in Eq. (13.6) by the best solution $\boldsymbol{g}^*$ found so far, it becomes the accelerated particle swarm optimization (APSO) [1].

- In the case of $\gamma \to +\infty$, the attraction term becomes zero, which means that its contribution becomes zero. If we impose $\alpha \neq 0$, the FA equation becomes a random walk, which is essentially the main mechanism in simulated annealing (SA).

This implies that DE, APSO, and SA algorithms can be considered as special cases of the firefly algorithm, and thus it is no surprise that FA can be very effective.

The above discussions also provide possible routes for improving the FA by modifying certain components or tuning parameters. In the rest of this chapter, we will briefly highlight a few such variants.

13.3   Variants of firefly algorithm

There are quite a few variants of the FA and many different modifications. It is not our intention to review most of them in this chapter. Instead, we just highlight a few so as to see the ideas of how different variants have been developed to suit different tasks.

13.3.1   Discrete FA

The standard FA is originally for optimization with continuous variables. For discrete and combinatorial optimization problems, the variables are discrete. Thus, we have to convert the standard FA into a discrete version. One way to discretize a continuous variable x is to use the sigmoidal or logistic function

$$S(x) = \frac{1}{1 + e^{-x}}, \tag{13.7}$$

which is an S-shaped function. It essentially converts a continuous variable x into a binary variable S when $|x|$ is large. When $x \to +\infty$, $S \to 1$. When $x \to -\infty$, $S \to 0$. However, this is not easy to implement in practice. So a random number $r \in [0, 1]$ is usually generated and used as a conditional switch. That is, if $S(x) > r$, $u = 1$, otherwise $u = 0$, which gives a binary variable $u \in \{0, 1\}$.

Once we have a binary variable $u \in \{0, 1\}$, we can convert it to binary variables with other discrete values. For example, we can use $y = 2u - 1 \in \{-1, +1\}$.

It is worth pointing out that a useful property of $S(x)$ is that its derivative can be computed by multiplication

$$\frac{dS}{dt} = -\frac{1}{(1 + e^{-x})^2}(-e^{-x}) = \frac{1}{1 + e^{-x}} \cdot \frac{e^{-x}}{1 + e^{-x}}$$

$$= \frac{1}{1 + e^{-x}} \cdot \frac{[(1 + e^{-x}) - 1]}{1 + e^{-x}} = \frac{1}{1 + e^{-x}} \cdot [1 - \frac{1}{1 + e^{-x}}] = S(1 - S). \tag{13.8}$$

Another way for discretization is to use round-up operations to get integer values. For example, we can use

$$y = \lfloor x \rfloor, \tag{13.9}$$

to convert x to integer y.

Alternatively, we can convert a continuous variable into m discrete integers by using a mod function

$$y = \lfloor x + k \rfloor \quad \mathrm{mod} \ m, \tag{13.10}$$

where k and $m > 0$ are integers.

Sometimes, a so-called random key can be used to generate a set of discrete values for nodal numbers (as in the travelling salesman problem) and job numbers as in job shop scheduling problems. For example, with a random key

$$x = [0.91, \ 1.1, \ 0.14, \ 0.09, \ 0.77, -0.23, 0.69], \tag{13.11}$$

which can be converted to

$$J = [6, \ 7, \ 3, \ 2, \ 5, \ 1, \ 4]. \tag{13.12}$$

This is done by ranking the real number vector x first, and then transforming them into labels of ranks. In some applications, such continuous numbers are drawn from uniformly distributed numbers in [0,1]. For example, we have

$$\begin{pmatrix} \text{Real numbers} & 0.65 & 0.25 & 0.37 & 0.04 & 0.89 \\ \downarrow & \downarrow & \downarrow & \downarrow & \downarrow & \downarrow \\ \text{Random keys} & 4 & 2 & 3 & 1 & 5 \end{pmatrix}. \tag{13.13}$$

Such random-keys-based approaches have been applied in many applications such as the travelling salesman problem (TSP), vehicle routing problems [4] and scheduling problems [9,21].

13.3.2 Chaos-based FA

One way to potentially enhance the performance of algorithms, including FA, is to use chaotic maps to replace the fixed values of algorithmic parameters. In the case of FA with chaos [14], they used 12 different chaotic maps such as the Chebyshev map, Gauss map and Logistic map to replace β and γ. For example, the Chebyshev map is based on Chebyshev polynomials and can be written as

$$u_{k+1} = \cos(k \cos^{-1}(u_k)), \tag{13.14}$$

which generates a chaotic sequence in the range of [-1, +1] (for $n = 1, 2, 3, ...$). The Logistic map is often written as

$$u_{k+1} = \lambda u_k (1 - u_k), \tag{13.15}$$

which gives a chaotic sequence in (0,1) for $\lambda = 4$ when $u_0 \neq 0$.

Extensive simulations carried out by Gandomi et al. [14] indicate that the best map for this purpose is the Gauss map

$$u_{k+1} = \frac{1}{u_k} - \left[\frac{1}{u_k} \right], \quad u_k \neq 0, \tag{13.16}$$

which essentially takes the mod operation to generate a chaotic sequence. The main reason for this enhancement is probably that chaotic maps tend to increase the mixing ability of different solutions/fireflies in the population, and thus increase the mobility and the probability of finding the global optimality of complex objective functions.

Other chaos-based improvement has been been investigated with promising results [13]. Similar enhancement can be also achieved by quantum-based approaches where the parameter values are usually replaced by the probability derived from quantum mechanics.

13.3.3   Randomly attracted FA with varying steps

Another type of variants may focus more on the variations of parameters related to iteration counter t. For example, in the randomly attracted FA variant, Wang et al. [6] used a formula for varying the attractiveness $\beta = \beta_0 e^{-\gamma r_{ij}^2}$ as

$$\beta = [\beta_{\min} + (\beta_{\max} - \beta_{\min})e^{-\gamma r_{ij}^2}]\frac{t}{G_{\max}}, \tag{13.17}$$

where $\beta_{\max}$ and $\beta_{\min}$ are the maximum and minimum of β, respectively, though $\beta_{\max} = 0.9$ and $\beta_{\min} = 0.3$ were used in their paper. Here, $G_{\max}$ is the maximum number of iterations or generations.

In addition, the full attraction model was used in the standard FA. That is, all the fireflies will be attracted to the best firefly. The advantage of this approach is that it can lead to very good convergence; however, if the attraction is too strong, premature convergence can occur under certain conditions. Thus, there is a need to reduce strong attraction. One modification is to use a random attraction approach as outlined by Wang et al. [6]. The main idea is that each better firefly can only be attracted to another firefly that is randomly selected. This is further enhanced by neighbourhood search in terms of one local operator and two global search operators. This FA variant showed a very good improvement in performance.

13.3.4   FA via Lévy flights

Another way of increasing the mobility of fireflies is to use other probability distributions. In the original FA, the random number vector ϵ_t^i in the perturbation term is drawn from either a uniform distribution or a Gaussian distribution. FA via Lévy flights [1,21] uses Lévy flights to replace the Gaussian distribution. Lévy flights are a random walk with step sizes s drawn from a Lévy distribution in the approximate power-law form

$$L(s) \sim \frac{1}{s^{1+\lambda}}, \quad 0 < \lambda \leq 2. \tag{13.18}$$

It is worth pointing out that this form is an approximation, because the Lévy distribution should strictly be defined in terms of an integral form, which makes it difficult to draw random numbers [1].

The advantage of Lévy flights is that such flights consist of a fraction of large step sizes in addition to many local steps. This leads to a phenomenon, called super-diffusion. As a result, the diversity of the firefly population is enhanced and the solutions generated can be sufficient far away from any local optima. This will ultimately increase the probability of finding the true global solution to the optimization problem under consideration [21].

13.3.5 FA with quaternion representation

In almost all metaheuristic algorithms (including FA), representations in terms of real numbers are used for formulating problems and representing the values of design variables. Such representations have rigorous mathematical foundations. In many applications such as electrical engineering, complex numbers (i.e., $a+bi$) are often used. In addition, there are other mathematical representations such as quaternions and octonions, which are the extensions of complex numbers. Such representations can have certain advantages. For example, in computational geometry, quaternion-based representations can be useful for manipulating 3D geometrical shapes.

Along this line, a quaternion-based FA variant was developed by Fister et al. [16]. The main idea was to use quaternions

$$q = a + bi + cj + dk, \tag{13.19}$$

where a, b, c, d are real numbers, while i, j, k satisfy Hamiltonian permutation conditions

$$ij = k, \quad jk = i, \quad ki = j, \tag{13.20}$$
$$ji = -k, \quad kj = -i, \quad ik = -j, \tag{13.21}$$

and

$$i^2 = j^2 = k^2 = -1. \tag{13.22}$$

Such quanternions obey standard quaternion algebra. For example, the norm of q is defined as

$$||q|| = \sqrt{a^2 + b^2 + c^2 + d^2}. \tag{13.23}$$

Though it seems that such representations may increase the computational efforts, the performance can indeed improve without any big increase in computation costs, as shown in the work by Fister et al. [16].

13.3.6 Multi-objective FA

The standard FA was initially designed for solving single objective optimization problems. However, most real-world problems are intrinsically multi-objective optimization. Thus, FA has been extended to solve multi-objective optimization by Yang in 2013 [3]. For example, even a simple bi-objective optimization problem such as Schaffer's min-min test function

$$f_1(x) = x^2, \quad f_2(x) = (x - 2)^2, \quad -1000 \le x \le 1000, \tag{13.24}$$

requires some modifications to algorithms for single objective optimization. This bi-objective optimization problem can have many solutions (in fact, an infinite number of solutions) forming a so-called Pareto front.

One simple approach is to use a weighted sum to combine the two functions into a single objective

$$f(x) = af_1 + bf_2, \quad a + b = 1, \quad a, b \in [0, 1].$$ (13.25)

Clearly, the optimal solution of $f(x)$ will depend on a (or $b = 1 - a$). As a varies from 0 to 1 (thus b from 1 to 0), a Pareto front can be traced. Thus, for a multi-objective optimization problem with m objective functions

$$\text{Minimize} \quad f_1(\boldsymbol{x}), f_2(\boldsymbol{x}), ..., f_m(\boldsymbol{x}), \quad \boldsymbol{x} \in \mathbb{R}^D,$$ (13.26)

we can convert it into a single objective problem

$$\text{Minimize} \quad F(\boldsymbol{x}), \quad \boldsymbol{x} \in \mathbb{R}^D,$$ (13.27)

where

$$F(\boldsymbol{x}) = \sum_{i=1}^{m} w_i f_i(\boldsymbol{x}), \quad \sum_{i=1}^{m} w_i = 1, \quad \forall w_i \in [0, 1].$$ (13.28)

Once this step is done properly, we can use the standard FA to solve it [3].

Numerical experiments show that the multi-objective firefly algorithm (MOFA) can be effective, in comparison with other algorithms such as multi-objective differential evolution (MODE), non-dominated sorting genetic algorithm (NSGA) and vector-evaluated genetic algorithm (VEGA) [3].

Though this approach is simple to implement and can be effective, it has some limitations because this approach is valid only if the Pareto front is convex. This is true for the above bi-objective problem and many problems, but many applications can have non-convex Pareto fronts. Therefore, some other approaches such as non-dominated sorting and ϵ-constraint methods should be used, in combination with the FA variants.

13.3.7 Other variants of FA

There are many other variants of FA and interested readers can refer to more advanced literature [2, 12, 15, 21]. The intention here is not to provide a comprehensive review, but to highlight ways to improve and modify the standard FA so as to inspire further research for designing more and better variants.

It is worth pointing out that the above ways of modifying the original FA to design different FA variants can also be used and extended to modify other algorithms such as the cuckoo search (CS), the bat algorithm (BA) and the flower pollination algorithm (FPA) as well as other algorithms such as particle swarm optimization (PSO).

In the rest of this chapter, we will review some applications and then conclude with some discussions.

13.4 Applications of FA and its variants

The applications of FA and its variants are very diverse, and it is again not our intention to review even a good fraction of them. Instead, we only highlight a few areas:

- Design optimization: A main class of application of the FA is design optimization in engineering and industry, such as pressure-vessel design, beam design and structure design [2, 3, 13]. Such design problems can be multimodal with multiple optimal solutions. One of the advantages of FA is that the overall swarm can subdivide into multiple subswarms automatically, which makes it naturally suitable for solving multimodal problems [21].

- Travelling salesman problem: Both the travelling salesman problem and vehicle routing problems are hard problems, and there are no efficient methods for solving such large-scale problems. Thus, some metaheuristics and approximations are needed. Preliminary studies [4] suggested that FA can obtain good results for such problems. Similarly, navigation problems for unmanned vehicles and path planning are also challenging problems, and recent studies show that FA can also be effective in this area [5].

- Scheduling: Scheduling is another class of combinatorial problems, which is not only difficult to solve, but also requires discretizing algorithms properly. For example, task scheduling for grid computing has been successfully carried out by FA [9]. Resource-constrained project scheduling has also been carried out by FA [18].

- Protein folding: The so-called protein folding problem is an NP-complete problem, and there are no efficient methods for tackling such problems. Due to its importance in computational biology and pharmaceutical applications, researchers have attempted various methods, including metaheuristic algorithms. For example, in a study by Maher et al. [8], they showed that the firefly-based approach can speed up the simulations.

- Feature selection and clustering: Feature selection is widely used in many applications. For example, detection of network intrusion can be carried out by FA [11], while a binary FA has been applied for return-cost-based feature selection application [19]. A study on clustering and satellite imaging has shown that FA can obtain the best results with the least computing efforts, among the 14 different algorithms and approaches [20].

- Machine learning: Machine learning has become popular in recent years. Though there are a class of specialized techniques for such applications, there are many problems and challenging issues that require alternative approaches. For example, optimization of granular neural networks has been carried out by FA with promising results, while a multi-objective approach based on the firefly algorithm and ant colony optimization has been success-fully used to carry out self-adaptive decision making for a swarm of robots without central control [10].

There are many other applications and the literature is rapidly expanding. Interested readers can refer to recent journal articles in these areas.

13.5   Conclusion

The firefly algorithm is simple, versatile and yet effective in solving many opti-mization problems in applications. This chapter introduces the basic form of the standard firefly algorithm and then highlights different ways of improving it and designing new variants. Applications have been briefly reviewed with different categories.

Despite the good performance of the FA and many other algorithms, it still lacks mathematical understanding why such metaheuristic algorithms can work well. Thus, some rigorous mathematical analyses in addition to extensive simulations are greatly needed to gain in-depth understanding of the meta-heuristic algorithms.

In addition, different variants can have different advantages and also dis-advantages. It would be useful to explore the further possibility of designing new variants by other approaches such as hybridization with other algorithms including traditional well-tested optimization algorithms.

Furthermore, most of these applications are small and moderate scales in the sense that the number of variables is typically a few dozens or a few hundred at most. Real-world applications can have thousands or even millions of design variables. As computers get faster and cheaper, a wide range of tools emerge, including grid computing and cloud computing. Thus, it would be useful to extend and explore how to use the FA and its variants to solve large-scale real-world applications.

References

1. X.S. Yang, *Nature-Inspired Metaheuristic Algorithms*, Luniver Press, UK, 2008.

2. I. Fister, I. Fister Jr., X.S. Yang, J. Brest, "A comprehensive review of firefly algorithms". *Swarm and Evolutionary Computation*, 13(1):34-46, 2013.

3. X.S. Yang, "Multiobjective firefly algorithm for continuous optimization", *Engineering with Computer*, 29(2): 175-184, 2013.

4. E. Osaba, X.S. Yang, F. Diaz, E. Onieva, A.D. Masegosa, A. Perallos, "A discrete firefly algorithm to solve a rich vehicle routing problem modelling a newspaper distribution system with recycling policy", *Soft Computing*, 21(18): 5295-5308, 2017.

5. Y. Ma, Y. Zhao, L. Wu, Y. He, X.S. Yang, "Navigability analysis of magnetic map with projecting pursuit-based selection method by using firefly algorithm", *Neurocomputing*, 159(1): 288-297, 2015.

6. H. Wang, Z. Cui, H. Sun, S. Rahnamayan, X.S. Yang, "Randomly attracted firefly algorithm with neighborhood search and dynamic parameter adjustment mechanism", *Soft Computing*, 21(18): 5325-5339, 2017.

7. X.S. Yang and X. S. He, "Why the firefly algorithm works", in: *Nature-Inspired Algorithms and Applied Optimization* (Edited by X.S. Yang), Springer, pp. 245-259, 2018.

8. B. Maher, A. Albrecht, M. Loomes, X.S. Yang, K. Steinhöfel, "A firefly-inspired method for protein structure prediction in lattice models", *Biomolecules*, 4(1): 56-75, 2014.

9. A. Yousif, A.H. Abdullah, S.M. Nor, A. Abdelaziz, "Scheduling jobs on grid computing using firefly algorithm", *J. Theor. Appl. Inform. Technol.*, 33(2): 155-164, 2011.

10. N. Palmieri, X.S. Yang, F. De Rango, A.F. Santamaria, "Self-adaptive decision-making mechanisms to balance the execution of multiple tasks for a multi-robots team", *Neurocomputing*, 306(1): 17-36, 2018.

11. B. Selvakumar, K. Muneeswaran, "Firefly algorithm based feature selection for network intrusion detection", *Computers & Security*, 81(1): 148-155, 2019.

12. Y. Mousavi, A. Alfi, "Fractional calculus-based firefly algorithm applied to parameter estimation of chaotic systems", *Chaos, Solitons & Fractals*, 114(1): 202-215, 2018.

13. A. Kaveh, S. M. Javadi, "Chaos-based firefly algorithms for optimization of cyclically large-size braced steel domes with multiple frequency constraints", *Computers & Structures*, 214(1): 28-39, 2019.

14. A.H. Gandomi, X.S. Yang, S. Talatahari, A. H. Alavi, "Firefly algorithm with chaos", *Commun. Nonlinear Sci. Numer. Simulation*, 18(1): 89-98, 2013.

15. L. Tighzert, C. Fonlupt, B. Mendil, "A set of new compact firefly algorithms", *Swarm and Evolutionary Computation*, 40 (1): 92-115, 2018.

16. I. Fister, X.S. Yang, J. Brest, I. Fister Jr., "Modified firefly algorithm using quaternion representation", *Expert Systems with Applications*, 40(18): 7220-7230, 2013.

17. D. Sánchez, P. Melin, O. Castillo, "Optimization of modular granular neural networks using a firefly algorithm for human recognition", *Engineering Applications of Artificial Intelligence*, 64(1): 172-186, 2017.

18. T. Kassandra, Rojaili, D. Suharono, "Resource-constrained project scheduling problem using firefly algorithm", *Procedia Computer Science*, 135: 534-543, 2018.

19. Y. Zhang, X.F. Song, D.W. Gong, "A return-cost-based binary firefly algorithm for feature selection", *Information Sciences*, 418-419: 561-574, 2017.

20. J. Senthilnath, S.N. Omkar, V. Mani, "Clustering using firefly algorithm: performance comparison", *Swarm and Evolutionary Computation*, 1(3): 164-171, 2011.

21. X.S. Yang, *Cuckoo Search and Firefly Algorithm: Theory and Applications*, Studies in Computational Intelligence, vol. 516, Springer, Heidelberg, 2014.

14

Glowworm Swarm Optimization - Modifications and Applications

Krishnanand Kaipa

Department of Mechanical and Aerospace Engineering
Old Dominion University, Norfolk, Virginia, USA

Debasish Ghose

Department of Aerospace Engineering
Indian Institute of Science, Bangalore, India

CONTENTS

14.1 Introduction

Glowworm Swarm Optimization (GSO) is a unique swarm intelligence algorithm that aims to capture all the local optima rather than just the global optimum as most other swarm intelligence algorithms do. There are several applications where this objective makes sense [1, 2, 3]. For instance, the problem of identifying multiple sources of signals in the environment needs each signal source to be identified. As it is more than likely that the location of each signal source will correspond to a local optimum, while the global optimum will merely correspond to the strongest signal source, searching for the local optimum will lead to a satisfactory solution. Another example where searching for local optima is desired is when the global optimum might be too costly to implement and a local optimum, which may be less optimum in terms of some performance measure could be a cheaper alternative. A third possibility is when the set of optima are contiguous points on the search space. This can happen when one is looking for the boundary of, say, a level set. In this chapter, many of these classes of engineering applications, which also require modifications in the basic GSO that become necessary due to the unique requirement of the application, are presented [4, 5, 6]. The chapter also presents a few modifications that show very innovative application of GSO to clustering and wireless network applications.

14.2   Brief description of GSO

GSO starts by placing a population of n glowworms randomly in the search space. Each cycle of the algorithm consists of a luciferin update phase, a movement phase, and a neighborhood range update phase. A detailed description of the original GSO algorithm was described in the chapter 14 of the companion tutorial book on swarm intelligence algorithms [19]. The equations representing the steps of GSO are repeated here.

1. Luciferin update:

$$\ell_i(t+1) \quad = \quad (1-\rho)\ell_i(t) + \gamma J(x_i(t+1)) \tag{14.1}$$

where $\ell_i(t)$ represents the luciferin level associated with glowworm i at time t, ρ is the luciferin decay constant $(0 < \rho < 1)$, γ is the luciferin enhancement constant and $J(x_i(t))$ represents the value of the objective function at agent i's location at time t.

2. Movement update: For each glowworm i, the probability of moving toward a neighbor j is given by

$$p_{ij}(t) \quad = \quad \frac{\ell_j(t) - \ell_i(t)}{\sum_{k \in N_i(t)} \ell_k(t) - \ell_i(t)} \tag{14.2}$$

where $j \in N_i(t)$ and

$$N_i(t) = \{j : d_{ij}(t) < r_d^i(t) \text{ and } \ell_i(t) < \ell_j(t)\} \tag{14.3}$$

is the set of neighbors of glowworm i at time t, $d_{ij}(t)$ represents the Euclidean distance between glowworms i and j at time t, and $r_d^i(t)$ represents the variable neighborhood range associated with glowworm i at time t. Let glowworm i select a glowworm $j \in N_i(t)$ with $p_{ij}(t)$ given by (14.2). Then, the movement update of each glowworm is given by

$$x_i(t+1) \quad = \quad x_i(t) + s \left(\frac{x_j(t) - x_i(t)}{\|x_j(t) - x_i(t)\|} \right) \qquad (14.4)$$

where $x_i(t) \in R^m$ is the location of glowworm i, at time t, in the m−dimensional real space R^m, $\| \cdot \|$ represents the Euclidean norm operator, and s (> 0) is the step size.

3. Neighborhood range update phase: To adaptively update the neighborhood range of each glowworm, the following rule is applied:

$$r_d^i(t+1) \quad = \quad \min\{r_s, \max\{0, r_d^i(t) + \beta(n_t - |N_i(t)|)\}\} \qquad (14.5)$$

where β is a constant parameter and n_t is a parameter used to control the number of neighbors.

14.3 Modifications to GSO formulation

14.3.1 Behavior switching modification

Localization of sources using mobile robot swarms has received considerable attention in the collective robotics community. Examples of such sources include sound, heat, light, leaks in pressurized systems, hazardous plumes/aerosols resulting from nuclear/chemical spills, fire-origins in forest fires, deep-sea hydrothermal vent plumes, hazardous chemical discharge in water bodies, oil spills, etc. This problem has also been recognized as one that involves significant risks to humans. Most research in this area has dealt with single sources, and relatively less research effort has been devoted to multiple source localization [7]. The problem is compounded when there are multiple sources. In all the above situations, there is an imperative need to simultaneously identify and neutralize all the sources before the emissions cause harm to the environment and people in the vicinity. In addition to this, mapping of the contaminant boundary facilitates a rapid planning effort to move people and valuable property out of the affected region [8].

The problem of identifying multiple source location should deal with the issue of how to automatically partition the robots into subgroups in order to ensure that each source is captured by one of the subgroups. Thomas and Ghose [9, 10] proposed a swarm algorithm that intelligently combines chemotactic, anemotactic, and spiralling behaviors in order to locate multiple odor sources. The chemotactic behavior was achieved by using GSO. Agents switch between the three behaviors based on the information available from the environment for optimal performance. The proposed algorithm was achieved by incorporating the following modifications into the GSO framework. In turbulent flows, the peak concentration value within a patch and the frequency of encountering a patch increase as the glowworm gets closer to the source. However, the instantaneous value might mislead the movement decision of the glowworms. Therefore, the authors defined the luciferin of each glowworm to be the maximum odor concentration encountered in the last N_{mem} seconds in its trajectory. This change was seen to improve algorithmic performance significantly. Accordingly, the luciferin update equation was modified as below:

$$\ell_i(t) \quad \leftarrow \quad \max\{C(x_i(t - N_{mem} + 1)), \ldots, C(x_i(t))\} \qquad (14.6)$$

where, $C(x_i(t))$ is the instantaneous odor concentration at glowworm i's location at time t.

A glowworm without a neighbor, but with a nonzero luciferin value switches to the anemotactic behavior: it takes a step in the upwind direction, as given by (14.7), only when the measured concentration is above a threshold. This condition prevents a glowworm from leaving the plume and proceeding upwind away from the source. In case the concentration measured at its current position is below the threshold value, the glowworm stays at its current position.

$$x_i(t) \quad \leftarrow \quad x_i(t) - sw \qquad (14.7)$$

where, w is the wind direction and s is the step size. A glowworm without a neighbor and with a zero luciferin value switches to a spiralling behavior until it either finds a neighbor or measures non-zero luciferin.

Most methods have addressed the problems of either (multiple) source localization or boundary mapping separately. However, [8] proposed a novel algorithm that enables a robotic swarm to achieve the following dual goals simultaneously: localization of multiple sources of contaminants spread in a region and mapping of the boundary of the affected region. The algorithm uses the basic GSO and modifies it considerably to make it suitable for both these tasks. Two types of agents, called the source localization agents (or S-agents) and boundary mapping agents (or B-agents) are used for this purpose. Whereas the S-agents behave according to the basic GSO, thereby achieving source localization, new behavior patterns are designed for the B-agents based on their terminal performance as well as interactions between them that help these agents to reach the boundary. The B-agents follow a luciferin update rule depending on the instantaneous point measurements of the level of contamination (a function value at a point in $x - y$ plane) made by the B-agent. Compared to the S-agents, the luciferin update rule is different for the B-agents. The acceptable level of contamination is defined as $\tilde{J}$. The instantaneous point measurement, which an i-th agent makes at a location $x_b^i(t)$ is

$J(x_b^i(t))$. If the value $J(x_b^i(t))$ is higher than $\tilde{J}$, the luciferin value of the i-th agent is updated as follows:

$$l_b^i(t) = (1 - \rho)l_b^i(t - 1) - \gamma J(x_b^i(t)); \quad \forall i = 1, \ldots, n_b \tag{14.8}$$

where, ρ and γ are the luciferin update scalar parameters in basic GSO and fixed at the same values as that of the S-agents. When $J(x_b^i(t)) \leq \tilde{J}$, the luciferin value associated with the ith agent is updated as

$$l_b^i(t) = 0 \tag{14.9}$$

By assigning the luciferin value to zero, those agents which are on the boundary are restricted from moving further to a safe region where the instantaneous point measurement could be less than $\tilde{J}$. By this method, the B-agents all converge on the boundary and mark the boundary by their presence.

14.3.2   Local optima mapping modification

The basic GSO was formulated to seek the local optima of a given multimodal function, where the fitness (luciferin level) of each glowworm is based on the objective function value at its current location. Standard benchmark multimodal functions were used to test the working of GSO. Aljarah and Ludwig [14] proposed Clustering-GSO, which involves a modification where the multimodal function value $J(g_j)$ of each glowworm g_j, $j = 1, \ldots, m$, where m is the swarm-size, is constructed as a function of its distances to data instances distributed in an d-dimensional space, where d is the size of each data instance and also the dimension of the position vector of each glowworm:

$$J(g_j) = \frac{InterDist \times \frac{1}{n}|cr_j|}{SSE \times \frac{intraD_j}{\max_j(intraD_j)}} \tag{14.10}$$

$$SSE = \sum_{j=1}^{k} \sum_{i=1}^{|C_j|} (Distance(x_i, c_j))^2 \tag{14.11}$$

$$InterDist = \sum_{i=1}^{k} \sum_{j=i}^{|C_j|} (Distance(c_i, c_j))^2 \tag{14.12}$$

$$IntraD_j = \sum_{i=1}^{|Cr_j|} Distance(cr_{ji}, g_j) \tag{14.13}$$

where SSE is the sum of squared errors, InterDist is a function of inter distance between centroids, IntraD$_j$ is a function of the distance of each glowworm g_j to the data instances covered by it, n is the number of data instances, x_j is the location of glowworm g_j, cr_j is the cluster of data instances covered by glowworm g_j, $C = \{C_1, C_2, \ldots, C_k\}$ is the set of clusters, $c = \{c_1, c_2, \ldots, c_k\}$

is the set of corresponding centroids of the clusters, and k is the number of clusters.

This modification enables the algorithm to identify different types of data instances, where each local optimum of the resulting multimodal function corresponds to the optimal centroid of a different cluster of data instances.

14.3.3   Coverage maximization modification

Wireless sensor networks (WSNs) are large collections of sensor nodes with capabilities of perception, computation, communication, and locomotion. They are usually deployed in outdoor fields to carry out tasks like climate monitoring, vehicle tracking, habitat monitoring, earthquake observation, and surveillance. The performance of a WSN is mainly influenced by its coverage of the service area. This problem deals with finding an efficient deployment of the sensor nodes so that every location in the region of interest is sampled by a minimum of one node.

Liao et al. [17] proposed a sensor deployment scheme based on GSO that maximizes the coverage of the sensors with limited movement after an initial random deployment. The decentralized nature of the GSO based approach leads to scalable WSNs. They presented simulation results to show that their approach outperforms the virtual force algorithm (VFA) in terms of coverage rate and sensor movement.

They modeled the sensor deployment problem in the framework of GSO as follows. Each sensor node is considered as a glowworm emitting luciferin whose intensity is a function of its distance from its neighbors. Each glowworm has a sensing range r_s and a communication radius r_c. In original GSO, a glowworm is attracted toward a neighbor of brighter luminescence. On the contrary, in the proposed approach, a glowworm is attracted toward its neighbors having dimmer luminescence and decides to move toward one of them. These local movement rules enable the glowworms to gradually distribute themselves within the sensing field so that coverage is maximized. The luciferin of each glowworm i at time t is computed as below:

$$\ell_i(t) \quad = \quad \ell_i(t-1) + \sum_{j=1}^{|N_i(t)|} \frac{\ell_j(t)}{d_{ij}^2(t)} \tag{14.14}$$

During the movement phase, each glowworm selects to move toward a neighbor with the following probability:

$$p_{ij}(t) \quad = \quad \frac{\ell_i(t) - \ell_j(t)}{\sum_{k \in N_i(t)} \ell_i(t) - \ell_k(t)} \tag{14.15}$$

$$j \quad \in \quad N_i(t) \neq \phi$$

$$N_i(t) \quad = \quad \{j : d_{ij}(t) < r_c \text{ and } \ell_j(t) < \ell_i(t)\} \tag{14.16}$$

where $N_i(t)$ is the set of neighbors of glowworm i at time t.

The optimal distance between neighboring sensors for maximum coverage is $\sqrt{3}r_s$ [18]. Therefore, the distance moved by the glowworm toward its neighbor is chosen as $\frac{\sqrt{3}r_s - d_{ij}(t)}{2}$. Therefore, the movement update for glowworm i is given by:

$$x_i(t+1) \quad = \quad x_i(t) + \left(\frac{\sqrt{3}r_s - d_{ij}(t)}{2} \right) \left(\frac{x_j(t) - x_i(t)}{||x_j(t) - x_i(t)||} \right) \quad (14.17)$$

14.3.4 Physical robot modification

It may be recalled that the original GSO was devised with the express intent to make it implementable on swarm robotic platforms. However, in order to implement GSO on a robotic platform it needs to be modified in order to make the implementation possible, without compromising with the working of the basic GSO. The swarm robotics based approach to source localization usually involves a swarm of mobile robots that search a given area for previously unknown signal-sources. These robots use cues such as their perception of signal-signatures at their current locations, and any information available from their neighbors, in order to guide their movements toward, and eventually to converge at, the signal-emitting sources. The GSO algorithm serves as an effective swarm robotics approach to the source localization problem.

Certain algorithmic aspects need modifications while implementing in a robotic swarm mainly because of the point-agent model of the basic GSO algorithm as against the physical dimensions and dynamics of a real robot. The modifications incorporated into the algorithm in order to make it suitable for a robotic implementation are the subject matter of this subsection.

There are mainly three issues that arise during a robotic implementation that are not taken into account in the algorithmic description of GSO:

1. The linear and angular movements of agents in the algorithm are instantaneous in nature. However, physical robots spend a finite time in performing linear and angular movements based on their prescribed speeds and turn rate capabilities.

2. In the algorithm, agents are considered as points and agent collisions are ignored. These are acceptable assumptions for numerical optimization problems. However, real robots have a physical shape and foot print of finite size, and cannot have intersecting trajectories. Thus, robots must avoid collisions with other robots in addition to avoiding obstacles in the environment.

3. In the point model implementation of GSO agents move over one another to perform a local search. Since this is not possible for real robots, they have to perform collision avoidance maneuvers around other robots.

The above issues are very important and need careful consideration as they call for changes in the agent movement models and alter the working of the basic GSO algorithm when it is implemented on a swarm of real robots.

Thus, each member of the mobile robot swarm used to implement GSO should possess the following capabilities:

1. Sensing and broadcasting of profile-value (luciferin level) at its location.

2. Detection of the number of its neighbors and their locations relative to its own location.

3. Reception of profile-values (luciferin levels) from its neighbors.

4. Selection of a neighbor using a probability distribution (based on the relative luciferin levels of its neighbors) and making a step-movement toward it.

5. Variation of the neighborhood range.

6. Avoiding collisions with obstacles and other robots in the environment.

As a real-robot simulation platform is used and perfect sensing and broadcast are assumed in the initial experiments, the first four robot capabilities are rather straightforward to implement. However, the mechanisms of implementing obstacle avoidance is non-trivial and are described below. They also serve to modify the GSO behavior.

The footprint of the robot is considered to be an octagon with a circumcircle radius r_{robot}. A pair of sonar based proximity-detection sensors with a range r_{sonar} and field of view θ_{fov} are mounted in the frontal region of the robot at a distance d_{sonar} from the robot-center. A simple obstacle avoidance rule is used where the robot decides to turn right (left) if the left (right) sensor detects the other robot. The robot performs the collision-avoidance maneuver by moving along an arc, whose radius of curvature is r_{cv}, and never reaching closer than a safe-distance d_{safe} to the other robot/obstacle. Using simple geometry, the radius of curvature r_{cv} can be calculated as below:

$$r_{cv} = \frac{(d_{sonar} + r_{sonar} + r_{robot})^2 - (2r_{robot} + d_{safe})^2}{2(r_{robot} + d_{safe})} \qquad (14.18)$$

14.4 Engineering applications of GSO

14.4.1 Application of behavior switching to multiple source localization and boundary mapping

In [9, 10] a number of experiments were conducted to validate the algorithm's ability to simultaneously capture multiple odor sources. The proposed approach was later tested on data obtained from a dye mixing experiment. It was also seen capable of locating an odor source under varying wind conditions. A detailed survey on robot algorithms for localization of multiple emission sources was presented by McGill and Taylor [11]. The survey recognized that GSO directly addresses the issue of automatically partitioning a swarm into subgroups as required by the multiple source localization problem. This is achieved by the adaptive local decision domain that facilitates the formation of subgroups in order to locate multiple sources simultaneously. In a comparative analysis, the authors reported that GSO can capture the highest number of sources (100 sources and swarm size of 1000), when compared to other state-of-the-art algorithms. Their experiments considered a gradient field consisting of ten Gaussian sources on a continuous 1000-1000 unit space [12]. Dead space was created by setting the field strength to zero if it was below a threshold value of 0.5. Three initial robot distributions were devised: uniform (agents are deployed at the node locations of a 2D grid spanning the search area), drop (all agents are deployed at a single location in the search area), and line (agents are deployed at one of the edges of the search area). In this, a glowworm employs original GSO once it acquires at least one neighbor.

For the boundary mapping problem [8], an area R in which there are three cumulative Gaussian distribution functions representing basically three sources of contamination. The simulations are done on this static function profile with these three contaminant sources with the assumption that the function values represent the time averaged contaminant intensities at that point. The contour plot of the contaminant spread is considered such that the outer contour corresponds to a value $\tilde{J} = 0.39$ representing an acceptable level of the contamination spread over the region. The objective of the problem is to determine simultaneously an approximation to the outer contour and the sources. The parameters for the modified GSO algorithm are set as follows: sensor range $r_s = 1$, local decision range the same as sensor range, repulsion distance is 0.5, which is less than the sensor range, and maximum number of iterations is 1500. For the experiment, 50 S-agents and 100 B-agents are randomly deployed initially over the region. Luciferin value is initialised with zero. The luciferin decay and enhancement constants are fixed at $\gamma = 0 : 6$ and $\rho = 0 : 4$. The number of desired neighbouring agents is 2. The decision range gain is defined as 0.08. Several cases were considered and it was shown that a good spread of B-agents was achieved while the S-agents converged to multiple sources.

Another very interesting engineering application of GSO to the boundary mapping problem is the implementation of the above algorithm for a boundary mapping problem and the development of a Swarm Algorithm Test-bed (SAT) [13]. The implementation involves the use of three robots, the SAT and a planner.

14.4.2 Application of local optima mapping modification to clustering

The problem of clustering deals with partitioning a set of objects into clusters so that the objects in the same cluster are more similar to each other than to those in other clusters. Clustering has important applications in exploratory data analysis, pattern recognition, machine learning, and other engineering fields. Aljarah and Ludwig [14] applied the local optima mapping modification to the clustering problem. The goal was to partition a set of data instances D into different clusters by having the glowworm swarm split into subswarms, and each subswarm converge to the centroid of each cluster. The GSO objective was modified to locate multiple optimal centroids such that each centroid represents a sub-solution and the combination of these sub-solutions formulate the global solution for the clustering problem. The proposed implementation consisted of four main phases: initialization phase, luciferin level update, glowworm movement, and candidate centroids set construction.

During initialization, a glowworm swarm of size m was created using uniform randomization within the given search space within the minimum and the maximum values that are calculated from the data set D. Then, the luciferin level was initialized using the initial luciferin level ℓ_0. The fitness function value $J(g_j)$ was initialized to zero. The local range r_s was set to an initial constant range r_0. Next, the set of data instances cr_j covered by g_j was extracted from D. The fitness function is evaluated for each glowworm g_j by using eq. (14.11). After this, the steps of the original GSO are applied.

The authors presented the results obtained using their modified algorithm on well-known data sets to conduct a reliable comparison. They presented experimental results to show that GSO with the proposed modification is efficient compared to other well-known clustering algorithms like K-Means clustering, average linkage agglomerative Hierarchical Clustering, Furthest First, and Learning Vector Quantization.

14.4.3 Application of coverage maximization modification to wireless networks

Liao et al. [17] evaluated the performance of the proposed GSO-based sensor deployment scheme by using simulations. The authors considered a variable number of sensor nodes ($n = 50$, 100, and 200) with two different initial deployments: center and random. A 2D obstacle-free environment was considered for deployment of sensors. The first result showed that the coverage rate increased with an increase in number of nodes and the GSO-based scheme achieved a higher coverage (96 % for n = 200) compared to VFA (40 % for n = 200) in all the cases. The second result showed that the coverage rate is higher in the case of random deployment than center deployment irrespective of the number of sensors in the network. The third result showed that the GSO-based scheme achieved a lower moving distance compared to VFA.

14.4.4 Application of physical robot modification to signal source localization

Several real-robot experiments are carried out on actual robot platforms. For this purpose, wheeled robots, called Kinbots [15], that were originally built for experiments related to robot formations, are used. By making necessary modifications to the Kinbot hardware, the robots are endowed with the capabilities required to implement the various behavioral primitives of GSO. These robots have been used for sound source localization [7] and light source localization experiments [16] demonstrate the potential of robots using GSO for localizing signal sources.

Sound source localization [7]: Each Glowworm is equipped with a sound pick up sensor in order to measure the intensity of sound at its location. A PIC16F877 microcontroller is used as the robot's processing unit.

The hardware used to implement the modules of luciferin broadcast/reception is as follows: The glow consists of an infrared light modulated by an 8-bit serial binary signal that is proportional to the Glowworm's luciferin value at the current sensing-decision-action cycle. Four emitters that are mounted vertically and symmetrically about the Glowworm's central axis cast the infrared-light onto a buffed aluminium conic reflector (with an azimuth of 45°) in order to obtain a near circular pattern of luciferin emission in the Glowworm's neighborhood. Two infrared receivers mounted on a sweep-platform are used as luciferin receptors. In order to avoid problems due to interference between data signals coming from different neighbors, the receiver sweeps and aligns along the line-of-sight with a neighbor before reading the luciferin data transmitted by it. Using the above scheme, a minimum threshold separation of 10 cm between neighbors was observed to be sufficient in order to distinguish data coming from different glowworm neighbors.

The hardware used to implement the modules of luciferin broadcast/reception and relative localization of neighbors is as follows: Two photodiodes mounted on the rotary platform perform a 180° sweep and record the intensity of the received infrared light as a function of the angle made by the direction of the source with the heading direction of a Glowworm. Local peaks in the intensity pattern indicate the presence and location of other glowworms. The received intensity of infrared light follows an inverse square law with respect to distance which is used to compute range information to other robots. Even though the Glowworm locates all others within the perception range of the distance-sensor (excepting those that are eclipsed by other glowworms), it identifies them as neighbors only when they are located within its current variable local-decision domain.

In the robotic platform experiment, localization of a single sound source was demonstrated. The sound source was a loud speaker activated by a square wave signal of frequency 28 Hz. A microphone based sound sensor enables each Glowworm to measure the sound intensity at its current location. The sound-intensity pattern in the workspace, is obtained by taking measurements at a sufficiently large number of locations. At first a Glowworm (A) is placed near the sound source and a dummy Glowworm (B) away from the source which is kept stationary but made to emit luciferin proportional to the intensity measurement at its location. A is already located at the source and doesn't get a direction to move and hence, remains stationary. Initially, since B is in the vicinity of C (while A is not), it moves towards B. However, as it reaches closer to B it senses A and hence, changes direction in order to move towards A. Since D is closer to A, it makes deterministic movements towards A at every step. In this manner, the glowworms localize the sound source eventually.

Light source localization [16]: A PIC16F877 microcontroller is used as the robot's processing unit. Experiments are conducted in which robots use GSO to localize a light source. A tethered power supply was used in these experiments.

As GSO is used in these experiments for locating a light source, a light pick-up sensor is used to measure the light intensity at the robot's location. The light sensor output is fed to an in-built analog-to-digital (A/D) convertor of the microcontroller. The output of the A/D module is converted into a luciferin glow that consists of an infrared emission modulated by an 8-bit serial binary signal that is proportional to the robot's luciferin value at the current sensing-decision-action cycle. Eight IR emitters that are mounted symmetrically about the robot's central axis are used to broadcast luciferin data in its neighborhood. An infrared receiver mounted on a sweep platform is used as a luciferin receptor. In order to avoid interference between data signals coming from different neighbors, the receiver sweeps and aligns along the line-of-sight with a neighbor to read the luciferin data broadcast by it.

For relative localization of neighbors, an improved neighbor localization hardware module is built with a circular array of sixteen infrared LEDs, placed radially outward. This array serves as a beacon to obtain a near circular emission pattern around the robot. These IR LEDs are actuated by a 1 KHz square wave signal. A photodiode, mounted on a rotary platform, performs a sweep and records the intensity of the received infrared light as a function of the angle made by the direction of the source with the heading direction of the robot. Local peaks in the intensity pattern indicate the presence of neighbors and serve to compute their location in polar coordinates. As in the sound localization problem, the received intensity of infrared light follows an inverse square law with respect to distance, which is used to compute range information to robot-neighbors and the angular position of the local peak is approximated as the relative angle to a neighbor. The photo-diode output is passed through a bandpass filter, centered around a frequency of 1 KHz, in order to make it sensitive only to the frequency of the actuation signal of the infrared-beacons and reject noise due to ambient light conditions and IR signals used to broadcast luciferin information.

The obstacle avoidance maneuver is achieved by the movement of the robot along an arc in physical simulations. However, in real-robot implementation, when a robot approaches closer than a safe distance $d_{safe} < s$ to another robot, it performs a simple collision-avoidance maneuver through a discrete sequence of point-turn and straight-line movements.

Localization of a single light source was demonstrated through an experiment, in which two Kinbots K_A and K_B implement GSO to detect, taxi toward, and co-locate at a light source. The robots provide simple light intensity measurements at their current locations as inputs to the GSO algorithm running on their onboard processors. The robots use a photodiode based light sensor for this purpose.

In this experiment, the robots are initially deployed in such a way that K_B is closer to the light source. Both the robots start scanning their respective neighborhoods and sense each other. As $\ell_A(0) < \ell_B(0)$, K_B remains stationary and K_A moves toward K_B until $t = 10$ sec. However, between $t = 10$ sec and $t = 30$ sec, K_A remains stationary. This can be attributed to the fact that the luciferin value received by K_A is corrupted. After $t = 30$ sec K_A resumes moving toward K_B until $t = 35$ sec when it switches to obstacle avoidance behavior. Now K_B starts moving toward K_A at $t= 45$ sec. However, at $t = 45$ sec, K_A is still relatively far from the light source than K_B. This can be attributed to the fact that when the robots are very close to each other, the difference in luciferin values interferes with the range of sensor noise. K_B stops moving at $t = 65$ sec). However, at $t= 110$ sec, K_A is closer to the source and $\ell_{K_A}(110) > \ell_{K_B}(110)$. Therefore, K_B moves toward K_A. This results in the robot pair moving closer to the light source.

The paths traced by the two robots, as they execute the modified GSO algorithm, show that toggling between the basic GSO and obstacle avoidance behaviors of the two robots, eventually leads to their localization at the light source. The robots remain idle during certain time intervals, which can be attributed to one of the following reasons:

1. A robot is isolated.

2. A robot is measuring the highest light intensity.

3. Both the robots are measuring almost the same light intensity (A robot is made to move toward another one only when $|\ell_A - \ell_B| > 3$).

4. Reception of luciferin data is corrupted.

From the experimental results it is also clear that the sensing-decision-action cycles of the robots are asynchronous with respect to each other.

14.5   Conclusions

This chapter presented a few modifications that researchers have suggested to the original GSO algorithm in order to make the algorithm suitable to the unique needs of that application. The applications themselves are also presented and discussed. These modifications and applications provide a glimpse into the diverse use of GSO in solving engineering problems. Apart from these, in the literature, there are a large number of other modifications of different kinds that combine GSO with other swarm intelligence algorithms and obtain superior results. These results are indicative of the usefulness of the basic philosophy of GSO to engineering applications and its possible application in many other practical problems.

References

1. K.N. Krishnanand and D. Ghose. Detection of multiple source locations using a glowworm metaphor with applications to collective robotics. In *Proceedings of IEEE Swarm Intelligence Symposium*, Pasadena, California, June 8−10, 2005, pp. 84−91.

2. K.N. Krishnanand and D. Ghose. Glowworm swarm optimization for simultaneous capture of multiple local optima of multimodal functions. *Swarm Intelligence*, Vol. 3, No. 2, pp. 87−124, 2009.

3. K.N. Kaipa and D. Ghose. *Glowworm Swarm Optimization: Theory, Algorithms, and Applications*. Studies in Computational Intelligence. Vol. 698. Springer, 2017.

4. Kalaiselvi, T., Nagaraja, P., & Basith, Z. A. A comprehensive study on Glowworm Swarm optimization. *Computational Methods, Communication Techniques and Informatics*, pp. 332-337.

5. Kalaiselvi, T., Nagaraja, P., & Basith, Z. A. A review on glowworm swarm optimization. *International Journal of Information Technology*, 3(2), 2017, pp. 49 56.

6. Singh A., Deep K. How Improvements in Glowworm Swarm Optimization Can Solve Real-Life Problems. In: *Das K., Deep K., Pant M., Bansal J., Nagar A. (eds) Proceedings of Fourth International Conference on Soft Computing for Problem Solving. Advances in Intelligent Systems and Computing*, vol. 336, 2015, Springer, New Delhi.

7. K.N. Krishnanand, P. Amruth, M.H. Guruprasad, S.V. Bidargaddi, and D. Ghose. Glowworm-inspired robot swarm for simultaneous taxis toward multiple radiation sources. In *Proceedings of IEEE International Conference on Robotics and Automation*, Orlando, Florida, May 15−19, 2006, pp. 958−963.

8. P.P. Menon and D. Ghose. Simultaneous source localization and boundary mapping for contaminants. *American Control Conference*, Montreal, Canada, June 27-29, 2012, pp. 4174-4179.

9. J. Thomas and D. Ghose. Strategies for locating multiple odor sources using glowworm swarm optimization. In *Proceedings of*

Indian International Conference on Artificial Intelligence, Tumkur, India, December 16−18, 2009, pp. 842−861.

10. J. Thomas and D. Ghose. A GSO-Based Swarm Algorithm for Odor Source Localization in Turbulent Environments. In *Handbook of Approximation Algorithms and Metaheuristics: Contemporary and Emerging Applications*, Vol. 2, pp. 711-738, Chapman & Hall/CRC Press, 2018.

11. K. McGill and S. Taylor. Robot algorithms for localization of multiple emission sources. *ACM Computing Surveys*, Vol. 43, No. 3, 2011.

12. K. McGill and S. Taylor. Comparing swarm algorithms for multisource localization. *IEEE International Workshop on Safety, Security, and Rescue Robotics*, Denver, Colorado, November 3-6, 2009.

13. A. Arvind, D. Ghose, P. Menon. Development of a Robotic Platform to Implement a Boundary Mapping Algorithm. International Conference on Advantages in Control and Optimization of Dynamical Systems, March 2014, Special Issue of IFAC Proceedings Volumes, 47(1), 2014, pp. 484–490.

14. I. Aljarah and S. Ludwig. A new clustering approach based on glowworm swarm optimization. *2013 IEEE Congress on Evolutionary Computation*, June 2013, pp. 2642-2649.

15. K.N. Krishnanand and D. Ghose. Formations of minimalist mobile robots using local templates and spatially distributed interactions. *Robotics and Autonomous Systems*, Vol. 53, No. 3−4, pp. 194−213, 2005.

16. K.N. Krishnanand and D. Ghose. A glowworm swarm optimization based multi-robot system for signal source localization. In *Design and Control of Intelligent Robotic Systems*, Springer-Verlag, Vol. 177, pp. 53−74, 2008.

17. W-H. Liao, Y. Kao, Y-S. Li. A sensor deployment approach using glowworm swarm optimization algorithm in wireless sensor networks. *Expert Systems with Applications*, Vol. 38, pp. 12180−12188, 2011.

18. H. Zhang and J.C. Hou. Maintaining sensing coverage and connectivity in large sensor networks. *Ad Hoc and Sensor Wireless Networks*, Vol. 1(1−2), 2005, pp. 89−124.

19. K.N. Krishnanand, D. Ghose. Glowworm Swarm Optimization - A Tutorial. In *Swarm Intelligence Algorithms: A Tutorial*, CRC Press, Taylor & Francis Group, 2020.

15

Grasshopper Optimization Algorithm - Modifications and Applications

Szymon Łukasik

Faculty of Physics and Applied Computer Science
AGH University of Science and Technology, Kraków, Poland

CONTENTS

15.1 Introduction

The optimization technique known as Grasshopper Optimization Algorithm (GOA) was introduced by Saremi, Mirjalili and Lewis in 2017 [24]. It uses the well known concept of intelligent swarms [10]. From the methodological side it includes both pairwise interactions between swarm members, as well as the influence of the best individual of the swarm. The first is reported to stem from the interaction of grasshoppers which demonstrates itself through slow movements (while in larvae stage) and dynamic motion (while in insect form). The second corresponds to the tendency to move towards the source of food, with deceleration observed on the approach path. The algorithm was positively evaluated both by the authors of the concept (and its original scheme) as well as researchers introducing the adaptation of the algorithm for a variety of the domains.

The rest of this paper is organized as follows: in Section 15.2 we provide a more detailed description of the basic scheme of the algorithm under-consideration, enclosing also the pseudo-code of its implementation. Section 15.3 reviews the modification proposed in the literature. Finally, in Section 15.4 we provide the results of our experiments with the clustering technique using GOA as the optimization engine.

15.2   Description of the original Grasshopper Optimization Algorithm

The original GOA, as presented in [24] constitutes a novel optimization algorithm dealing with continuous optimization, i.e. the task of finding a value of x – within the feasible search space $S \subset R^D$ – denoted as x^* such as $x^* = \text{argmin}_{x \in S} f(x)$, assuming that the goal is to minimize cost function f. GOA belongs to a class of population based metaheuristics [27]. Thus it uses the population comprising P agents to tackle the aforementioned problem. Each of them is represented as a solution vector x_p, $p = 1, ..., P$ and corresponds to exactly one solution in the domain of considered function f [15].

The movement of individual p in iteration k is described with the following equation:

$$x_{pd} = c \left(\sum_{q=1, q \neq p}^{P} c \frac{UB_d - LB_d}{2} s(|x_{qd} - x_{pd}|) \frac{x_{qd} - x_{pd}}{dist(x_q, x_p)} \right) \\ + x_d^* \qquad (15.1)$$

where $d = 1, 2, ..., D$ represents dimensionality of the search space, and $dist(a, b)$ Euclidean distance between a and b. The equation (15.1) contains two components: the first line corresponds to the pairwise social interactions between grasshoppers, the second – to the movement attributed to the wind. GOA implement it by a shift towards the best individual. In contrast to the behavior of real grasshopper swarms [6] the algorithm does not contain a gravitation factor.

Function s represents the strength of social forces, and was originally defined as:

$$s(r) = f e^{\frac{-r}{l}} - e^{-r} \qquad (15.2)$$

with default values of $l = 1.5$ and $f = 0.5$.

It divides the space between two considered grasshoppers into three separate zones. Individuals being very close are found in the so called repulsion zone. On the other hand distant grasshoppers are located in the zone of attraction. The zone (or equilibrium state) between them – called the comfort zone – is characterized by the lack of social interactions. The first factor is additionally normalized by the upper (UB_d) and lower bounds (LB_d) of the feasible search space.

Parameter c – occurring twice in formula (15.1) – is decreased according to the following equation:

$$c = c_{max} - k\frac{c_{max} - c_{min}}{K_{max}} \qquad (15.3)$$

with maximum and minimum values given by c_{max}, c_{min} respectively, and maximum number of iterations denoted as K_{max}. The first occurrence of c in (15.1) reduces the movements of grasshoppers around the target – balancing between exploration and exploitation of the swarm around it. The same concept is used in the Particle Swarm Optimization Algorithm and it is known as inertia weight [5]. The component $c\frac{UB_d - LB_d}{2}$ on the other hand linearly decreases the space that the grasshoppers should explore and exploit.

The pseudo-code for the generic version of GOA technique is presented in Algorithm 17.

Algorithm 17 Grasshopper Optimization Algorithm.

1: $k \leftarrow 1$, $f(x^*(0)) \leftarrow \infty$ ▷ initialization
2: **for** $p = 1$ to P **do**
3:     $x_p(k) \leftarrow$ Generate_Solution(LB, UB)
4: **end for**
5:   ▷ find best
6: **for** $p = 1$ to P **do**
7:     $f(x_p(k)) \leftarrow$ Evaluate_quality($x_p(k)$)
8:     **if** $f(x_p(k)) < f(x^*(k-1))$ **then**
9:         $x^*(k) \leftarrow x_p(k)$
10:     **else**
11:         $x^*(k) \leftarrow x^*(k-1)$
12:     **end if**
13: **end for**
14: **repeat**
15:     $c \leftarrow$ Update_c($c_{max}, c_{min}, k, K_{max}$)
16:     **for** $p = 1$ to P **do**
17:                                 ▷ move according to formula (15.1)
18:         $x_p(k) \leftarrow$ Move_Grasshopper($c, UB, LB, x^*(k)$)
19:                                     ▷ correct if out of bounds
20:         $x_p(k) \leftarrow$ Correct_Solution($x_p(k), UB, LB$)
21:         $f(x_p(k)) \leftarrow$ Evaluate_quality($x_p(k)$)
22:         **if** $f(x_p(k)) < f(x^*k)$ **then**
23:            $x^*(k) \leftarrow x_p(k)$, $f(x^*k) \leftarrow f(x_p(k))$
24:         **end if**

25:    **end for**
26:    **for** $p = 1$ to P **do**
27:        $f(x_p(k+1)) \leftarrow f(x_pk), \ x_p(k+1) \leftarrow x_p(k)$
28:    **end for**
29:    $f(x^*(k+1)) \leftarrow f(x^*k), \ x^*(k+1) \leftarrow x^*(k)$
30:    $k \leftarrow k + 1$
31: **until** $k < K_{max}$ **return** $f(x^*(k)), \ x^*(k)$

For more details regarding the scheme of the original algorithm, as well as its practical implementation one could refer to [18].

15.3   Modifications of the GOA technique

Since its introduction in 2017 the Grasshopper Optimization Algorithm has attracted significant attention indicated by 221 citations in Scopus of the original paper introducing the algorithm up to this day (Sept. 4th 2019). Besides GOA applications – in its standard variant as described in the previous section – in a variety of fields numerous improvements and modifications have been suggested. It will be briefly presented in the subsequent parts of this section.

15.3.1   Adaptation to other optimization domains

As was observed in the enclosed description of GOA – it was originally designed for continuous optimization tasks. However the algorithm can be easily modified to be used for binary optimization. As an example two variants of GOA – presented in [19] – could be named. They involve using either transfer functions or stochastic mutation as an attraction operator. Both approaches proved to be highly competitive for feature selection problems considered in the aforementioned publication.

In addition to that the GOA scheme was also modified to allow the algorithm to tackle multi-objective optimization tasks [20]. In [21] a concept of archive – storing Pareto optimal solutions with additional elimination of solutions lying within a crowded neighbourhood – were introduced to form Multi-objective Grasshopper Optimization Algorithm (MOGOA). Its performance was tested for well known ZDT and CEC 2009 test suites. During those tests the algorithm was also compared with Multi-objective PSO, Multi-objective Dragonfly Algorithm, Multi-objective Ant Lion Algorithm and NSGA-II technique. It was concluded therein that MOGOA offers relatively high distribution across all objectives and quick convergence.

15.3.2   Structural modifications

Several modifications to the basic algorithm scheme were also proposed. They were primarily aimed to further improve its optimization capabilities.

First of all experiments involving Levy flight in the movement towards the best individual were conducted in [13]. Each solution vector is obtained by:

$$x_{pd}^{new} = x_{pd} * R * f_L(\beta) \tag{15.4}$$

with R representing a random value in the $[0, 1]$ interval and $f_L(\beta)$ random value for Levy flight with parameter β. It was demonstrated – for the image thresholding problem – that Levy flight can maximize the diversity of population exploring the search domain bringing a positive outcome in terms of final optimization result. Similarly an additional random component, in the form of random mutation was considered in [26] for the optimal power flow problem, also proving to be beneficial.

More fundamental modification was under investigation in [30] – for general continuous optimization tasks. The paper considers the introduction of dynamic weight modifying the impact of the first component of eq. (15.1) and creating three phases of search space exploration. In addition to that random jumps, for stagnating solutions are introduced. These two components significantly improve GOA performance in aspects of accuracy, stability, and convergence rate.

So called Adaptive GOA – proposed in [29] – seems to represent the most invasive modification of the basic algorithm's scheme. It encompasses natural selection strategy, three elite agents and dynamic adjustment of c, similar to the one presented in [30]. The authors demonstrated that such an approach significantly over-performs standard GOA over a set of target tracking problems.

Finally, the trend to use chaotic maps in metaheuristics finds its representation also for the Grasshopper Optimization Algorithm. Ten different chaotic maps were considered in [25] to form a class of Enhanced Chaotic Grasshopper Optimization Algorithms (ECGOAs). They are used to modify the value of parameter c . No definitive suggestions concerning the choice of the maps were derived, however the paper concludes with the general observation that such chaotic formulation of c brings very positive results. Similar – though more general – experiments were also conducted in [4]. Their outcome is a recommendation to use circle maps to improve GOA search capabilities. Such a map is obtained in this case with:

$$c^{new} = c + b - \frac{P}{2\pi} \sin(2\pi c) \quad \mod (1). \tag{15.5}$$

for fixed values of parameters $P = 0.5$ and $b = 0.2$.

15.3.3   Hybrid algorithms

GOA was also considered as a component of hybrid algorithms – including the use of two and more heuristic approaches. The list of representative contributions in this field is not long however. In this context GOA was successfully hybridized with the Genetic Algorithm, to deal with generation of security keys [2]. Solutions from both GA and GOA are merged to improve search capabilities of the combined algorithm. Similarly a hybrid with Differential Evolution (DE) was proposed in [11]. It was aimed at solving the Multilevel Satellite Image Segmentation problem. DE is supposed to supplement GOA in the latter stages of optimization preserving the population diversity.

15.4   Application example: GOA-based clustering

15.4.1   Clustering and optimization

Clustering represents a task of dataset division into a set of C disjoint clusters (groups). Its goal is to find elements resembling each other – and to form clusters composed of them. Cluster analysis is used in diverse applications like automatic control [14], text analysis [1], agriculture [7] or marketing [22]. Several clustering algorithms were already presented in the literature of the subject [1]. Here we will demonstrate the use of the Grasshopper Optimization Algorithm to tackle the aforementioned problem.

Let us represent a dataset by data matrix Y of $M \times N$ dimensionality. Each of its N columns represents a feature characterizing data sample. Samples correspond to a row of the matrix, commonly known as the dataset's elements or cases.

Clustering remains an unsupervised learning procedure, frequently with a known number of clusters C being the only information available. Its task is to assign dataset elements $y_1, ..., y_M$ to clusters $CL_1, CL_2, ..., CL_C$. To validate the clustering solution one could compare its results with correct cluster labels. If they are unavailable so called internal validation assessing if the obtained solution reflects the structure of the data and natural groups which can be identified within its records [9] can be employed.

Using any heuristic optimization algorithm requires choosing proper solution representation. In the case of clustering it is natural to represent the solution as a vector of cluster centers $x_p = [u_1, u_2, ..., u_C]$. Consequently the dimensionality D used in the description of GOA, in the case of the data clustering problem, is equal to $C * N$ as each cluster center u_i is a vector of N elements.

Another important aspect is choosing the proper tool for assessing the quality of generated solutions. Here an idea already presented in [16] was implemented. After assigning each data element y_i to the closest cluster center the solution x_p (representing those centers) is evaluated according to the formula:

$$f(x_p) = \frac{1}{I_{CH,p}} + \#_{CL_{i,p}=\emptyset,\, i=1,...,C} \tag{15.6}$$

where the value $I_{CH,p}$ represents Calinski-Harabasz index calculated for solution x_p. It is generally obtained via:

$$I_{CH} = \frac{N-C}{C-1} \frac{\sum_{i=1}^{C} dist(u_i, U)}{\sum_{i=1}^{C} \sum_{x_j \in CL_i} dist(x_j, u_i)} \qquad (15.7)$$

wheras $u_i \in R^N$ for non-empty cluster CL_i corresponds to the cluster center defined by:

$$u_i = \frac{1}{M_i} \sum_{y_j \in CL_i} y_j, \quad i = 1, ..., C \qquad (15.8)$$

with M_i being cardinality of cluster i and – likewise – U corresponding to the center of gravity of the dataset:

$$U = \frac{1}{M} \sum_{j=1}^{M} y_j \qquad (15.9)$$

To penalize solutions which do not include the desired number of clusters a penalty – equal to a number of empty clusters identified in the x_p clustering solution written in (15.6) as $\#CL_{i,p=\emptyset}, i=1,...,C$ – is added.

The choice of this index was motivated by our successful experiments on other heuristic algorithms using I_{CH} value [12] as a key component and similar positive results presented in [3].

15.4.2   Experimental setting and results

Clustering as such is an unsupervised learning problem, therefore its evaluation requires an assumption about the availability of correct cluster labels. In practice a labeled dataset containing the information about assignment of data elements to classes can be used as a substitute.

A clustering solution understood as a set of cluster indexes provided for all data points should be then compared with a set of class labels. Such a comparison can be done with the use of the Rand index [23], external validation index which measures similarity between cluster analysis solutions. It is characterized by a value between 0 and 1. A low value of R suggests that the two clusterings are different and 1 indicates that they represent exactly the same solution – even when the formal indexes of clusters are mixed.

For computational experiments a set of standard synthetic clustering benchmark instances known as S-sets was used [8]. Figure 15.1 demonstrates the structure of those two dimensional data sets. It can be seen that they are characterized by different degree of cluster overlapping.

To make the study more representative six additional real-world problems, taken from UCI Machine Learning Repository were taken into consideration [28]. Table 15.1 characterizes all datasets used in this paper for experimental evaluation. For each instance it contains properties like dataset size M,

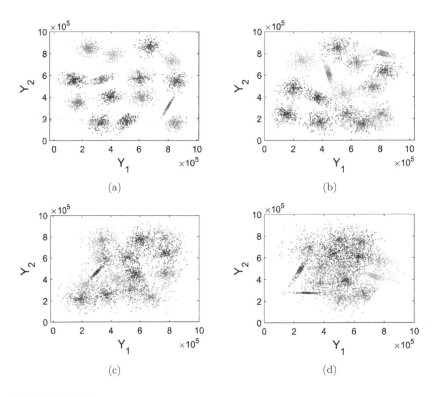

FIGURE 15.1
Scatter plots for *s1* (a), *s2* (b), *s3* (c) and *s4* (d) datasets.

TABLE 15.1
Experimental datasets description.

Dataset	Sample size	Dimensionality	No of clusters
s1	5000	2	15
s2	5000	2	15
s3	5000	2	15
s4	5000	2	15
glass	214	9	6
wine	178	13	3
iris	150	4	3
seeds	210	7	3
heart	270	13	2
yeast	1484	8	10

TABLE 15.2

K-means vs. GOA-based clustering.

	K-means clustering		GOA clustering	
	$\overline{R}$	$\sigma(R)$	$\overline{R}$	$\sigma(R)$
s1	0.980	0.009	0.990	0.006
s2	0.974	0.010	0.984	0.006
s3	0.954	0.006	0.960	0.005
s4	0.944	0.006	0.951	0.003
glass	0.619	0.061	0.643	0.035
wine	0.711	0.014	0.730	0.000
iris	0.882	0.029	0.892	0.008
seeds	0.877	0.027	0.883	0.004
heart	0.522	0.000	0.523	0.000
yeast	0.686	0.033	0.676	0.034

dimensionality N and the number of classes C – used as desired number of clusters for the grouping algorithms. For non-synthetic problems it was naturally assumed that each class manifests itself as a single cluster.

As a point of reference for evaluating performance of clustering methods classic K-means algorithm was used. It is also the case of this contribution.

To evaluate clustering methods they were run 30 times with mean and standard deviation values of Rand index – $\overline{R}$ and $\sigma(R)$ – being recorded. For the GOA-based algorithm a population of $P = 20$ swarm members was used. The algorithm terminates when $C * N * 1000$ cost function evaluations were performed. It is a standard strategy for evaluating metaheuristics – making the length of the search process dependent on data dimensionality.

We used default values of all GOA parameters, with $c = 0.00001$. It means that c quickly approaches values close to zero. The summary of obtained results for this case is provided in Table 15.2. It can be observed that GOA-based clustering outperforms K-means on the majority of the datasets – it is also less prone to getting stuck in local minima. It is indicated by the values of standard deviations, which in the case of K-means algorithm are significantly higher.

For more extensive study on the GOA application in clustering – including alternative values of c and chaotic variant of the algorithm – one could refer to [17].

15.5 Conclusion

This chapter has modified variants of of Grasshopper Optimization Algorithm. We have studied alternative GOA schemes including different optimization setting, structural modifications and hybridization with other nature-inspired techniques. In addition to that the application of GOA for clustering task was also presented.

The algorithm in its basic form was more extensively discussed in the chapter: Grasshopper optimization algorithm [18]. Along with more precise description of its mechanics, the chapter also provides technical details of GOA implementation and walk-through GOA structural components.

References

1. C. Aggarwal and C. Zhai, "A survey of text clustering algorithms," in *Mining Text Data*, C. C. Aggarwal and C. Zhai, Eds. Springer US, 2012, pp. 77–128.

2. A. Alphonsa, M. M. Mohana, N. Sundaram, "A reformed grasshopper optimization with genetic principle for securing medical data". *Journal of Information Security and Applications*, vol. 47, pp. 410-420, 2019.

3. O. Arbelaitz, I. Gurrutxaga, J. Muguerza, J. M. Pérez, and I. Perona, "An extensive comparative study of cluster validity indices," *Pattern Recognition*, vol. 46, no. 1, pp. 243 – 256, 2013.

4. S. Arora, P. Anand, "Chaotic grasshopper optimization algorithm for global optimization". *Neural Computing and Applications*, doi: 10.1007/s00521-018-3343-2, 2018.

5. J. C. Bansal, P. K. Singh, M. Saraswat, A. Verma, S. S. Jadon, and A. Abraham. "Inertia weight strategies in particle swarm optimization", *2011 Third World Congress on Nature and Biologically Inspired Computing*, Oct. 2011, pp. 633-640.

6. R.F. Chapman and A. Joern. "Biology of grasshoppers", A Wiley-Interscience publication, Wiley, 1990.

7. M. Charytanowicz, J. Niewczas, P. Kulczycki, P. A. Kowalski, S. Łukasik, and S. Żak, "Complete gradient clustering algorithm for features analysis of X-Ray images," in *Information Technologies in Biomedicine*, ser. Advances in Intelligent and Soft Computing, E. Pietka and J. Kawa, Eds. Springer Berlin Heidelberg, 2010, vol. 69, pp. 15–24.

8. P. Fränti and O. Virmajoki, "Iterative shrinking method for clustering problems," *Pattern Recognition*, vol. 39, no. 5, pp. 761 – 775, 2006.

9. M. Halkidi, Y. Batistakis, and M. Vazirgiannis, "On clustering validation techniques," *Journal of Intelligent Information Systems*, vol. 17, no. 2-3, pp. 107–145, 2001.

10. A. E. Hassanien and E. Emary. *Swarm Intelligence: Principles, Advances, and Applications*, CRC Press, 2018.

11. H. Jia, C. Lang, D. Oliva, W. Song, X. Peng, "Hybrid grasshopper optimization algorithm and differential evolution for multilevel satellite image segmentation". *Remote Sensing*, vol. 11, paper no 1134, 2019.

12. P. A. Kowalski, S. Łukasik, M. Charytanowicz, and P. Kulczycki, "Clustering based on the krill herd algorithm with selected validity measures," in *2016 Federated Conference on Computer Science and Information Systems (FedCSIS)*, Sept. 2016, pp. 79–87.

13. H. Liang, H. Jia, Z. Xing, J. Ma and X. Peng, "Modified Grasshopper Algorithm-Based Multilevel Thresholding for Color Image Segmentation," *IEEE Access*, vol. 7, pp. 11258–11295, 2019.

14. S. Łukasik, P. Kowalski, M. Charytanowicz, and P. Kulczycki, "Fuzzy models synthesis with kernel-density-based clustering algorithm," in *Fuzzy Systems and Knowledge Discovery, 2008. FSKD '08. Fifth International Conference on*, vol. 3, Oct. 2008, pp. 449–453.

15. S. Łukasik and P. A. Kowalski. "Study of flower pollination algorithm for continuous optimization", in *Intelligent Systems 2014* (P. Angelov, K. T. Atanassov, L. Doukovska, M. Hadjiski, V. Jotsov, J. Kacprzyk, N. Kasabov, S. Sotirov, E. Szmidt, and S. Zadrozny, eds.), Springer International Publishing, 2015, pp. 451-459.

16. S. Łukasik, P. A. Kowalski, M. Charytanowicz, and P. Kulczycki, "Clustering using flower pollination algorithm and calinski-harabasz index," in *2016 IEEE Congress on Evolutionary Computation (CEC)*, July 2016, pp. 2724–2728.

17. S. Łukasik, P. A. Kowalski, M. Charytanowicz, and P. Kulczycki. "Data clustering with grasshopper optimization algorithm", *2017 Federated Conference on Computer Science and Information Systems (FedCSIS)*, pp. 71-74, 2017.

18. S. Łukasik. "Grasshopper Optimization Algorithm", *Swarm Intelligence Algorithms: A Tutorial* (A. Slowik, ed.), CRC Press, 2020.

19. M. Mafarja, I. Aljarah, H. Faris, A.I. Hammouri, A.M. Al-Zoubi, S. Mirjalili, "Binary grasshopper optimisation algorithm approaches for feature selection problems", *Expert Systems with Applications*, vol. 117, pp. 267-286, 2019.

20. S. Z. Mirjalili, S. Mirjalili, S. Saremi, H. Faris, and I. Aljarah. "Grasshopper optimization algorithm for multi-objective optimization problems", *Applied Intelligence*, vol. 48, no. 4, pp. 805–820, 2018.

21. S. Mirjalili, S. Mirjalili, S. Saremi, H. Faris, I. Aljarah, "Grasshopper optimization algorithm for multi-objective optimization problems", *Applied Intelligence*, vol. 48, pp. 805–820, 2018.

22. H. Müller and U. Hamm, "Stability of market segmentation with cluster analysis - a methodological approach," *Food Quality and Preference*, vol. 34, pp. 70 – 78, 2014.

23. H. Parvin, H. Alizadeh, and B. Minati, "Objective criteria for the evaluation of clustering methods," *Journal of the American Statistical Association*, vol. 66, pp. 846?–850, 1971.

24. S. Saremi, S. Mirjalili, and A. Lewis. "Grasshopper optimisation algorithm: Theory and application", *Advances in Engineering Software*, vol. 105, pp. 30-47, 2017.

25. A.Saxena, R. Kumar, "Chaotic Variants of Grasshopper Optimization Algorithm and Their Application to Protein Structure Prediction". in: *Applied Nature-Inspired Computing: Algorithms and Case Studies*, N. Dey, A. Ashour, S. Bhattacharyya, (eds) Springer (Singapore), pp. 151–175, 2020.

26. M.A. Taher, S. Kamel, F. Jurado, M. Ebeed, "Modified grasshopper optimization framework for optimal power flow solution". *Electrical Engineering*, doi: 0.1007/s00202-019-00762-4, 2019.

27. X. S. Yang. *Nature-Inspired Optimization Algorithms*, Elsevier, London, 2014.

28. UCI Machine Learning Repository, `http://archive.ics.uci.edu/ml/`, access Sept 4, 2019.

29. J. Wu, H. Wang, N. Li, P. Yao, Y. Huang, Z. Su, Y. Yu, "Distributed trajectory optimization for multiple solar-powered UAVs target tracking in urban environment by Adaptive Grasshopper Optimization Algorithm". *Aerospace Science and Technology*, vol. 70, pp. 497–510, 2017.

30. R. Zhao, H. Ni, H. Feng, X. Zhu, "A Dynamic Weight Grasshopper Optimization Algorithm with Random Jumping". In: S. Bhatia, S. Tiwari, K. Mishra, M. Trivedi (eds.) *Advances in Computer Communication and Computational Sciences*. Springer, Singapore, pp. 401–413, 2019.

16

Grey Wolf Optimizer – Modifications and Applications

Ahmed F. Ali
Department of Computer Science
Suez Canal University, Ismaillia, Egypt

Mohamed A. Tawhid
Department of Mathematics and Statistics
Faculty of Science, Thompson Rivers University, Kamloops, Canada

CONTENTS

16.1 Introduction

The grey wolves live in groups and they have a special dominant social structure. In 2014, Mirjalili et al. proposed a swarm intelligence algorithm [4] which was called grey wolf optimizer (GWO). The GWO algorithm mimics the social behavior of the grey wolves. Many researchers have applied the GWO algorithm due to its efficiency in various applications, for example, for CT liver segmentation [1], minimizing potential energy function [8], feature selection [2], [11], minimax and integer programming problems [9], global optimization problem [10], flow shop scheduling problem [3], optimal reactive power dispatch problem [7], and the casting production scheduling [12]. The rest of the organization of our chapter is as follows. We present the original GWO algorithm in Section 16.2. We modify the GWO algorithm in Section 16.3. In Section 16.4, we solve the engineering optimization problem by the chaotic grey wolf optimization (CGWO) algorithm. Finally, we summarize the conclusion in Section 16.5.

16.2 Original GWO algorithm in brief

In this section, we present the main steps of the GWO algorithm. These steps are shown in Algorithm 18.

Algorithm 18 Grey wolf optimizer algorithm.

1: Set the initial values of the population size n, parameter a and the maximum number of iterations Max_{itr}
2: Set $t := 0$
3: **for** $(i = 1 : i \leq n)$ **do**
4:     Generate an initial population $\vec{X_i}(t)$ randomly
5:     Evaluate the fitness function of each search agent (solution) $f(\vec{X_i})$
6: **end for**
7: Assign the values of the first, second and the third best solutions $\vec{X_\alpha}$, $\vec{X_\beta}$ and $\vec{X_\delta}$, respectively
8: **repeat**
9:     **for** $(i = 1 : i \leq n)$ **do**
10:         Decrease the parameter a from 2 to 0
11:         Update the coefficients $\vec{A}$ and $\vec{C}$ as shown in Equations (16.4)–(16.5)
12:         Update each search agent in the population as shown in Equations (16.1)–(16.3)
13:         Evaluate the fitness function of each search agent $f(\vec{X_i})$

14: **end for**
15: Update the vectors $\vec{X}_\alpha$, $\vec{X}_\beta$ and $\vec{X}_\delta$.
16: Set $t = t + 1$
17: **until** $(t \geq Max_{itr})$ ▷ Termination criteria are satisfied
18: Produce the best solution $\vec{X}_\alpha$

16.2.1 Description of the original GWO algorithm

The standard GWO starts by setting the initial parameters of the population size n, the parameter a and the maximum number of iterations Max_{itr}. The iteration counter t is initialized where $t = 0$. The initial population n is randomly generated and each search agent (solution) $\vec{X}_i$ is evaluated by calculating its fitness function $f(\vec{X}_i)$. The overall best three solutions are assigned according to their fitness values which are alpha α, beta β and the delta δ solutions $\vec{X}_\alpha$, $\vec{X}_\beta$ and $\vec{X}_\delta$, respectively. The main loop is repeated until the termination criterion is satisfied. Each search agent (solution) in the population is updated according to the position of the α, β and δ solutions as shown in Equations 16.1–16.3.

$$\begin{aligned}
\vec{D}_\alpha &= |\vec{C}_1 \cdot \vec{X}_\alpha - \vec{X}|, \\
\vec{D}_\beta &= |\vec{C}_2 \cdot \vec{X}_\beta - \vec{X}|, \\
\vec{D}_\delta &= |\vec{C}_3 \cdot \vec{X}_\delta - \vec{X}|,
\end{aligned} \tag{16.1}$$

$$\begin{aligned}
\vec{X}_1 &= \vec{X}_\alpha - \vec{A}_1 \cdot (\vec{D}_\alpha), \\
\vec{X}_2 &= \vec{X}_\beta - \vec{A}_2 \cdot (\vec{D}_\beta), \\
\vec{X}_3 &= \vec{X}_\delta - \vec{A}_3 \cdot (\vec{D}_\delta),
\end{aligned} \tag{16.2}$$

$$\vec{X}(t+1) = \frac{\vec{X}_1 + \vec{X}_2 + \vec{X}_3}{3}, \tag{16.3}$$

The parameter a is gradually decreased from 2 to 0 and the coefficients $\vec{A}$ and $\vec{C}$ are updated as shown in Equations 16.4–16.5.

$$\vec{A} = 2\vec{a} \cdot \vec{r}_1 \cdot \vec{a} \tag{16.4}$$

$$\vec{C} = 2 \cdot \vec{r}_2 \tag{16.5}$$

where components of $\vec{a}$ are linearly decreased from 2 to 0 over the course of iterations and $\vec{r}_1$, $\vec{r}_2$ are random vectors in $[0, 1]$.

Each search agent (solution) in the population is evaluated by calculating its fitness function $f(\vec{X}_i)$. The first three best solutions are updated $\vec{X}_\alpha$, $\vec{X}_\beta$ and $\vec{X}_\delta$, respectively. The iteration counter is increased where $t = t + 1$. Once the termination criteria are satisfied, the algorithm is terminated and the overall best solution $\vec{X}_\alpha$ is produced.

16.3   Modifications of the GWO algorithm

Recently, many metaheuristic algorithms have been developed and suggested for global search. These developed algorithms can improve computational efficiency and solve various optimization problems [1], [8], [2], [11], [9], [10], [3], [5], [7], [12]. Among these algorithms, grey wolf optimizer (GWO) was developed in 2014 by Mirjalili et al. as a swarm intelligence algorithm [4]. The GWO algorithm simulates the social behavior of the grey wolves. On the other hand recent studies in nonlinear dynamics such as chaos, have attracted more attention in many areas [16] . One of these areas is the integration of chaos into optimization algorithms to improve and enhance certain algorithm-dependent parameters [17]. Researchers have used chaotic sequences to tune up parameters in metaheuristic optimization algorithms such as genetic algorithms [18], particle swarm optimization [19], ant and bee colony optimization [20], and simulated annealing [21]. Empirical studies show that chaos can improve the capability of standard metaheuristics when a chaotic map exchanges a fixed parameter. It turns out the solutions generated from such exchange may have higher diversity. For this reason, it is desirable to perform more studies by introducing chaos to recent metaheuristic algorithms. In this section, we replace the random vectors $\vec{r_1}$, $\vec{r_2}$ in Equations 16.4–16.5 by the chaotic maps $C1, C2$. The proposed algorithm is called Chaotic Grey Wolf Optimization (CGWO) algorithm. Invoking the two chaotic map in the standard GWO algorithm can help it to escape from being trapped in local minima due to increasing the range of the random numbers from [0,1] to [-1,1].

16.3.1   Chaotic maps

Although the chaotic maps have no random variables, they show a random behavior. In Table 16.1, we present the mathematical forms of these maps. The GWO suffers from being trapped in local minima like other meta-heuristics algorithms. We invoke two chaotic maps in GWO instead of using the standard random parameters to increase the diversity of the search and avoid being trapped in local minima. We use the initial point $x^0 = 0.7$ as chosen in [6].

16.3.2   Chaotic grey wolf operator

We replace the random vectors $\vec{r_1}$, $\vec{r_2}$ in Equations 16.4–16.5 by two chaotic maps with range [-1,1] as shown in Table 16.1. The two new equations are

TABLE 16.1

Chaotic maps.

No	Name	Chaotic map	Range
C1	Chebyshev [13]	$x_{i+1} = \cos(i \cos^{-1}(x_i))$	(-1,1)
C2	Iterative [15]	$x_{i+1} = \sin(\frac{a\pi}{x_i}), a = 0.7$	(-1,1)

presented as follows.

$$\vec{A} = 2\vec{a} \cdot \vec{C}_1 \cdot \vec{a} \tag{16.6}$$

$$\vec{C} = 2 \cdot \vec{C}_2 \tag{16.7}$$

16.4 Application of GWO algorithm for engineering optimization problems

In this section, we test the efficiency of the proposed CGWO by solving five engineering optimization problems as follows.

16.4.1 Engineering optimization problems

In this subsection, we give a brief description for five engineering optimization problems. These problems are the welded beam design problem, pressure vessel design problem, speed reducer design problem and three-bar truss design problem.

16.4.1.1 Welded beam design problem

The main objective of a welded beam design problem is minimizing the cost of its structure which consists of a beam A and the required weld to hold it to member B subject to constraints on shear stress τ, beam bending stress θ, buckling load on the bar P_c, beam end deflection δ. The four design variables $h(x_1), l(x_2), t(x_3)$ and $b(x_4)$ are shown in Figure 16.1. The mathematical form of the welded beam problem is shown in Equation 16.8.

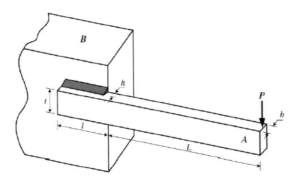

FIGURE 16.1
Welded beam design problem.

$$Minimize \quad f(x) = \quad 1.10471x_1^2x_2 + 0.04811x_3x_4(14.0 + x_2) \quad (16.8)$$

Subject to

$$g_1(x) = \quad \tau(x) - \tau_{max} \le 0$$
$$g_2(x) = \quad \sigma(x) - \sigma_{max} \le 0$$
$$g_3(x) = \quad x_1 - x_4 \le 0$$
$$g_4(x) = \quad 0.10471x_1^2 + 0.04811x_3x_4(14.0 + x_2) - 0.5 \le 0$$
$$g_5(x) = \quad 0.125 - x_1 \le 0$$
$$g_6(x) = \quad \delta(x) - \delta_{max} \le 0$$
$$g_7(x) = \quad P - P_c(x) \le 0$$

where the other parameters are defined as follows.

$$\tau(x) = \sqrt{((\tau')^2 + (\tau'')^2 + \frac{2\tau'\tau''x_2}{2R}}, \quad \tau' = \frac{p}{\sqrt{2}x_1x_2}$$

$$\tau'' = \frac{MR}{J}, \quad M = P(L + \frac{x_2}{2}), \quad R = \sqrt{\left(\frac{x_1 + x_3}{2}\right)^2 + \frac{x_2^2}{4}}$$

$$J = 2\left\{\frac{x_1x_2}{\sqrt{2}}\left[\frac{x_2^2}{12} + \left(\frac{x_1 + x_3}{2}\right)^2\right]\right\}, \quad \sigma(x) = \frac{6PL}{x_4x_3^2}$$

$$\delta(x) = \frac{4PL^3}{Ex_4x_3^3}, \quad P_c(x) = \frac{4.013\sqrt{EGx_3^2x_4^6/36}}{L^2}\left(1 - \frac{x^3}{2L}\sqrt{\frac{E}{4G}}\right)(16.9)$$

where $P = 6000$lb, $L = 14$, $\delta_{max} = 0.25$,in, $E = 30,106$ psi, $G = 12,106$ psi, $\tau_{max} = 13,600$psi, $\sigma_{max} = 30,000$ psi and $0.1 \le x_i \le 10.0$ ($i = 1, 2, 3, 4$).

16.4.1.2   Pressure vessel design problem

The objective of this problem is to minimize the total cost of material, forming and welding. Figure 16.2 shows the following four parameters, T_s (x_1, thickness

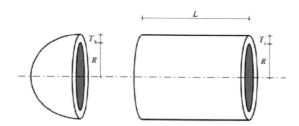

FIGURE 16.2
Pressure vessel design problem.

of the shell), T_h (x_2, thickness of the head), R (x_3, inner radius) and L. (x_4, length of the cylindrical section of the vessel without the head). T_s and T_h are integer multiples of 0.0625in, R and L are continuous variables. The mathematical form of this problem is shown in Equation 16.10.

$$\begin{aligned} Minimize \quad f(x) = \quad & 0.6224x_1x_3x_4 + 1.7781x_2x_3^2 \\ & +3.1661x_1^2x_4 + 19.84x_1^2x_3 \end{aligned} \quad (16.10)$$

$$\begin{aligned} Subject \ to \\ g_1(x) = \quad & -x_1 + 0.0193x_3 \\ g_2(x) = \quad & -x_2 + 0.00954x_3 \leq 0 \\ g_3(x) = \quad & -\pi x_3^2 x_4 - 4/3\pi x_3^3 + 1296000 \leq 0 \\ g_4(x) = \quad & x_4 - 240 \leq 0 \end{aligned}$$

where $1 \leq x_1 \leq 99$, $1 \leq x_2 \leq 99$, $10 \leq x_3 \leq 200$ and $10 \leq x_4 \leq 200$.

16.4.1.3 Speed reducer design problem

In this problem, we minimize a gear box volume and weight subject to the following constraints as shown in Equation 16.11. In this problem, we have seven design variables $x_1 - x_7$, which can be described as the following. x_1 is a width of the gear face (cm), x_2 teeth module (cm), x_3 number of pinion teeth, x_4 shaft 1 length between bearings (cm), x_5 shaft 2 length between bearing (cm), x_6 diameter of shaft 1 (cm) and x_7 diameter of shaft 2 (cm). The speed reducer design problem is shown in Figure 16.3.

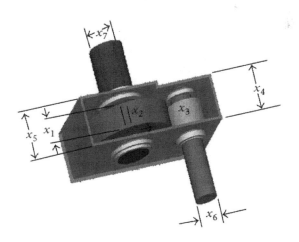

FIGURE 16.3
Speed reducer design problem.

$$Minimize \quad f(x) = \quad 0.7854x_1x_2^2(3.3333x_3^2 \tag{16.11}$$
$$+14.9334x_3 - 43.0934) - 1.508x_1(x_6^2 + x_7^2)$$
$$+7.4777(x_6^3 + x_7^3) + 0.7854(x_4x_6^2 + x_5x_7^2)$$

Subject to

$$g_1(x) = \quad \frac{27}{x_1x_2^2x_3} - 1 \leq 0$$

$$g_2(x) = \quad \frac{397.5}{x_1x_2^2x_3^2} - 1 \leq 0$$

$$g_3(x) = \quad \frac{1.93x_4^3}{x_2x_6^4x_3} - 1 \leq 0$$

$$g_4(x) = \quad \frac{1.93x_5^3}{x_2x_7^4x_3} - 1 \leq 0$$

$$g_5(x) = \quad \frac{[(745x_4/x_2x_3)^2 + 16.9 \times 10^6]^0.5}{110.0x_6^3} - 1 \leq 0$$

$$g_6(x) = \quad \frac{[(745x_5/x_2x_3)^2 + 157.5.9 \times 10^6]^0.5}{85.0x_7^3} - 1 \leq 0$$

$$g_7(x) = \quad \frac{x_2x_3}{40} - 1 \leq 0$$

$$g_8(x) = \quad \frac{5x_2}{x_1} - 1 \leq 0$$

$$g_9(x) = \quad \frac{x_1}{12x_2} - 1 \leq 0$$

$$g_{10}(x) = \quad \frac{1.5x_6 + 1.9}{x_4} - 1 \leq 0$$

$$g_{11}(x) = \quad \frac{1.1x_7 + 1.9}{x_5} - 1 \leq 0 \tag{16.12}$$

where $2.6 \leq x_1 \leq 3.6$, $0.7 \leq x_2 \leq 0.8 \leq$, $17 \leq x_3 \leq 28$, $7.3 \leq x_4 \leq 8.3$, $7.3 \leq x_5 \leq 8.3$, $2.9 \leq x_6 \leq 3.9$, $5 \leq x_7 \leq 5.5$

16.4.1.4   Three-bar truss design problem

In the three-bar truss design problem, we try to minimize the weight $f(x)$ of three bar trusses subject to some constraints as shown in Equation 16.13.

$$Minimize \quad f(x) = \quad (2\sqrt{2}x_1 + x_2) \times l \tag{16.13}$$

Subject to

$$g_1(x) = \quad \frac{\sqrt{2}x_1 + x_2}{\sqrt{2}x_1^2 + 2x_1x_2}P - \sigma \leq 0$$

$$g_2(x) = \quad \frac{x_2}{\sqrt{2}x_1^2 + 2x_1x_2}P - \sigma \leq 0$$

$$g_3(x) = \quad \frac{1}{\sqrt{2}x_2 + x_1}P - \sigma \leq 0$$

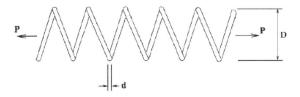

FIGURE 16.4
Tension compression spring problem.

where $0 \leq x_1 \leq 1$ and $0 \leq x_2 \leq 1$, $l = 100$ cm, $P = 2kN/cm^2$, and $\sigma = 2kN/cm^2$. The minimum weighted structure should be achieved by determining the optimal cross-sectional areas x_1, and x_2.

16.4.1.5   Tension compression spring problem

The last problem is the tension compression spring problem. In this problem, we need to minimize the weight $f(x)$ of a tension compression spring design subject to three constraints as shown in Equation 16.14. The mean coil diameter $D(x_2)$, the wire diameter $d(x_1)$ and the number of active coils $P(x_3)$ are the design variables as shown in Figure 16.4.

$$Minimize \quad f(x) = \quad (x_3 + 2)x_2 x_1^2 \qquad (16.14)$$
$$Subject \ to$$

$$g_1(x) = \quad 1 - \frac{x_2^3 x_3}{71785 x_1^4} \leq 0$$

$$g_2(x) = \quad \frac{4x_2^2 - x_1 x_2}{12566(x_2 x_1^3 - x_1^4)} + \frac{1}{5108 x_1^2} - 1 \leq 0$$

$$g_3(x) = \quad 1 - \frac{140.45 x_1}{x_2^2 x_3} \leq 0$$

where $0.05 \leq x_1 \leq 2$, $0.25 \leq x_2 \leq 1.3$ and $2 \leq x_3 \leq 15$.

16.4.2   Description of experiments

The proposed algorithm was programmed in Matlab. We set the population size $n = 20$ and the $Max_{itr} = 1000$. We handled the constraints by transforming the constrained optimization problems to an unconstrained optimization problem as shown in [14].

16.4.3   Convergence curve of CGWO with engineering optimization problems

In Figure 16.5, we show the performance and the convergence curve of the proposed CGWO by plotting the number of iterations versus the function

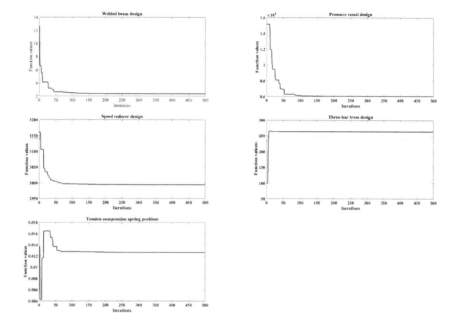

FIGURE 16.5
Convergence curve of the proposed CGWO algorithm with engineering optimization problems.

values for the engineering optimization problems. We can see that the CGWO can reach to near optimal global minimum for all problems in a small number of iterations.

16.4.4   Comparison between CGWO and GWO with engineering optimization problems

We verify the efficiency of the proposed CGWO algorithm by comparing it with the GWO algorithm. We report the results in Table 16.2 and the best

TABLE 16.2
Comparison between CGWO and GWO with engineering optimization problems.

Design problem	GWO	CGWO
Welded beam design	2.382447652	**2.380956951**
Pressure vessel design	6076.292065	**6059.714335**
Speed reducer design	3000.385887	**2994.471066**
Three-bar truss design	263.896704	**263.8958434**
A tension/compression	0.012717623	**0.012665233**

results in **bold text**. The results in Table 16.2 show that the proposed CGWO outperforms the GWO which means that invoking the chaotic maps in CGWO can enhance and obtain better results than the standard GWO.

16.5   Conclusions

In this chapter, we present a proposed algorithm to solve five engineering optimization problems. The proposed algorithm is called Chaotic Grey Wolf Optimization CGWO. Replacing the random vectors in the standard GWO by chaotic maps with range [-1,1] increases the diversity of the proposed algorithm and avoids being trapped at the local minima. In order to verify the efficiency of CGWO, we compare it with the standard GWO algorithm. The results show that the proposed CGWO is better than GWO in all cases.

References

1. A.F. Ali, A. Mostafa, G.I, Sayed, M.A. Elfattah and A.E. Hassanien. "Nature inspired optimization algorithms for ct liver segmentation". In *Medical Imaging in Clinical Applications* (pp. 431–460), 2016. Springer.

2. E.Emary, H.M. Zawbaa and A.E. Hassanien. "Binary grey wolf optimization approaches for feature selection". *Neurocomputing*, vol. 172, pp. 371–381, 2016.

3. G. M. Komaki and V. Kayvanfar. "Grey Wolf Optimizer algorithm for the two-stage assembly flow shop scheduling problem with release time". *Journal of Computational Science*, vol. 8, pp. 109-120, 2015.

4. S. Mirjalili, S. M. Mirjalili, and A. Lewis. "Grey Wolf Optimizer". *Advances in Engineering Software*, vol. 69, pp. 46–61, 2014.

5. A. Pasquet. "Cooperation and prey capture efficiency in a social spider, Anelosimus eximius (Araneae, Theridiidae)". *Ethology*, vol. 90, pp. 121—133, 1991.

6. S. Saremi,S. Mirjalili, and A. Lewis. "Grasshopper optimisation algorithm: Theory and application". *Advances in Engineering Software*, vol. 105, pp. 30–47, 2017.

7. M. H. Sulaiman, Z. Mustaffa, M. R. Mohamed, and O. Aliman. "Using the gray wolf optimizer for solving optimal reactive power

dispatch problem". *Applied Soft Computing*, vol. 32, pp. 286-292, 2015.

8. M.A. Tawhid and A.F. Ali. "A hybrid grey wolf optimizer and genetic algorithm for minimizing potential energy function". *Memetic Computing*, vol. 9, no. 4, pp. 347–359, 2017.

9. M.A. Tawhid and A.F. Ali. "A simplex grey wolf optimizer for solving integer programming and minimax problems". *Numerical Algebra, Control & Optimization*, vol. 7, no. 3, pp. 301–323, 2017.

10. M.A. Tawhid and A.F. Ali. "Multidirectional Grey Wolf Optimizer Algorithm for solving global optimization problems". *International Journal of Computational Intelligence and Applications*, pp. 1850022, 2018.

11. Q. Tu, X. Chen, and X. Liu. "Multi-strategy ensemble grey wolf optimizer and its application to feature selection". *Applied Soft Computing*, vol. 76, pp. 16–30, 2019.

12. H. Qin, P. Fan, H. Tang, P. Huang, B. Fang and S. Pan. "An effective hybrid discrete grey wolf optimizer for the casting production scheduling problem with multi-objective and multi-constraint". *Computers & Industrial Engineering*, vol. 128, pp. 458–476, 2019.

13. N. Wang, L. Liu. "Genetic algorithm in chaos". *OR Trans*, vol. 5, pp. 1–10, 2001.

14. J.M. Yang, Y.P. Chen, J.T. Horng and C.Y. Kao. "Applying family competition to evolution strategies for constrained optimization". In *Lecture Notes in Mathematics* vol. 1213, pp. 201–211, New York, Springer, 1997.

15. G. Zhenyu, C. Bo, Y. Min, C. Binggang. "Self-adaptive chaos differential evolution Zhenyu". *Advances in Natural Computation*, pp. 972–975, 2006.

16. L.T. Pecora. "Carroll, synchronization in chaotic system". *Physical Review Letters* vol. 64, no. 8, pp. 821–824, 1990.

17. D. Yang, G. Li, G. Cheng. "On the efficiency of chaos optimization algorithms for global optimization". *Chaos, Solitons & Fractals*, vol. 34, pp. 136–1375, 2007.

18. G. Gharooni-fard, F. Moein-darbari, H. Deldari, A. Morvaridi. "Scheduling of scientific workflows using a chaos-genetic algorithm". *Procedia Computer Science*, vol. 1, pp. 1445–1454, 2010.

19. A.H. Gandomi, G.J. Yun, X.S. Yang, S. Talatahari. "Chaos-enhanced accelerated particle swarm algorithm". *Communications in Nonlinear Science and Numerical Simulation*, vol. 18, no. 2, pp. 327–340, 2013.

20. B. Alatas. "Chaotic bee colony algorithms for global numerical optimization". *Expert Systems with Applications*, vol. 37, pp. 5682–5687, 2010.

21. J. Mingjun, T. Huanwen. "Application of chaos in simulated annealing". *Chaos, Solitons & Fractals*, vol. 21, pp. 933–941, 2004.

17

Hunting Search Optimization Modification and Application

Ferhat Erdal

Department of Civil Engineering
Akdeniz University, Turkey

Osman Tunca

Department of Civil Engineering
Karamanoglu Mehmetbey University, Turkey

Erkan Dogan

Department of Civil Engineering
Celal Bayar University, Turkey

CONTENTS

17.1 Introduction

Hunting search algorithm (HuS) is one of the recent additions to the meta-heuristic search techniques of combinatorial optimization problems, introduced by Oftadeh et al. [1]. This approach is based on the group hunting of animals such as lions, wolves and dolphins. The commonality in the way these animals hunt is that they all hunt in a group. They encircle the prey and gradually tighten the ring of siege until they catch the prey. Each member of the group corrects its position based on its own position and the position of other members during this action. If a prey escapes from the ring, hunters reorganize the group to siege the prey again. The hunting search algorithm is based on the way wolves hunt. The procedure involves a number of hunters which represents the hunting group which are initialized randomly in the search space of an objective function. Each hunter represents a candidate solution of the optimum design problem. This chapter is organized as follows. In Section 17.2 a brief introduction of the simple hunting search algorithm is presented. In Section 17.3 improvements are given. The Levy flight procedure is introduced. In Section 17.4 a numerical example, the welded beam design problem, is taken into consideration and application of the hunting search algorithm is demonstrated.

17.2   Original HuS algorithm in brief

The original hunting search algorithm can be presented using pseudo-code of the algorithm. Steps of the method are also listed in the following.

17.2.1   Description of the original hunting search algorithm

In Algorithm 19 the pseudo-code of the original hunting search algorithm is given, which summarizes the routine.

Algorithm 19 Pseudo code of original HuS.

1: determine the D-th dimensional objective function $\mathbf{F}$
2: determine the range of variability for each j-th dimension $[X_{i,j}^{min}, X_{i,j}^{max}]$
3: determine the HuS algorithm parameters such as hunting group size, maximum movement toward the leader, hunting group consideration rate, maximum and minimum values of arbitrary distance radius, convergence rate parameters and number of iterations per epoch.
4: randomly create hunting group which consists of N hunters.
5: **for** each i-th hunter X_i from group **do**
6:     Evaluate objective function
7:     Determine the leader hunter X_i which has the minimum objective function
8: **end for**
9: **while** termination condition not met **do**
10:     **for** each i hunter in the group **do**

11:　　　　**for** each dimension **do**

12:　　　　　　update the position $\mathbf{X_{i,j}}$ using formula

13:　　　　　　$X_{i,j} = X_{i,j} + r \times (MML)(X_{i,j}^L - X_{i,j})$

14:　　　　　　check if the newly created value $X_{i,j}$ is within the range $[X_{i,j}^{min}, X_{i,j}^{max}]$ if not correct it

15:　　　　**end for**

16:　　　　**evaluate** the hunter X_i with objective function

17:　　　　**if** objective function of X_i better than leader **then**

18:　　　　　　assign the hunter X_i to the leader X_i^L

19:　　　　**end if**

20:　　**end for**

21:　　**for** each i hunter in the group **do**

22:　　　　**for** each dimension **do**

23:　　　　　　apply position correction-cooperation between members with $\mathbf{X_{i,j}}$ using formula

24:　$x_i^{j'} \leftarrow \begin{cases} x_i^{j'} \in \{x_i^1, x_i^2, i = 1, \ldots, x_i^{HGS}\} & i = 1, \ldots, N \\ x_i^{j'} = x_i^j \pm Ra \text{ with probability } (1 - HGCR) & j = 1, \ldots, HGS \end{cases}$

25:　　　　　　check if the newly created value $X_{i,j}$ is within the range $[X_{i,j}^{min}, X_{i,j}^{max}]$ if not correct it

26:　　　　**end for**

27:　　　　**evaluate** the hunter X_i with objective function

28:　　　　**if** objective function of X_i better than leader **then**

29:　　　　　　assign the hunter X_i to the leader X_i^L

30:　　　　**end if**

31:　　**end for**

32:　　**for** each i hunter in the group **do**

33:　　　　**for** each dimension **do**

34:　　　　　　apply reorganizing the group with $\mathbf{X_{i,j}}$ using formula

35:　　　　　　$x_i' = x_i^L \pm r \times (x_i^{max} - x_i^{min}) \times \alpha(-\beta \times EN)$

36:　　　　　　check if the newly created value $X_{i,j}$ is within the range $[X_{i,j}^{min}, X_{i,j}^{max}]$ if not correct it

37:　　　　**end for**

38:　　　　**evaluate** the hunter X_i with objective function

39:　　　　**if** objective function of X_i better than leader **then**

40:　　　　　　assign the hunter X_i to the leader X_i^L

41:　　　　**end if**

42:　　**end for**

43: **end while**

In step 1, parameters of the algorithm are initialized. The algorithm has eight parameters that require initial values to be assigned. These are hunting group size (number of solution vectors in the hunting group, HGS), maximum movement toward the leader (MML), hunting group consideration rate (HGCR) which varies between 0 and 1, maximum and minimum values of arbitrary distance radius (Ra^{max} and Ra^{min}), convergence rate parameters (α *and* β) and number of iterations per epoch (IE). Then, in step 2, hunting group is initialized. Based on the number of hunters (HGS), the hunting group matrix is filled with feasible randomly generated solution vectors. The values of objective function are computed for each solution vector and the leader is defined depending on these values. Step 3 is devoted to the generation of the new hunters' positions. New solution vectors $x' = \{x'_1, x'_2, \ldots, x'_n\}$ are generated by moving toward the leader (the hunter that has the best position in the group) as follows.

$$x'_i = x_i + r(MML)(x^L_i - x_i) \qquad i = 1, \ldots, n \qquad (17.1)$$

where MML is the maximum movement toward the leader, r is a uniform random number [0,1] and x'_i is the position value of the leader for the *ith* variable. If the movement of a hunter toward the leader is successful, it stays in its new position. However, if the movement is not successful, i.e., its previous position is better than its new position, it comes back to the previous position. This results in two advantages. First, the hunter is not compared with the worst hunter in the group to allow the weak members to search for other solutions. They may find better solutions. The second advantage is that, for prevention of rapid convergence of the group the hunter compares its current position with its previous position; therefore, good positions will not be eliminated. In step 4, position correction-cooperation between members is performed. In order to conduct the hunt more efficiently, the cooperation among hunters should be modeled. After moving toward the leader, hunters tend to choose another position in order to conduct the 'hunt' more efficiently, i.e. better solutions. Positions of the hunters can be corrected in two ways; real value correction and digital value correction. In real value correction which is considered in the present study, the new hunter's position $x' = \{x'_1, x'_2, \ldots, x'_n\}$ is generated from HG, on the basis of hunting group considerations or position corrections, which is expressed as;

$$x^{j'}_i \leftarrow \begin{cases} x^{j'}_i \in \{x^1_i, x^2_i, i = 1, \ldots, x^{HGS}_i\} & i = 1, \ldots, N \\ x^{j'}_i = x^j_i \pm Ra \ with \ probability \ (1 - HGCR) & j = 1, \ldots, HGS \end{cases}$$

$$(17.2)$$

For instance, the value of the first design variable for the *jth* hunter $x^{j'}_1$ for the new vector can be selected as a real number from the specified $HG(x^1_i, x^2_i, \ldots, x^{HGS}_i)$ or corrected using the HGCR parameter (chosen between 0 and 1).

In the above equation, HGCR is the probability of choosing one value from the hunting group stored in the HG. It is reported that values of this parameter between 0.1 and 0.4 produce better results. Ra is referred to as an arbitrary distance radius for the continuous design variable, which can be reduced or fixed during the optimization process. Through the former assumption, Ra can be reduced by use of the following exponential function.

$$Ra(it) = Ra_{min}(x_i^{max} - x_i^{min}) \exp \left(\dfrac{\ln\left(\dfrac{Ra_{max}}{Ra_{min}}\right) \times it}{itm} \right) \qquad (17.3)$$

where it represents the iteration number, x_i^{min} and x_i^{max} are the maximum and minimum possible values for x_i. Ra^{max} and Ra^{min} denote the maximum and minimum of relative search radius of the hunter, respectively and itm is the maximum number of iterations in the optimization process.

In digital value correction, instead of using real values of each variable, the hunters communicate with each other by the digits of each solution variable. For example, the solution variable with the value of 23.4356 has six meaningful digits. For this solution variable, the hunter chooses a value for the first digit (i.e. 2) based on hunting group considerations or position correction. After the quality of the new hunter position is determined by evaluating the objective function, the hunter moves to this new position; otherwise it keeps its previous position. Next in step 5 the hunting group is reorganized. In order to prevent being trapped in a local optimum they must reorganize themselves to get another opportunity to find the optimum point. The algorithm does this in two independent conditions. If the difference between the values of the objective function for the leader and the worst hunter in the group becomes smaller than a preset constant $\varepsilon 1$ and the termination criterion is not satisfied, then the algorithm reorganizes the hunting group for each hunter. Alternatively, after a certain number of searches the hunters reorganize themselves. The reorganization is carried out as follows: the leader keeps its position and the other hunters randomly choose their position in the design space.

$$x_i' = x_i^L \pm rand \times (max(x_i) - min(x_i)) \times \alpha \exp(-\beta \times EN) \qquad (17.4)$$

where, x_i^L is the position value of the leader for the ith variable, r represents the random number between 0 and 1, x_i^{min} and x_i^{max} are the maximum and minimum possible values for x_i, respectively. EN counts the number of times that the hunting group has trapped until this step. As the algorithm goes on, the solution gradually converges to the optimum point. Parameters α and β are positive real values which determine the convergence rate of the algorithm. Finally, the process is terminated if the predetermined value of iterations is reached.

17.3 Improvements in the hunting search algorithm

In order to increase the chance of obtaining the best solution, the Levy flight procedure [2] is adopted to the simple hunting search algorithm. How we adopt this to the algorithm is that, this step is added to the iteration process as a new search for the better position of the hunters. A Levy flight is a random walk in which the steps are defined in terms of the step-lengths which have a certain probability distribution, with the directions of the steps being isotropic and random. Hence Levy flights necessitate selection of a random direction and generation of steps under the chosen Levy distribution. The Mantegna algorithm [2] is one of the fast and accurate algorithms which generate a stochastic variable whose probability density is close to Levy stable distribution characterized by arbitrary chosen control parameter α ($0.3 \leq \alpha \leq 1.99$). Using the Mantegna algorithm, the step size λ is calculated as;

$$\lambda = \frac{x}{|y|^{1/\alpha}} \tag{17.5}$$

where x and y are two normal stochastic variables with standard deviation σ_x and σ_y which are given as;

$$\sigma_x(\alpha) = \left[\frac{\Gamma(1+\alpha)sin(\pi\alpha/2)}{\Gamma((1+\alpha)/2)\alpha 2^{(\alpha-1)/2}} \right]^{1/\alpha} and \ \sigma_y(\alpha) = 1 \ for \ \alpha = 1.5 \tag{17.6}$$

in which the capital Greek letter Γ represents the gamma function that is the extension of the factorial function with its argument shifted down by 1 to real and complex numbers. That is if k is a positive integer. $\Gamma(k) = (k-1)!$ Once the step size λ is determined according to random walk with Levy flights, the hunters' new positions are generated as follows;

$$x_i' = x_i \pm \beta\lambda r(x_i - x_i^L) \tag{17.7}$$

where, $\beta > 1$ is the step size which is selected according to the design problem under consideration, r is random number from standard normal distribution.

17.4 Applications of the algorithm to the welded beam design problem

17.4.1 Problem description

A carbon steel rectangular cantilever beam design problem [3] is taken as an optimization problem. The geometric view and the dimensions of the beam

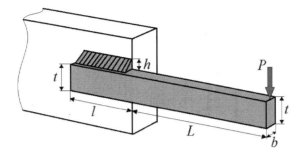

FIGURE 17.1
Welded beam.

are illustrated in Figure 17.1. The beam is designed to carry a certain P load acting at the free tip with minimum overall cost of fabrication. Geometric properties of the beam can be listed as the following: h the thickness of the weld, l is the length of the welded joint, t and b are the width and the thickness of the beam, respectively.

17.4.2 How can the hunting search algorithm be used for this problem?

We want to create an optimization problem to obtain the best dimensions for the welded beam. To do this, the objective function of the problem should be constructed first. This can be performed by use of the weight function of the beam. Then this weight function will be the objective function to be optimized or minimized. This function is given in the following.

$$f(x) = 1.10471x_1^2 x_2 + 0.04811x_3 x_4 (14.0 + x_2) \qquad (17.8)$$

Once we have constructed this, now we can proceed with the design variables which are to be changed during the process. In fact, these are the dimensions of the beam listed above. We can determine the vector of variables as follows;

$x_1 : h$: the thickness of the weld
$x_2 : l$: the length of the welded joint
$x_3 : t$: the width of the beam
$x_4 : b$: the thickness of the beam

In the optimization applications, problems generally have constraints. Hence, an optimally designed welded beam should satisfy the below constraints for a correct design. These can be either side constraints or behavioral constraints. Below the first seven constraints are treated as behavioral constraints for this problem. And side constraints are given at the end.

$g_1(x) = \tau(x) - \tau_{max} \leqslant 0$: shear stress

$g_2(x) = \sigma(x) - \sigma_{max} \leqslant 0$: bending stress in the beam

$g_3(x) = x_1 - x_4 \leqslant 0$: side constraint

$g_4(x) = 0.10471x_1^2 + 0.04811x_3x_4(14.0 + x_2) - 5 \leqslant 0$: side constraint

$g_5(x) = 0.125 - x_1 \leqslant 0$: side constraint

$g_6(x) = \delta(x) - \delta_{max} \leqslant 0$: end deflection of the beam

$g_7(x) = P - P_c(x) \leqslant 0$: buckling load on the bar

where

$$\tau(x) = \sqrt{(\tau')^2 + 2\tau'\tau''\frac{x_2}{2R} + (\tau'')^2}$$

$$\tau' = \frac{P}{\sqrt{2}\,x_1\,x_2}$$

$$\tau'' = \frac{MR}{J}$$

$$M = P\left(L + \frac{x_2}{2}\right)$$

$$R = \sqrt{\frac{x_2^2}{4} + \left(\frac{x_1+x_2}{2}\right)}$$

$$J = 2\left\{\sqrt{2}x_1x_2\left[\frac{x_2^2}{12} + \left(\frac{x_1+x_3}{2}\right)^2\right]\right\}$$

$$\delta(x) = \frac{4PL^3}{Ex_3^3 x_4}, \sigma(x) = \frac{6PL}{x_4 x_3^2}$$

$$P_c(x) = \frac{4.013\ E\sqrt{\frac{x_3^2 x_4^6}{36}}}{L^2}\left(1 - \frac{x_2}{2L}\sqrt{\frac{E}{4G}}\right)$$

$P = 600\ lb, L = 14\ in,$

$E = 30 \times 10^6\ psi, G = 12 \times 10^6\ psi$

$\tau_{max} = 13.600\ psi, \sigma_{max} = 30.000\ psi,$

$\delta_{max} = 0.25\ in.$

The side constraints for the design variables are given as follows:

$0.1 \leqslant x_1 \leqslant 2.0, \quad\quad 0.1 \leqslant x_2 \leqslant 10$

$0.1 \leqslant x_3 \leqslant 10, \quad\quad 0.1 \leqslant x_4 \leqslant 2.0$

The iteration process is initialized by selecting the parameters. Then hunters are generated randomly. This means coordinates of the positions of each hunter are assigned. After that, it is observed whether the designs satisfy the constraints or not. If yes, then this design can be treated as a candidate. Then the objective function of each design is determined. Afterwards, the positions, namely the dimensions of the welded beam, are updated through the leader. That means new designs are related with the previous one of each hunter. Once again, it is observed whether the new designs are feasible or not. If yes, and weights are less than the previous one, then positions are updated and a new leader is determined. In order to obtain better solutions, positions of each hunter are corrected after moving towards the leader. While doing these changes, the algorithm generates random numbers. As conducted in the previous step, the leader is determined, which is also treated as the current best. In the same manner, the update of the objective function for each hunter will be performed in the next two steps as well. Finally the design with the minimum objective function will be treated as the optimum solution of the problem.

17.4.3   Description of experiments

The method presented above was tested with the well-known benchmark 'welded beam design problem', the description of which is given in the previous part. Design pool was constructed by taking into account the interval of each variable. At the beginning of the solution process, we did not consider the value of the objective function of the problem. What is important at this stage is the generation of a predetermined number of feasible solutions. A flyback mechanism [4] was utilized in handling the constraints. While performing this, error is assumed to be 0. However, when the algorithm was in the iteration stage, an adaptive error strategy was applied. This strategy works in this way: If some hunters are slightly infeasible, then such hunters are kept in the solution. These hunters having one or more slightly infeasible constraints are utilized in the design process that might provide a new hunter that may be feasible. This is achieved by using larger error values initially for the acceptability of the new design vectors and then reducing this value gradually during the design cycles using finally an error value of 0.001 or whatever necessary value is required to be selected for the permissible error term towards the end of iterations. As expected, the algorithm produced a number of slightly feasible solutions and these solutions were kept in the memory as current best for the corresponding cycle. Afterwards, while iterations were proceeding, error was gradually reduced and finally feasible solutions were obtained in this neighborhood. The value of algorithm parameters are as follows: Number of hunters = 40, MML = 0.005, HGCR = 0.3, Ramax= 0.01, Ramin = 0.0000001, α = 1.2, β = 0.02, maximum number of iterations in one epoch = 25.

17.4.4   Result obtained

Optimum welded beam design was achieved after 119000 iterations with the objective function value of 1.724941. Design variables, e.g., the dimensions of optimum beam are 0.205731, 3.47112, 9.036624 and 0.205730. (Table 17.1)

The design history curve for these solutions is also plotted in Fig. 17.2, which displays the variation of the feasible best design obtained so far during the search versus the number of designs sampled. It is clear from this figure that the hunting search with levy flights algorithm performs the best convergence rate toward the optimum solution.

This solution was determined by the improved hunting search algorithm. On the other hand, the one attained by the simple hunting search method is

TABLE 17.1

Optimum designs obtained by simple and improved versions of hunting search algorithm.

Design Variable	Method	
	Improved HuS	Simple HuS
x_1	0.205731	0.20573
x_2	3.47112	3.47112
x_3	9.036624	9.03663
x_4	0.205730	0.20573
f	**1.724941**	**1.724944**
Number of iterations	**119000**	**119500**

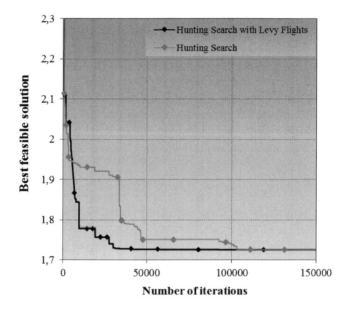

FIGURE 17.2

Design history graph for welded beam design problem.

stored as 1.724944. Although the difference between these two results seems to be negligibly small, this means a lot for the comparison. Therefore, in view of the results one can conclude that the Levy flight extension improved the performance of the algorithm.

17.5 Conclusions

In this subsection, improved and original version of hunting search optimum design algorithm presented in the previous sections is used to design well-known welded beam benchmark problem. Problem descriptions and the steps of the working process of the algorithm were given in detail. Dimensions of welded beam are treated as design variables of the optimum design problem and weight of the beam is assumed to be the objective function to be minimized. This problem was solved with both algorithms a number of times and results were compared. Accordingly, as compared to the solution of the standard hunting search algorithm, which is 1.724944, a much better final design value of 1.724941 is located by the hunting search with levy flights (Table 17.1). It is observed that Levy flights increases the performance of the algorithm. One common drawback of metaheuristic search methods is that the iteration process generally gets stuck in the local optima. In this study, it is observed that solutions with a constant error strategy experience this fact. However, it is observed in this study that an adaptive error strategy for constraint handling prevents this problem and provides an efficient search for each algorithm.

References

1. R. Oftadeh, M.J. Mahjoob, M. Shariatpanahi. "A novel meta-heuristic optimization algorithm inspired by group hunting of animals: Hunting search". *Computers and Mathematics with Applications*, vol. 60(7), pp. 2087-2098, 2010.

2. R.N. Mantegna. "Fast, accurate algorithm for numerical simulation of Levy stable stochastic processes". *Physical Review E.*, vol. 49(5), pp. 4677-4683, 1994.

3. K.M. Ragsdell, D.T. Phillips. "Optimal design of a class of welded structures using geometric programming". *Journal of Engineering for Industry*, vol. 98(3), pp. 1021-1025, 1976.

4. K. Deb. "Optimal design of a welded beam via genetic algorithms". *AIAA journal*, vol. 29(11), pp. 2013-2015, 1991.

18

Krill Herd Algorithm – Modifications and Applications

Ali R. Kashani

Department of Civil Engineering
University of Memphis, Memphis, United States

Charles V. Camp

Department of Civil Engineering
University of Memphis, Memphis, United States

Hamed Tohidi

Department of Civil Engineering
University of Memphis, Memphis, United States

Adam Slowik

Department of Electronics and Computer Science
Koszalin University of Technology, Koszalin, Poland

CONTENTS

18.1 Introduction

The krill herd (KH) algorithm is one of the swarm intelligence algorithms developed by Gandomi and Alavi [1]. KH as a bio-inspired algorithm imitates the herding behavior of krill in the seas using a Lagrangian model and evolutionary operators. Basically, krill individuals tend to be gathered in large swarms for finding food resources. Therefore, herding of the krill addresses two main objectives: (i) increasing krill density, and (ii) reaching food. Considering this fact, the global optimal solution is represented by higher density of krill swarm and shorter distance from the food source. KH algorithm explores the solution space using a swarm of krill as potential solutions. The global optimum solution is the closest krill to the food resource. A krill's movements within the search space is governed by three basic rules: (1) movements induced by other krill; (2) foraging activity; and (3) random diffusion. Moreover, genetic operators i.e., crossover and mutation are utilized in this algorithm to improve the algorithm's performance. The remainder of this chapter is organized accordingly. In Section 18.2 a summary of the fundamentals of KH algorithm is presented. In Section 18.3 some modifications of KH are described. Finally, in Section 18.4 the application of KH algorithm to the optimization of a shallow foundation is discussed.

18.2 Original KH algorithm in brief

In this section the fundamentals of the original KH algorithm are discussed using a step by step pseudo-code.

Algorithm 20 Pseudo-code of the original KH.

1: determine the $D-th$ dimensional objective function $OF(.)$
2: determine the range of variability for each $j-th$ dimension $\left[K_{i,j}^{min}, K_{i,j}^{max}\right]$
3: determine the KH algorithm parameter values such as NK – number of krill, MI – maximum iteration, V_f – foraging speed, D_{max} – maximum diffusion, N_{max} – maximum induced speed
4: randomly create swarm P which consists of NK krill individuals (each krill individual is a D-dimensional vector)
5: finding the best krill and its relevant vector
6: $Iter = 0$
7: **while** termination condition not met (here is reaching MI) **do**
8:    evaluate $X^{food} = \frac{\sum_{i=1}^{NK} \frac{1}{K_i} \cdot X_i}{\sum_{i=1}^{NK} \frac{1}{K_i}}$
9:    **for** each i-th krill in swarm P **do**
10:        $\alpha_i^{target} = 2 \cdot \left(rand + \frac{Iter}{MI}\right) \cdot \widehat{K}_{i,best} \cdot \widehat{X}_{i,best}$

11: $\quad\quad R_{z,i} = \sum_{j=1}^{N} \|X_i - X_j\|$ and $d_{z,i} = \frac{1}{5 \cdot N} \cdot R_{z,i}$

12: $\quad\quad$ **if** $R_{z,i} < d_{z,i}$ and $K(i) \neq K(n)$ **then**

13: $\quad\quad\quad \alpha_i^{local} = \sum_{j=1}^{NN} \frac{K_i - K_j}{K^{worst} - K^{best}} \times \frac{X_j - X_i}{\|X_j - X_i\| + \epsilon}$

14: $\quad\quad$ **end if**

15: $\quad\quad \omega = 0.1 + 0.8 \times \left(1 - \frac{1}{MI}\right)$

16: $\quad\quad N_i^{new} = N^{max} \cdot \left(\alpha_i^{target} + \alpha_i^{local}\right) + \omega \cdot N_i^{old}$

17: $\quad\quad \beta_i^{food} = 2 \cdot \left(1 - \frac{Iter}{MI}\right) \cdot \widehat{K}_{i,food} \cdot \widehat{X}_{i,food}$

18: $\quad\quad \beta_i^{best} = \widehat{K}_{i,best} \cdot \widehat{X}_{i,best}$

19: $\quad\quad F_i^{new} = V_f \cdot \left(\beta_i^{food} + \beta_i^{best}\right) + \omega \cdot F_i^{old}$

20: $\quad\quad D_i = D^{max} \cdot \left(1 - \frac{Iter}{MI}\right) \cdot \delta$

21: $\quad\quad \frac{dX_i}{dt} = N_i^{new} + F_i^{new} + D_i$

22: $\quad\quad X_i = crossover\,(X_i, X_c)$

23: $\quad\quad X_i = mutation\,(X_i, X_{best}, M_u)$

24: $\quad\quad X_i^{new} = X_i + \frac{dX_i}{dt}$

25: $\quad$ **end for**

26: $\quad$ update best-found solution

27: $\quad$ $Iter = Iter + 1$

28: **end while**

29: post-processing the results

Here, KH algorithm deals with a D-dimensional objective function $OF(.)$ as demonstrated in Algorithm 20. At the second step the permitted j-th variable's domain for i-th krill is defined as $\left[K_{i,j}^{min}, K_{i,j}^{max}\right]$. In step 3, the necessary parameters for KH algorithm are initialized accordingly; NK is number of krill, MI is the maximum iteration, V_f is the foraging speed, D^{max} is the maximum diffusion speed, N_{max} is the maximum induced speed. In the original paper, D^{max} varies between 0.002 and 0.010 (ms^{-1}), V_f and N_{max} are recommended to be 0.02 and 0.01 (ms^{-1}), respectively. In step 4, an initial swarm of NK number of krill is reproduced randomly within the valid solution domain. Each krill is represented by a D-dimensional vector as follows:

$$K_i = \{k_{i,1}, k_{i,2}, ..., k_{i,D}\} \tag{18.1}$$

where $k_{i,1}$ to $k_{i,D}$ are decision variables varying between $K_{i,j}^{min}$ and $K_{i,j}^{max}$.

In step 5, the global best solution and its relevant vector are determined. After that, the current iteration counter is initialized as zero. Now, the main loop of KH algorithm will be started. In step 8, the virtual food location is evaluated. In the original paper [1], the center of food is proposed to be evaluated based on distribution of the krill individuals' fitness. It is somehow like evaluating "*center of mass*". As mentioned previously, the motion of krill individuals within the solution space can be characterized by three basic rules: I. Movement induced by the other krill individuals (N_i); II. Foraging action; and III. Random diffusion.

In KH the following time-dependent Lagrangian model is developed to explore our D-dimensional search space:

$$\frac{dX_i}{dt} = F_i + N_i + D_i \qquad (18.2)$$

To address Equation (18.2), step 9 in Algorithm 20 is devoted to illustrating krill individuals' pace through an iterative loop. The first action is to evaluate N_i at first. To this end, in step 10, α_i^{target} results from the best-found solution to define the target direction effect. Another important effect is imposed by the neighbors through local effect α_i^{local}. Before that, the sensing distance for each i-th particle $(d_{s,i})$ is computed based on the distance between this particle and all of the others as shown in step 11. In step 13, α_i^{local} is determined to consider the effect of neighbors. In KH algorithm two different inertia weights are defined; inertia weight for induced motion by other krill individuals (ω_n) and inertia weight of the foraging motion (ω_f). Both of ω_n and ω_f vary in the range of $[0, 1]$, though in this study a time-dependent value is considered for both as shown in step 15. In step 16, the movement resulting from other krill individuals (N_i^{new}) is calculated. To examine foraging motion (step 19), we need two important factors, food attractiveness (β_i^{food}) and the effect of the best fitness of the i-th krill (β_i^{best}). We evaluate β_i^{food} and β_i^{best} in steps 17 and 18, respectively. In step 20, the final movement caused by physical diffusion of the krill individuals is estimated where δ is the random directional vector. The next position of the krill individuals in an interval of $\frac{dX_i}{dt}$ is evaluated in step 21. However, before updating the solutions two evolutionary operators, crossover and mutation, are applied to the current generation in steps 22 and 23, respectively. In step 24, new positions are generated by applying the term of $\frac{dX_i}{dt}$ to the current solutions. The best-found solution is updated in step 26. Finally, the best-found solution is proposed when the termination criteria are satisfied.

18.3 Modifications of the KH algorithm

18.3.1 Chaotic KH

In original KH, tuning inertia weights (ω_n, ω_f) play an important role in the final performance of this algorithm. Higher values of inertia weights push the algorithm toward exploration and lower values encourage the KH algorithm to focus on exploitation. Wang et al. [2] proposed a chaotic KH (CKH) using a chaotic-based pattern for adjusting those parameters to reach faster convergence. In this way, the authors normalized the range of the chaotic map between 0 and 1. CKH updates the values of inertia weights in each iteration using 12 different chaotic maps collected in Table 18.1. Wang et al. [2] applied

TABLE 18.1

Different chaos maps.

ID	Name	Formulation
1.	Chebyshev	$X_{i+1} = cos\left(k cos^{-1}\left(X_i\right)\right)$
2.	Circle	$X_{i+1} = mod\left(X_i + b - \left(\frac{a}{2\pi}\right) sin\left(2\pi X_i\right), 1\right)$
3.	Gaussian	$X_{i+1} = \begin{cases} 0 & X_i = 0 \\ \frac{1}{X_i mod(1)} & \text{otherwise}, \end{cases} \frac{1}{X_i mod(1)} = \frac{1}{X_i} - \left[\frac{1}{X_i}\right]$
4.	Intermittency	$X_{i+1} = \begin{cases} \epsilon + X_i + C X_i^n & 0 < X_i \le p \\ \frac{X_i - p}{1-p} & p < X_i < 1 \end{cases}$
5.	Iterative	$X_{i+1} = sin\left(\frac{a\pi}{X_i}\right)$
6.	Liebovitch	$X_{i+1} = \begin{cases} \alpha X_i & 0 < X_i \le p_1 \\ \frac{p - X_i}{p_2 - p_1} & p_1 < X_i \le p_2 \\ 1 - \beta\left(1 - X_i\right) & p_2 < X_i \le 1 \end{cases}$
7.	Logistic	$X_{i+1} = \alpha X_i \left(1 - X_i\right)$
8.	Piecewise	$X_{i+1} = \begin{cases} \frac{X_i}{p} & 0 \le X_i < p \\ \frac{X_i - p}{0.5 - p} & p \le X_i < 0.5 \\ \frac{1 - p - X_i}{0.5 - p} & 0.5 \le X_i < 1 - p \\ \frac{1 - X_i}{p} & 1 - p \le X_i < 1 \end{cases}$
9.	Sine	$X_{i+1} = \frac{a}{4} sin\left(\pi X_i\right)$
10.	Singer	$X_{i+1} = \mu(7.86 X_i - 23.31 X_i^2 + 28.75 X_i^3 - 13.302875 X_i^4), \mu = 1.07$
11.	Sinusoidal	$X_{i+1} = a X_i^2 sin\left(\pi X_i\right)$
12.	Tent	$X_{i+1} = \begin{cases} \frac{X_i}{0.7} & X_i < 0.7 \\ \frac{10}{3}\left(1 - X_i\right) & X_i \ge 0.7 \end{cases}$

their proposed algorithm to 14 benchmark functions. It was mentioned in their study that the results recorded considerable superiority of CKH over original KH algorithm.

18.3.2 Levy-flight KH

Metaheuristic optimization algorithms deal with objective function based on two important strategies: exploration and exploitation. Any attempt toward finding an appropriate balance between them improves the algorithm's performance effectively. Wang et al. [3] proposed levy-flight krill herd (LKH) using local levy-flight operator to strengthen the exploitation ability of original KH. Wang et al. [3] utilized the following random walking steps:

$$dX_i^{t+1} = X_i^{t+1} + \beta L\left(s, \lambda\right) \qquad (18.3)$$

$$L\left(s, \lambda\right) = \frac{\lambda \Gamma\left(\lambda\right) sin\left(\frac{\pi \lambda}{2}\right)}{\pi} \times \frac{1}{s^{1+\lambda}}, \left(s, s_0 > 0\right)$$

where β is the step size scaling factor and should have a positive value. In LKH, a local Levy flight strengthens the exploitation ability of the KH algorithm by providing much longer step length for searching around the best-found solution.

In the proposed local Levy flight, a time-dependent step size is proposed by the following equation:

$$\alpha = \frac{A}{t^{\Omega}} \qquad (18.4)$$

where A is the maximum step size and *Iter* is the current iteration number.

For updating the new position of the krill, a vector of random numbers *(random)* with the size of problem dimension (D) will be produced and each d-th dimension updated as follows:

$$X_i^{t+1} = \begin{cases} \alpha \times \frac{dX_i}{dt} + X_{D-r+1} & random\,(d) \leq 0.5 \\ \alpha \times \frac{dX_i}{dt} - X_{D-r+1} & random\,(d) > 0.5 \end{cases} \qquad (18.5)$$

where r is a random number. The current solution would be replaced by V_i in case of improvement.

This LKH was tested through 14 benchmark functions and the results are compared to several optimization algorithms. The authors claimed that their proposed modified algorithm achieved better performance rather than the original KH algorithm.

18.3.3   Multi-stage KH

In another effort, Wang et al. [4] tried to boost the KH algorithm by balancing between exploration and exploitation. Original KH proposed a satisfying performance during the search of solution space (exploration). Therefore, any attempt toward strengthening the exploitation ability of KH can be helpful to increase its performance. Wang et al. [4] proposed the multi-stage KH (MSKH) by mounting the original KH with a local mutation and crossover (LMC) operator to intensify the local search concentration. In this way the MSKH algorithm searches the solution space based on different stages. In the first stage, the original KH shrinks the search space as much as possible and in the second stage LMC operator is enlisted to search around the limited space resulting from the first stage more accurately. To be more exact, in the first stage, the position of krill in a herd will be updated using the step size $\frac{dX_i}{dt}$ (line 21 in Algorithm 20). After that, LMC operator modifies the d-th element of each i-th krill's position for all the krill in a herd following the below equation:

$$X_i^{t+1} = \begin{cases} X_{best}\,(d) & random\,(d) \leq 0.5 \\ X_{best}\,(r) & random\,(d) > 0.5 \end{cases} \qquad (18.6)$$

where X_i^{t+1} is the current position of the i-th krill, $X_{best}(d)$ is the d-th element of the best solution and $X_{best}(r)$ is the r-th element of the best solution. To validate the proposed algorithm, MSKH is tackled for 25 benchmark functions and compared to several optimization algorithms. As the authors in [4] mentioned, their proposed algorithm, MSKH, performed very well especially in dealing with complex multimodal problems.

18.3.4   Stud KH

To compensate for the weakness of original KH in exploitation, Wang et al. [5] proposed the stud krill herd (SKH) algorithm. In fact, SKH by imitating the stud genetic algorithm tends to use the best-found solution for crossover in each iteration. To be more exact, Wang et al. [5] by incorporating the stud selection and crossover (SSC) operator to the KH algorithm improved the ability of this algorithm to evade local minima. Based on this methodology, for updating the positions of all the krill two different attitudes may be selected. First, the step size defined in the original KH, $\frac{dX_i}{dt}$, will be evaluated. After that, by applying the crossover operator the best krill modifies one of the selected solutions by the selection operator. Next, the quality of the reproduced offspring will be evaluated and in case of improvement the krill position will be updated. Otherwise, the krill individual will move to the next position by adding the term of step size to the current position. This algorithm was tested using 22 benchmark functions and based on the authors' assertion; this simple modification resulted in better performance of the algorithm and increasing accuracy of the global optimality.

18.3.5   KH with linear decreasing step

Gandomi and Alavi [1], developed the basic KH method and they defined $\frac{X_i}{dt}$ which works as a scale factor of the speed vector and it depends on search space and a constant parameter C_t. Li et al. [6] tried to improve the basic method of Gandomi and Alavi [1] by analyzing the parameter C_t. They have conducted various experiments on KH method to investigate the effect of C_t size on search space. They figured that the C_t is large enough that the KH searches entire space, but it cannot search local space accurately. On the other hand when the C_t is small the step will be smaller and the KH can search local space, but it misses the global area. Li et al. [6] claimed that if the step size of C_t can be controlled by increasing the number of iterations, KH can search both the local and global area carefully. By increasing iterations in number, the C_t becomes linearly smaller in each iteration until it covers both the local and global area. The proposed new C_t is provided in Equation (18.7):

$$Ct(t) = Ct_{max} - \frac{Ct_{max} - Ct_{min}}{MI} \cdot t \tag{18.7}$$

where Ct_{max} and Ct_{min} are the maximum and minimum values of Ct, and MI is the maximum number of iterations and t is the current number of iterations.

18.3.6 Biography-based krill herd

Since the basic KH method cannot produce an appropriate convergence all the time, in 2012 Gandomi and Alavi [1] tried to improve the performance of basic KH convergence by adding genetic reproduction mechanisms to the basic KH algorithm. However, sometimes the exploitation ability of KH is still not satisfactory. Wang et al. [7] combined the krill migration (KM) operator with KH to improve the exploitation ability of KH. This method is known as biography-based krill herd (BBKH). Algorithm 21 shows the KM operator steps used in the BBKH.

Algorithm 21 Krill migration (KM) operator.

1: Select i-th krill (its position X_i) without probability based on k_i
2: **if** $rand\,(0,1) < k_i$ **then**
3:     **for** $j = 1$ to d (all elements) **do**
4:         Select X_j with probability based on l_j
5:         **if** $rand\,(0,1) < l_j)$ **then** Randomly select an element r from X_j Replace a random element in X_i with r
6:         **end if**
7:     **end for**
8: **end if**

To keep the krill optimal all the time, Wang et al. [7] added a kind of elitism to the algorithm. Therefore, the BBKH algorithm was developed by combining KM and concentrated elitism in the KM algorithm as can be seen in Algorithm 22.

Algorithm 22 Biogeography-based KH method.

1: Initialization. Set the generation counter $t = 1$; initialize the population P of NP krill randomly; set V_f, D^{max}, N^{max}, S_{max}, and p_{mod}.
2: Fitness evaluation. Evaluate each krill.
3: **while** $t < MI$ **do**
4:     Sort the krill from best to worst
5:     Store the best krill
6:     **for** $i = 1$ to NP (all krill) **do**
7:         Perform the three-motion calculation
8:         Update the krill position by $X_i\,(t + \Delta t) = X_i\,(t) + \Delta t \frac{dX_i}{dt}$
9:         Fine-tune $X_i + 1$ by performing KM operator in Algorithm 20
10:        Evaluate each krill by $X_i + 1$
11:    **end for**
12:    Replace the worst krill with the best krill

13:     Sort the krill and find the current best

14:     $t = t + 1$

15: **end while**

16: Output the best solutions

18.4   Application of KH algorithm for optimum design of retaining walls

18.4.1   Problem description

In nature there are many cases in which a soil slope (natural or artificial) is not stable due to the inherent angle of inclination. A retaining wall is one of the structures that are used to stabilize slope. Since this type of structure is used for various applications such as highways, railways, bridges, etc., optimum cost and weight are very important factors that affect the design process of retaining walls. Besides the economical aspect of the design procedure of retaining walls, also a retaining wall must be designed to meet the geotechnical and structural requirements. There are 12 variables (please see Figure 18.1) that should be considered in order to reach an optimal design of a retaining wall; eight of those variables ($X1$ to $X8$) are wall geometry related and are expressed as: width of the base ($X1$), toe width ($X2$), footing thickness ($X3$), thickness at the top of the stem ($X4$), base thickness ($X5$), the distance from the toe to the front of shear key ($X6$), shear key width ($X7$), shear key depth ($X8$). The other four variables ($R1$ to $R4$) are steel; reinforcement dependent in which the vertical steel reinforcement is in the stem ($R1$), the horizontal steel reinforcement of the toe and heel ($R2$ and $R3$, respectively), and the vertical reinforcement of the shear key ($R4$).

Each phase of retaining wall design (geotechnical and structural) is investigated for various failure modes. In the geotechnical phase, the resistance of the wall is evaluated against overturning, sliding, and bearing using the factor of safety. The structural design phase is assessed for shear and moment failure of stem, heel, toe, and shear key. The main step of reaching an optimal design is defining an appropriate objective function. The objective function that is going to be introduced in this chapter, was presented by Saribas and Erbatur [8] for minimizing cost and weight of the wall as provided in Equations (18.8) and (18.9):

$$f_{cost} = C_s W_{st} + C_c V_c \tag{18.8}$$

$$f_{weight} = W_{st} + 100 V_c \gamma_c \tag{18.9}$$

where C_s is the unit cost of steel, C_c is the unit cost of concrete, W_{st} is the weight of reinforcing steel, V_c is the volume of concrete, and γ_c is the unit weight of concrete.

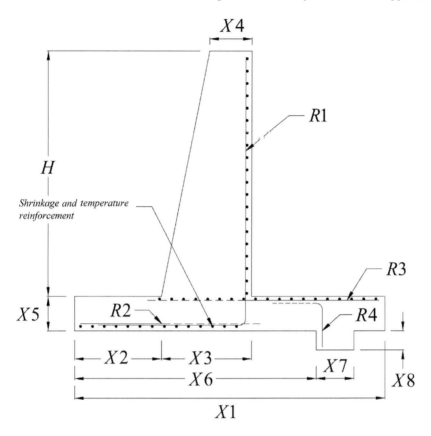

FIGURE 18.1
Reinforced cantilever retaining wall typical geometry and input parameters.

The other predominant factor of each optimization problem is constraints which define satisfying certain conditions to design a safe and stable retaining wall. The constraints can be divided into geometrical, structural, and geotechnical which are included in boundary inequality constraints. According to Saribas and Erbatur [8] and Gandomi et al. [9,10], the boundary limitations contain the variables that will be satisfied by changing wall geometry. Inequality constraints check the final structural and geotechnical strength and stability. All the structural strength and geotechnical stability limitations are defined as the constraints summarized in Table 18.2.

18.4.2 How can KH algorithm be used for this problem?

In this sub-section we plan to introduce an automatic process for optimum design of retaining walls based on KH algorithm. In the first step a population

TABLE 18.2

Inequality constraints.

Constraint	Function
$g_1(x)$	$\frac{FS_{Odesign}}{FS_O} - 1 \leq 0$
$g_2(x)$	$\frac{FS_{Sdesign}}{FS_S} - 1 \leq 0$
$g_3(x)$	$\frac{FS_{Bdesign}}{FS_B} - 1 \leq 0$
$g_4(x)$	$q_{min} \geq 0$
$g_{[5-8]}(x)$	$\frac{M_u}{M_n} - 1 \leq 0$
$g_{[9-12]}(x)$	$\frac{V_u}{V_n} - 1 \leq 0$
$g_{[13-16]}(x)$	$\frac{A_{smin}}{A_s} - 1 \leq 0$
$g_{[17-20]}(x)$	$\frac{A_s}{A_{smax}} - 1 \leq 0$
$g_{21}(x)$	$\frac{X_2+X_3}{X_1} - 1 \leq 0$
$g_{22}(x)$	$\frac{X_6+X_7}{X_1} - 1 \leq 0$
$g_{23}(x)$	$\frac{l_{dbstem}}{X_5-cover} - 1 \leq 0$ or $\frac{l_{dbstem}}{X_5-cover} - 1 \leq 0$
$g_{24}(x)$	$\frac{l_{dbtoe}}{X_1-X_2-cover} - 1 \leq 0$ or $\frac{12d_{btoe}}{X_5-cover} - 1 \leq 0$
$g_{25}(x)$	$\frac{l_{dbheel}}{X_2+X_3-cover} - 1 \leq 0$ or $\frac{12d_{bheel}}{X_5-cover} - 1 \leq 0$
$g_{26}(x)$	$\frac{l_{dbkey}}{X_5-cover} - 1 \leq 0$ or $\frac{l_{dhkey}}{X_5-cover} - 1 \leq 0$

of krill with the size of NK is initialized. Each krill is a randomly created 12-dimensional vector. Before estimating the objective functions, we need to evaluate several constraints defined by standard codes for the whole swarm. The number of violations will be added to the objective values as a penalty term. Then, we find the best krill individual with the lowest cost value.

Next, the time-dependent additive for updating the position of the krill swarm is evaluated. To this end, we need to evaluate three motion types according to the rules mentioned in Section 18.2: movement caused by other krill (N_i); foraging activity (F_i); random diffusion (D_i). To evaluate N_i, two main factors, α_i^{target} (the effect of the best krill) and α_i^{local} (the effect of neighbors), are required. Foraging motion (F_i) results from the food attractiveness and effect of the best-found solution. Food attractiveness is directly related to the location of the food resources which is proposed to be the center of food. Now, we evaluate D_i to provide random movements of the krill individuals. This randomness follows an annealing-based rule which is decreased as the generation proceeds. Before updating the position of the swarm, we need to apply two genetic operators, crossover and mutation, to the current population. At the final step, we evaluate the penalized objective function and determine the best-found solution. This procedure will be iterated until satisfying the termination criteria.

18.4.3   Description of experiments

In this section our proposed methodology is tested by means of a set of numerical retaining wall models. To evaluate the performance of the presented algorithm, two case studies utilized from the study were done by Saribas and Erbatur [8]. ACI 318-05 [11] requirements are considered in the design for steel reinforcement. To set up the experiments, each algorithm was run 101 times, and population size and number of iterations are inserted as 50 and 1000, respectively. In the first example, the retaining wall was designed twice for both objective functions of cost and weight; also, the sensitivity of designs was investigated for different surcharge load, backfill, slope, and friction angle. Since the first example did not include base shear key, the second example designed a retaining wall for two cases of base shear key, one with base shear key and one without base shear key by optimizing cost and weight. The structural and geotechnical design parameters are summarized in Table 18.3. Moreover, boundaries limitations for the input parameters are gathered in Table 18.4. In both examples the best, the worst, mean, and standard deviation (SD) values were obtained and will be discussed in results section.

18.4.4   Results obtained

In this chapter the optimum design of retaining wall optimization is tackled for further discussion. Due to the stochastic basis of metaheuristic algorithms

TABLE 18.3
Utilized soil parameters for the case study.

Input parameters	Symbol	Value Example 1	Value Example 2	Unit
Stem height	H	3.0	4.5	m
Reinforcing steel yield strength	f_y	400	400	MPa
Concrete compressive strength	f_c	21	21	MPa
Concrete cover	C_c	7	7	cm
Shrinkage and temperature reinforcement percentage	ρ_{st}	0.002	0.002	-
Surcharge load	q	20	30	kPa
Backfill slope	β	10	0	°
Internal friction angle of retained soil	ϕ	36	36	°
Internal friction angle of base soil	ϕ'	0	34	°
Unit weight of retained soil	γ_s	17.5	17.5	kN/m^3
Unit weight of base soil	γ'_s	18.5	18.5	kN/m^3
Unit weight of concrete	γ_c	23.5	23.5	kN/m^3
Cohesion of base soil	c	125	0	kPa
Depth of soil in front of wall	D	0.5	0.75	m
Cost of steel	C_s	0.4	0.4	\$/kg
Cost of concrete	C_c	40	40	\$/m^3
Factor of safety for overturning stability	$SF_{Odesign}$	1.5	1.5	-
Factor of safety for sliding	$SF_{Sdesign}$	1.5	1.5	-
Factor of safety for bearing capacity	$SF_{Bdesign}$	3	3	-

TABLE 18.4

Design variables (DVs) permitted domain.

DVs	Unit	Example 1		Example 2	
		Lower Bound	Upper Bound	Lower Bound	Upper Bound
$X1$	m	1.3090	2.3333	1.96	5.5
$X2$	m	0.4363	0.7777	0.65	1.16
$X3$	m	0.2000	0.3333	0.25	0.5
$X4$	m	0.2000	0.3333	0.25	0.5
$X5$	m	0.2722	0.3333	0.4	0.5
$X6$	m	-	-	1.96	5.5
$X7$	m	-	-	0.2	0.5
$X8$	m	-	-	0.2	0.5
$R1$	-	1	223	1	223
$R2$	-	1	223	1	223
$R3$	-	1	223	1	223
$R4$	-	-	-	1	223

TABLE 18.5

Design cost values for numerical case studies.

KH optimization algorithm	Cost (\$/m)			
	Best	Worst	Mean	SD
Example 1	73.17	79.93	74.17	1.12
Example 2 (Case I)	164.62	187.24	170.31	4.15
Example 2 (Case II)	162.46	176.62	165.92	2.57
KH optimization algorithm	Weight (kg/m)			
	Best	Worst	Mean	SD
Example 1	2667.23	2738.08	2684.66	11.83
Example 2 (Case I)	5661.94	6007.72	5719.46	50.54
Example 2 (Case II)	5556.63	5704.27	5603.84	30.67

the algorithm is run 101 times and the results for both examples one and two are reported based on best, worst, mean, standard deviation and median in Table 18.5. The total number of function evaluations is 50,000 for all the cases. In the first case, a 3 m-tall retaining wall design without a base shear key was considered for both low-cost and low-weight design. The results have been kept unchanged after 817-th and 728-th generations for low-cost and low-weight designs, respectively. For the second case study, a 4.5 m-tall retaining wall is considered. Convergence history proved that the results remained unchanged after 872-nd and 958-th generations. In the third numerical simulation, the second case is reconsidered by an additional base shear key. As expected, this additive element causes lower cost and weight design. The final best solution of the KH algorithm in this experienced no improvement after the 932nd and

985th iterations for low-cost and low-weight design, respectively. The convergence history of KH algorithm demonstrated that it reached valid designs which satisfy all the geotechnical and structural limitations since the first iteration for both low-cost and low-weight designs over all the case studies.

18.5   Conclusions

In this chapter, the krill herd algorithm with a step-by-step description was presented in detail. Next, several modifications and improvements of this algorithm were demonstrated. Finally, KH algorithm was applied to one real-world engineering problem. The retaining wall design procedure as one of the important and complicated tasks in geotechnical engineering is examined here. In this study, first we clarify the necessary steps in using KH algorithm to automate the design procedure. Then, we explain all the effective parameters and limitations for handling the retaining wall problems which come from standard codes. As can be seen in Section 18.4, KH deals with this problem successfully and reached solutions.

References

1. A.H. Gandomi, A.H. Alavi. "Krill herd: a new bio-inspired optimization algorithm". *Communications in Nonlinear Science and Numerical Simulation*, vol. 17(12), pp. 4831-4845, 2012.

2. G.G Wang, L. Guo, A.H. Gandomi, G.S. Hao, H. Wang. "Chaotic krill herd algorithm". *Information Sciences*, vol. 274, pp. 17-34, 2014.

3. G. Wang, L. Guo, A.H. Gandomi, L. Cao, A.H. Alavi, H. Duan, J. Li. "Lévy-flight krill herd algorithm". *Mathematical Problems in Engineering*, Volume 2013, Article ID 682073, 14 pages, Hindawi 2013.

4. G.G. Wang, A.H. Gandomi, A.H. Alavi, S. Deb. "A multi-stage krill herd algorithm for global numerical optimization". *International Journal on Artificial Intelligence Tools*, vol. 25(2), 1550030, 2016.

5. G.G Wang, A.H. Gandomi, A.H. Alavi. "Stud krill herd algorithm". *Neurocomputing*, vol. 128, pp. 363-370, 2014.

6. J. Li, Y. Tang, C. Hua, X. Guan. "An improved krill herd algorithm: Krill herd with linear decreasing step". *Applied Mathematics and Computation*, vol. 234, pp. 356-367, 2014.

7. G.G. Wang, A.H. Gandomi, A.H. Alavi. "An effective krill herd algorithm with migration operator in biogeography-based optimization". *Applied Mathematical Modelling*, vol. 38(9-10), pp. 2454-2462, 2014.

8. A. Saribas, F. Erbatur. "Optimization and sensitivity of retaining structures". *Journal of Geotechnical Engineering*, vol. 122(8), pp. 649-656, 1996.

9. A.H. Gandomi, A.R. Kashani, D.A. Roke, M. Mousavi. "Optimization of retaining wall design using recent swarm intelligence techniques". *Engineering Structures*, vol. 103, pp. 72-84, 2015.

10. A.H. Gandomi, A.R. Kashani, D.A. Roke, M. Mousavi. "Optimization of retaining wall design using evolutionary algorithms". *Structural and Multidisciplinary Optimization*, vol. 55(3), pp. 809-825, 2017.

11. American Concrete Institute. Building code requirements for structural concrete and commentary (ACI 318-05), Detroit, 2005.

19

Modified Monarch Butterfly Optimization and Real-life Applications

Pushpendra Singh

Department of Electrical Engineering
Govt. Women Engineering College, Ajmer, India

Nand K. Meena

School of Engineering and Applied Science
Aston University, Birmingham, United Kingdom

Jin Yang

School of Engineering and Applied Science
Aston University, Birmingham, United Kingdom

CONTENTS

19.1   Introduction

Nowadays, the complexity of engineering optimization problems has increased due to involvement of multiple objectives and constraints. Knowing this fact, many researchers are contentiously working to find acceptable solutions of such problems. In this regard, various new optimization problems have been introduced in recent years with better problem solving capabilities such as the squirrel search algorithm (SSA) [1], water cycle algorithm (WCA) [2] etc. However, some of the researchers have upgraded the standard version of well-known optimization techniques and their inspirational improvements are demonstrated on benchmark functions, e.g., improved elephant herding optimization [3], novel monarch butterfly optimization with greedy strategy and self adaptive crossover operation (GCMBO) [4] etc.

The monarch butterfly optimization (MBO) technique is recently proposed by Gai-Ge Wang et al. [5], inspired by the monarch butterfly of the North American region which migrate from region 1 to region 2 and vice-versa, in particular months. The moth flutter upgrades its population by means of two reproduction operators, i.e., the migration operator and butterfly adjusting operators. In [5], the MBO was tested on various benchmark functions to demonstrate the promising problem solving abilities of the method. As identified in [4], the standard version of MBO shows some limitations regarding real-life optimization problems such as worst average and standard fitness values.

In order to overcome such limitations, Gai-Ge Wang et al. proposed a new version of MBO, known as GCMBO. The amendment is done by employing two strategies in the standard version of MBO. First is self-adaptive crossover (SAC), an improved crossover operator, integrated with the butterfly adjusting operator in the standard version of the method. This is proposed to increase the diversity of the monarch flutter during the later searching stage. This SAC operator can also utilize the information of the whole flutter. The second advancement is introduced by incorporating a greedy strategy with migration and butterfly adjusting operators. This operator is designed to accept the monarch individual which shows better fitness than its parents. This operator also helps in accelerating the convergence of the method. Rest condition and operators remain the same as the basic version.

This chapter is organised in five sections. Section 19.1 presents the introduction of this chapter and then the brief description of the basic version of MBO is presented in Section 19.2. In Section 19.3, some suggested improvements in MBO are discussed followed by its implementation on a real-life complex optimization problem of distributed generation (DG) allocations in distribution systems. The conclusions of this chapter are presented in Section 19.7.

19.2 Monarch butterfly optimization

In the standard version of MBO, some sets of rules are defined in order to find the global or near global solution of an optimization problem. The migration of a monarch butterfly flutter is described below by help of some basic rules [5].

- The whole population of the monarch flutter are found in region 1 and 2.

- Every offspring, produced by the butterfly is in region 1 or region 2.

- The size of the monarch flutter will remain unchanged during the optimization process.

- A certain number of fittest butterfly flutter individuals can't be updated by butterfly adjustment operators or migration operators.

19.2.1 Migration operator

The population of the monarch flutter can be bifurcated as, sub-population 1 (P_{n1}) and sub-population 2 (P_{n2}) on the basis of their stay in region 1 and region 2. The (P_{n1}) is equal to the ceiling value of $P_r \times P_n$. Whereas, P_{n2} is expressed as P_n - P_{n1}, where, P_r is the ratio of monarch flutter halting in region 1 and P_n is the absolute number of monarch butterfly population. The migration process is mathematically expressed as

$$z_{j,l}^{i+1} = z_{r_1,l}^i \tag{19.1}$$

where, $z_{j,l}^{i+1}$ is the lth element of z_j at the generation level $i + 1$ and j is the position of the monarch butterfly. Likewise, $z_{r_1,l}^i$ is the lth element of z_{r_1} which is a newly generated position of monarch butterfly r_1 and i is the current stage of generation. Here r_1 is a butterfly randomly selected from the monarch butterflies of sub-population 1. When, $r \leq P_r$, then the component of l in the newly born monarch butterfly is produced by (19.1). The value of r is determined as

$$r = rnd * \psi \tag{19.2}$$

where, rnd is a random number drawn from a uniform distribution and ψ is a constant considered to be 1.2 for the 12 month period. On other hand, if $r > P_r$ then newborn monarch butterflies will be produced as

$$z_{j,l}^{i+1} = z_{r_2,l}^i \tag{19.3}$$

The r_2 is a position of monarch butterfly offspring, randomly obtained from the monarch butterflies of sub-population 2. It is to be noted that by adjusting the value of P_r, the direction of the migration operator can be balanced. For example, if the value of P_r is larger or bigger then more members of monarch flutter P_{n1} will participate. On other side, if the value of P_r is smaller then more chances will be given to the members of the monarch flutter from sub-population 2, P_{n2}. Therefore, P_{n2}, the sub-population of region 2 is an important factor in producing new monarch butterflies. In this chapter, the value of P_r is considered to be $5/12$.

19.2.2   Butterfly adjusting operator

For all the elements in the monarch flutter in k, $rand \leq P_r$ is updated as

$$z_{k,l}^{i+1} = z_{best,l}^i \qquad (19.4)$$

where, $z_{k,l}^{i+1}$ represents the lth element of z_k in generation $i + 1$. Similarly, $z_{best,l}^i$ is the lth element of the fittest monarch butterfly z_{best}. However, if $P_r \leq rand$ then updated as follows.

$$z_{k,l}^{i+1} = z_{r3,l}^i \qquad (19.5)$$

The $z_{r3,l}^i$ represents lth element of z_{r3}. For $r_3 \in \{1, 2, ..., P_{n2}\}$. Under this condition, if $rand > BAR$ then it is further updated as

$$z_{k,l}^{i+1} = z_{k,l}^{i+1} + \alpha * (dz_l - 0.5) \qquad (19.6)$$

where, dz is the Levy flight calculated walk step size of monarch butterfly k, determined as

$$dz = Levy(z_k^t) \qquad (19.7)$$

In (19.6), α is the weighting factor calculated as

$$\alpha = S_{max}/i^2 \qquad (19.8)$$

where, S_{max} is the maximum walk step taken by a butterfly in a single move step.

19.3   Modified monarch butterfly optimization method

MBO is a robust nature-inspired optimization technique tested, for promising results, on various benchmark systems. However, it may fail to obtain effective results of standard and average fitness value for some benchmark function [4]. To overcome this deficiency, some strategic improvements are incorporated in basic MBO. The improved MBO called as GCMBO is discussed in the following sections.

19.3.1 Modified migration operator

In the standard version of MBO, all new offspring are accepted and transferred to the next generation while in the upgraded version of MBO, a greedy strategy is adopted. In this strategy, the new butterfly individuals with better fitness value than their respective parents are only moved to the next generation; otherwise the new butterflies are rejected and their parents move to the next generation. In this strategy, the position of the jth butterfly updated in (19.1) is again modified as

$$z_{j,new}^{i+1} = \begin{cases} z_{j,l}^{i+1}; & \text{if } f(z_{j,l}^{i+1}) < f(z_{r_1/r_2,l}^i) \\ z_{r_1/r_2,l}^i; & \text{else} \end{cases} \tag{19.9}$$

where, $z_{j,new}^{i+1}$ is a newly produced offspring of butterfly individual j moving to next generation, $i+1$.

19.3.2 Modified butterfly adjustment operator

In this modification, the monarch butterfly individual of sub-population 2, P_{n2} is updated as per (19.4)–(19.8). In this modification, an adaptive crossover and greedy strategy operators are discussed. The generated monarch butterfly individuals updated in (19.4)–(19.8) are referred to as z_{k1}^{i+1}. To utilize the complete benefits of monarch flutter information, a crossover operator is implemented in the butterfly adjustment operators, expressed as

$$z_{k2}^{i+1} = z_{k1}^{i+1} \times (1 - C_r) + z_k^i \times C_r \tag{19.10}$$

where, z_{k2}^{i+1} is newly produced butterfly offspring by deploying z_{k1}^{i+1}, z_k^i and crossover rate C_r is calculated as

$$C_r = 0.8 + 0.2 \times \frac{f(z_k^i) - f(z_{best})}{f(z_{worst}) - f(z_{best})} \tag{19.11}$$

where $f(z_k^i)$ represents the fitness value of butterfly k from sub-population 2 P_{n2}. $f(z_{worst})$ and $f(z_{best})$ are the fitnesses of z_{worst} and z_{best} monarch butterfly individuals. It is observed that the value obtained for C_r, from (19.11) is in the range of [0.2 0.8]. The new butterfly individual is determined by deploying the greedy strategy as

$$z_{k,new}^{i+1} = \begin{cases} z_{k1}^{i+1}; & \text{if } f(z_{k1}^{i+1}) < f(z_{k2}^i) \\ z_{k2}^{i+1}; & \text{if } f(z_{k2}^{i+1}) < f(z_{k1}^i) \end{cases} \tag{19.12}$$

where, $f(z_{k1}^{i+1})$ and $f(z_{k2}^{i+1})$ are fitnesses of the butterflies z_{k1}^{i+1} and z_{k2}^{i+1} respectively.

19.4   Algorithm of modified MBO

In this section, the algorithm of GCMBO, discussed in previous sections, is point-wise described here. The flow chart of GCMBO is presented in Fig. 19.1.

Modified monarch butterfly optimization

Step-1: Initialize random but feasible population of P_n monarch butterflies and set the values of required parameters, constants and maximum number of iterations, i.e., $S_{Max.}$, BAR, t, etc.

Step-2: calculate the fitness values of each monarch butterfly generated in previous step

Step-3: start the iteration $t = 1$

Step-4: sort all monarch butterfly individuals with respect to their fitness values and then divide these monarch butterflies into two populations, P_{n1} and P_{n2}, for the region 1 and region 2 respectively

Step-5: the sub-population P_{n1} of region 1 generates the new sub-population 1 by using the modified migration operator expressed in (19.9). Similarly, the sub-population P_{n2} of region 2 is updated by the modified adjusting operator succeeded by adaptive crossover and greedy strategy operators in (19.10)–(19.12).

Step-6: a correction algorithm may be applied to correct the infeasible individuals, if any

Step-7: repeat steps 4 to 6 until convergence criteria or maximum number of iterations is reached

Step-8: Print the best solution.

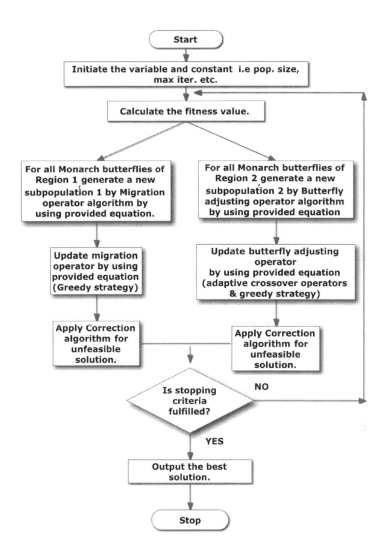

FIGURE 19.1
Flow chart of monarch butterfly optimization with greedy strategy and self adaptive crossover operation.

19.5   Matlab source-code of GCMBO

For basic understanding of GCMBO, the Matlab source-code is presented in Listing 19.1.

```matlab
LB=[];UB=[]; %set upper & lower limits of each variable
iter=0;Max_iter=10;     %initial & maximum number of iterations
Pn = 8 ;      % population size
nFlutr =length(UB);  % number of variables
nBF1=ceil(Pn*5/12);  % butterfly of region1 & 2
nBF2=Pn - nBF1;
Lnd1=zeros(nBF1,nFlutr); %flutter at region 1 & 2
Lnd2=zeros(nBF2,nFlutr);
%==========MSO Const======
Stepmax=1.0;  partt=5/12;  period=1.5;
Keep=2;  partition=5/12;  BAR=partition;
for  j=1:Pn % initialization %rand variables
for  k= 1:nFlutr
x(j,k)=LB(k)+rand*(UB(k)-LB(k));
end
x(j,:)=Correction_Algo(x(j,:),LB,UB); % Apply Correction
[fitness(j)]=OF(x(j,:)); %Fitness_Function(x(j,:));
end
best_fit=min(fitness); worst_fit=max(fitness); nn_best=find(min(
    fitness)==fitness);
nn_worst=find(max(fitness)==fitness); x_best=x(nn_best(1),:);
xx=x_best; x_worst=x(nn_worst(1),:);
while  iter <=Max_iter
iter = iter + 1;COMB_POP=[x, fitness '];
x_srt=sortrows(COMB_POP(:,nFlutr));
x_keep = x_srt(1:Keep,:); % Elitism Strategy
x_srt((Pn-1):Pn,:)=x_keep; x=COMB_POP(:,1:nFlutr);
fitness=(COMB_POP(:,nFlutr+1)) ';
for  pii=1:Pn %%Divide the whole population into two parts
if  pii<= nBF1
pop1(pii,:)=x(pii,:);
else
pop2(pii,:)=x(pii,:);
end
end
for  j=1:nBF1 %migration operator
for  k=1:nFlutr
rr1=rand*1.2;
if  rr1<=partition
rr2=round(nBF1*rand + 0.5);
Lnd1(j,k)=pop1(rr2,k);
else
rr3=round(nBF2*rand + 0.5);
Lnd1(j,k)=pop2(rr3,k);
end
end
x_x(2,:)=Lnd1(j,:); %greedy strategy
for  ryn=1:2
x_x(ryn,:)=Correction_Algo(x_x(ryn,:), LB, UB);
arn(ryn,:)=OF(x_x(ryn,:)); %Fitness_Function(x_new(j,:));
end
x_x(2,:);arn(2,:);
if  arn(1)<arn(2)
Lnd1(j,:)=x_x(1,:);
else
Lnd1(j,:)=x_x(2,:);
end
end
for  j=nBF2 %butterfly adjustment operator%%%%%
```

```
59  scale=Stepmax/(iter)^2; stepsize=ceil(exprnd(2*Max_iter,1,1)); deltaX=
        LevyFlight(stepsize,nFlutr);
60  for k=1:nFlutr
61  if (rand>=partition)
62  Lnd2(j,k)=x_best(1,1);
63  else
64  rr4=round(nBF2*rand + 0.5);        Lnd2(j,ci)=pop2(rr4,k);
65  if (rand> BAR)
66  Lnd2(j,k)=Lnd2(j,k)+ scale*(deltaX(k)-0.5);
67  end
68  end
69  end
70  %%%%%%%% Self adaptive crossover operator %%%%%%%%%
71  Lnd2(j,:)=Correction_Algo(Lnd2(j,:), LB, UB);
72  x_L(1,:)=OF(Lnd2(j,:));        x_B(1,:) = OF(x_best(1,:));
73  x_W(1,:)=OF(x_worst(1,:));  %worst & best
74  C_rate=0.8+0.2.*((x_L(1,:) - x_B(1,:))/(x_W(1,:) - x_B(1,:)));
75  Cr= rand(nFlutr,1) < C_rate;%%%Cross Rate%%%
76  new_Lnd2=Lnd2(j,k).*(1-Cr') + (x(1,:).*Cr');
77  x_x_L1(1,:)=Lnd2(j,:);x_x_L2(1,:)=new_Lnd2(1,:);
78  x_x_L1(1,:)=Correction_Algo(x_x_L1(1,:), LB, UB);
79  x_x_L2(1,:)=Correction_Algo(x_x_L2(1,:), LB, UB);
80  fx_x_L1(1,:)=OF(x_x_L1(1,:));  fx_x_L2(1,:) = OF(x_x_L2(1,:));
81  if fx_x_L1(1,:) <fx_x_L2(1,:)    %%%%%%%greedy strategy%%%%%%%%%
82  Lnd2(j,:)=x_x_L1(1,:);
83  else
84  Lnd2(j,:)=x_x_L2(1,:);
85  end
86  end
87  x_new=[Lnd1;Lnd2];        % fitness calculation
88  for j=1:Pn
89  x_new(j,:)=Correction_Algo(x_new(j,:), LB, UB);
90  [fitness(j)]=OF(x_new(j,:));
91  end
92  best_fit1=min(fitness); nn_best=find(min(fitness)==fitness);
93  x_best1=x(nn_best(1),:);
94  if best_fit1<best_fit
95  best_fit=best_fit1;        x_best=x_best1;
96  end
97  best_fit; best=x_best; x=x_new;
98  m(iter)=mean(fitness);b(iter)=best_fit;%storing mean& iter
99  end
100 best_fit; %Best Fitness
101 best=x_best; %Best sting obtain
```

Listing 19.1
Source-code of GCMBO in Matlab.

19.6   Application of GCMBO for optimal allocation of distributed generations

19.6.1   Problem statement

The optimal allocation of distributed generations (DGs) in a distribution system is found to be a complex mixed-integer, non-linear, and non-convex optimization problem [3], which therefore needs an effective optimization method. In this chapter, a benchmark 33-bus distribution system is considered [6]. The objective is to determine the optimal sites (nodes) and sizes (DG capacities) of 3 DGs for minimum real power loss, P_{Loss} of 33-bus distribution system. The expression of real power loss is expressed as

$$P_{Loss} = \sum_{i=1}^{N} \sum_{j=1}^{N} \alpha_{ij}(P_iP_j + Q_iQ_j) + \beta_{ij}(Q_iP_j - P_iQ_j) \qquad (19.13)$$

$$\alpha_{ij} = \frac{r_{ij}\cos(\delta_i - \delta_j)}{V_iV_j} \qquad (19.14)$$

$$\beta_{ij} = \frac{r_{ij}\sin(\delta_i - \delta_j)}{V_iV_j} \qquad (19.15)$$

s. t.

1. *Nodal power balance constraints*

$$P_i = P_{G_i} - P_{D_i} = V_i \sum_{j=1}^{N} V_j Y_{ij}\cos(\phi_{ij} + \delta_j - \delta_i) \quad \forall\, i \quad (19.16)$$

$$Q_i = Q_{G_i} - Q_{D_i} = -V_i \sum_{j=1}^{N} V_j Y_{ij}\sin(\phi_{ij} + \delta_j - \delta_i) \;\; \forall\, i \quad (19.17)$$

2. *Voltage limits constraints*

$$V^{Min} \leq V_i \leq V^{Max} \qquad \forall\, i \qquad (19.18)$$

3. *Feeder capacity limits constraint*

$$0 \leq I_{ij} \leq I_{ij}^{Max} \qquad \forall\, i,j \qquad (19.19)$$

4. *Maximum penetration limit of each bus*

$$0 \leq P_{DG_i} \leq P_{DG}^{Max} \qquad \forall\, i \qquad (19.20)$$

5. *Maximum DG penetration limits of the system*

$$0 \leq \sum_{i=1}^{N} \sigma_i P_{DG_i} \leq P^{Peak} \qquad \forall\, i \qquad (19.21)$$

where, P_i, Q_i, P_{G_i}, Q_{G_i}, P_{D_i}, Q_{D_i}, σ_i, V_i, and δ_i are representing the real power injection, reactive power injection, real power generation, reactive power generation, real power load, reactive power load, binary decision variable of DG installation, magnitude of node voltage and voltage angles of node i respectively. Furthermore, r_{ij}, ϕ_{ij}, Y_{ij}, I_{ij}, I_{ij}^{Max}, N, P_{DG}^{Max}, P^{Peak} denote the branch resistance, impedance angle, Y-Bus element, feeder current, maximum allowed current limit of the branch connected between node i and j, and number of system nodes, allowed DG capacity limit at a node, and in the system respectively.

19.6.2 Optimization framework for optimal DG allocation

In this section, we discuss the implementation of GCMBO on optimal DG allocation problems. As stated in the problem statement, we need to determine the sites and sizes of three DGs in 33-bus distribution system. Therefore, the number of variables used in a butterfly individual will be 6, i.e., 3 sites + 3 sizes. The lower and upper limits of the first three variables will be [1 33]; whereas, it will be [0 2000] for the remaining 3 variables, if $P_{DG}^{Max} = 2000$kW and $P^{Peak} = 6000$kW. Fig. 19.2 presents the contour plot of power loss of a 33-bus distribution system with respect to nodes and DG sizes up to 2000kW.

It may be observed that DG placement is a non-convex and non-linear optimization problem which has multiple solutions. The solution variables are mixed integers as nodes are integer whereas, DG sizes are non-integers. Now, the DG allocation problem discussed in the previous section is solved by using GCMBO. The Matlab source-code of objective function $OF(.)$, i.e., power loss minimization, is presented in Listing 19.2.

The optimal nodes and sizes obtained are

DG-1: 811 kW at node number 13

DG-2: 1487 kW at node number 24

DG-3: 1001 kW at node number 30

The minimum values of the power loss obtained is, $P_{Loss} = 75.7282$ kW.

The convergence characteristics of MBO and GCMBO are compared and presented in Fig. 19.3. The figure shows that the suggested improvements significantly improve the solution searching abilities of the method. The GCMBO continuously searches solutions in each iteration. The other performance parameters of the MBO and GCMBO algorithms, such as best fitness, worst fitness, mean fitness and standard deviation, for 50 independent trials, are also compared and then presented in Table 19.1. The table shows that GCMBO outperformed in the global optimal solution search.

```
1 function [OBJ]=OF(X)
2 X=floor(X);
3 % System data
4 Linedata=[]; Busdata=[];
5 d=1; %slack bus
6 Niter=4; % No. of iterations
7 Base_Kv=12.66; % Base KV
8 Base_MVA=100; % 100MVA Base;
9 Z_Base=(((Base_Kv)^2)/(Base_MVA)); fb=Linedata(:,1);
10 tb=Linedata(:,2); N=max(max(fb),max(tb));
11 Nb=length(fb); R=Linedata(:,3); X=Linedata(:,4);
12 Z_Line=complex(R,X)/Z_Base; PL=Busdata(:,7);
13 QL=Busdata(:,8); SL=((complex(PL,QL))/(1000*Base_MVA));
14 PG=Busdata(:,5); QG=Busdata(:,6);
15 %Find terminal and internal buses
16 Terminalbuses=zeros(N,1); Intermediatebuses=zeros(N,1);
17 for i=1:N-1
18 Terminalbuses(i,1)=0; Intermediatebuses(i,1)=0;
19 end
20 do=0;in=0;
21 for i=1:Nb
22 for j=1:Nb
23 if (i~=j)&&(tb(i)==fb(j))
24 do=do+1;
```

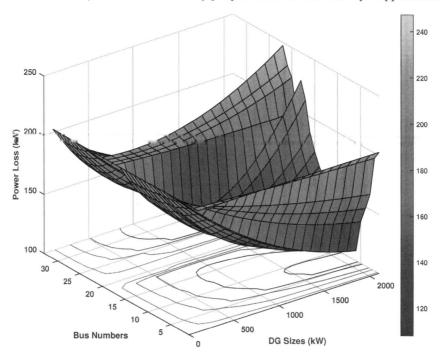

FIGURE 19.2
Contour plot of power loss with respect to nodes and DG sizes.

```
25 end
26 end
27 if do==0
28 in=in+1;
29 Terminalbuses(in,1)=tb(i);
30 else
31 in=in+1;
32 Intermediatebuses(in,1)=tb(i);
33 end
34 do=0;
35 end
36 T1=Terminalbuses; In=Intermediatebuses;
37 T2=find(T1); In2=flipud(find(In));
38 PG=zeros(N,1); QG=zeros(N,1);
39 %DG placement from X
40 for opj=1:length(X)/2
41 PG(X(opj))=PG(X(opj))+X(opj+length(X)/2);
42 end
43 SG=((complex(PG,QG))/(1000*Base_MVA)); Sinj=SG − SL;
44 V=Busdata(:,3); Ang=Busdata(:,4);
45 %iteration starts..................
46 for iter=1:Niter
47 % Calculation of current injection Matrix
48 I_injection1=zeros(N,N);
49 for i=1:N
50 I_injection1(i,i)=conj((Sinj(i))/(V(i)));%Diagonal elements of
      injection matrix
51 end
52 %Current Calculations in Terminal Buses.
```

```
53 I_injection2=zeros(N,N);
54 for i=1:length(T2)
55 I_injection2(fb(T2(i)),tb(T2(i)))=-I_injection1(tb(T2(i)),tb(T2(i)));%
       Current Calculations from Sub Terminal to Terminal Buses.
56 end
57 % Current Calculations between Intermediate & Terminal Buses.
58 for j=1:length(In2)
59 I_injection2(fb(In2(j)),tb(In2(j)))=-I_injection1(tb(In2(j)),tb(In2(j)
       ))+sum(I_injection2(tb(In2(j)),:));
60 I_injection2(tb(In2(j)),fb(In2(j)))=-I_injection2(fb(In2(j)),tb(In2(j)
       ));
61 end
62 for i=1:length(T2)
63 I_injection2(tb(T2(i)),fb(T2(i)))=-I_injection2(fb(T2(i)),tb(T2(i)));%
       Current Calculations from Sub Terminal to Terminal Buses.
64 end
65 %voltage update using ohms law.
66 V(d)=1.0;
67 for k=1:Nb
68 V(tb(k))=(V(fb(k)))-Z_Line(k)*(I_injection2(fb(k),tb(k)));
69 end
70 V_mag=abs(V);
71 end
72 %Branch Power Flows in the system.
73 S_Flow=zeros(N,N);
74 %Power Loss calculation in the system..
75 S_Loss = zeros(Nb,1);
76 for m = 1:Nb
77 S_Flow(fb(m),tb(m)) = (V(fb(m))*conj(I_injection2(fb(m),tb(m))))*1000;
78 S_Flow(tb(m),fb(m)) = -(V(tb(m))*conj(I_injection2(fb(m),tb(m))))
       *1000;
79 S_Loss(m) = S_Flow(fb(m),tb(m)) + S_Flow(tb(m),fb(m));
80 end
81 S_Loss=S_Loss*Base_MVA;
82 P_Loss = real(S_Loss);
83 OBJ=sum(P_Loss);
```

Listing 19.2

Source-code of objective function OF() for power loss minimization.

TABLE 19.1

Performance comparison of MBO and GCMBO for 50 independent trials.

Method	Best fitness	Worst fitness	Mean fitness	Standard deviation	CPU time(s)
MBO	0.0753	0.0880	0.0827	0.0034	7.04
GCMBO	0.0734	0.0793	0.0741	0.0023	7.06

19.7 Conclusion

In this chapter, a new nature inspired meta-heuristic optimization technique is developed. It is inspired by the migration behaviour of monarch butterflies found in the continent of North America. It is found that MBO provides

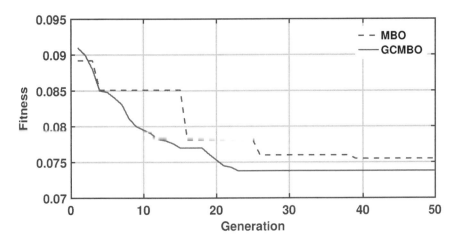

FIGURE 19.3
Best Fitness comparison of MBO & GCMBO.

promising results for various benchmark functions; however, it generates poor standard deviation and worst mean fitness for some bench functions. In order to improve the performance of the standard version of MBO, some modifications are suggested in migration operators, known as GCMBO. In GCMBO, a greedy and self adaptive crossover operator is injected into the migration and butterfly operators. A real-life DG allocation problem of power loss minimization is then solved and it is demonstrated that the greedy operators significantly accelerated the convergence speed and efficiency.

Acknowledgement

This work was supported by the Engineering and Physical Sciences Research Council (EPSRC) of United Kingdom (Reference Nos.: EP/R001456/1 and EP/S001778/1).

References

1. M. Jain, V. Singh and A. Rani. "A novel nature-inspired algorithm for optimization: Squirrel search algorithm". *Swarm and Evolutionary Computation*, vol. 44, 2019, pp.148-175.

2. Eskandar, H., Sadollah, A., Bahreininejad, A. and Hamdi, M. Water cycle algorithm–A novel metaheuristic optimization method for solving constrained engineering optimization problems. *Computers & Structures*, 110, pp.151-166, 2012.

3. N. K. Meena, S. Parashar, A. Swarnkar, N. Gupta and K. R. Niazi. "Improved elephant herding optimization for multiobjective DER accommodation in distribution systems". *IEEE Transactions on Industrial Informatics*, vol. 14(3), 2018, pp.1029-1039.

4. Wang, G.G., Deb, S., Zhao, X. and Cui, Z. A new monarch butterfly optimization with an improved crossover operator. *Operational Research*, 18(3), pp.731-755, 2018.

5. G G. Wang, S. Deb and Z. Cui. "Monarch butterfly optimization". *Neural Comput Appl.*, 2015, pp. 1433-3058. doi:10.1007/s00521-015-1923-y.

6. M. E. Baran and F. F. Wu. "Network reconfiguration in distribution systems for loss reduction and load balancing". *IEEE Transactions on Power Delivery*, vol. 4(2), 1989, pp. 1401-1407.

20

Particle Swarm Optimization – Modifications and Application

Adam Slowik

Department of Electronics and Computer Science
Koszalin University of Technology, Koszalin, Poland

CONTENTS

20.1   Introduction

The particle swarm optimization (PSO) [1] algorithm and ant colony optimization (ACO) [2] algorithm are the two first algorithms which began the new era of research called swarm intelligence [7]. The PSO algorithm is a global optimization technique which is inspired by the cooperation behavior (social behavior) of flocks of birds. This algorithm has many common features with evolutionary computing techniques [8]. The PSO algorithm also operates on a randomly created population (called a swarm) of potential solutions (called particles). The optimal solution searches by creating subsequent generations of solutions. However, the genetic operators such as crossover and mutation do not exist in this algorithm. In the PSO algorithm the particles move towards the actual optimal solution/solutions in the search space. The main idea of optimization using the PSO algorithm is based on changes of velocity (acceleration) for each particle moving toward the *Pbest* position and *Lbest* position for the so-called local version of the PSO algorithm or toward the *Pbest* position and *Gbest* position for the so-called global version of the PSO algorithm. The acceleration is randomly modified independently for each direction. This chapter is organized as follows. In Section 20.2 a short introduction to the PSO algorithm is given. The global version of the PSO algorithm (GPSO) and local version of the PSO algorithm (LPSO) are presented and discussed in detail. In Section 20.3, some selected modifications of the PSO algorithm such as velocity clamping, inertia weight, constriction coefficient and acceleration coefficient are described. In Section 20.4, the application of the PSO algorithm to the design of IIR digital filters with non-standard amplitude characteristics is demonstrated.

20.2   Original PSO algorithm in brief

The original PSO algorithm in global version (GPSO) and local version (LPSO) can be presented using pseudo-code in Algorithm 23. The code which is used only for the local version of the algorithm is underlined with a dotted line, and the code which is only for the global version is underlined with a solid line.

20.2.1   Description of the original PSO algorithm

In Algorithm 23 the pseudo-code of the PSO algorithm was presented. Now in this section we will discuss this algorithm in detail. Before we start, we should determine the $D-th$ dimensional objective function $OF(.)$ which is carried out in step 1. In step 2, we determine the range of the variability $[P_{i,j}^{min}, P_{i,j}^{max}]$ for each $j-th$ decision variable ($j \in [1, D]$). In step 3, we determine such PSO algorithm parameters as number of particles in the swarm (P), values of learning factors (c_1 and c_2), and for the LPSO algorithm only we must additionally determine the number of returned results ($Re \in [1, N]$), the type of the neighborhood topology and the number of the nearest neighbors (Ne). In step 4, we randomly create the swarm P which consists of N particles. Each particle P_i is a $D-th$ dimensional vector and is presented as follows:

$$P_i = \{P_{i,1}, P_{i,2}, P_{i,3}, ..., P_{i,D-1}, P_{i,D}\}$$

where: $P_{i,1}, P_{i,2}$, and so on, are the values of the particular decision variables (from the previously determined range $[P_j^{min}, P_j^{max}]$).

Algorithm 23 Pseudo-code of the original PSO.

1: determine the $D - th$ dimensional objective function $OF(.)$
2: determine the range of variability for each $j - th$ dimension $\left[P_{i,j}^{min}, P_{i,j}^{max} \right]$
3: determine the PSO algorithm parameter values such as N – number of particles in the swarm, c_1, c_2 – learning factor, number of returned results $Re \in [1, N]$, the topology and number of nearest neighbors Ne
4: randomly create swarm P which consists of N particles (each particle is a D-dimensional vector)
5: create the D-dimensional *Gbest* vector
6: **for** each i-th particle P_i from swarm P **do**
7:     create a *Pbest$_i$* particle equal to i-th particle P_i
8:     create a *Lbest$_i$* particle for each particle P_i
9:     create a velocity V_i for each particle P_i
10:     evaluate the particle P_i using $OF(.)$ function
11: **end for**
12: assign the best particle P_i to the *Gbest*; for each particle *Lbest$_i$* assign the best particle among Ne nearest neighbors of particle P_i
13: **while** termination condition not met **do**
14:     **for** each i particle in the swarm P **do**
15:         **for** each j dimension **do**
16:             update the velocity $V_{i,j}$ using formula
17:             $V_{i,j} = V_{i,j} + c_1 \cdot r_1 \cdot (Pbest_{i,j} - P_{i,j}) + c_2 \cdot r_2 \cdot (Gbest_j - P_{i,j})$
18:             $V_{i,j} = V_{i,j} + c_1 \cdot r_1 \cdot (Pbest_{i,j} - P_{i,j}) + c_2 \cdot r_2 \cdot (Lbest_{i,j} - P_{i,j})$
19:             update the particle $P_{i,j}$ using formula $P_{i,j} = P_{i,j} + V_{i,j}$
20:             check if newly created value $P_{i,j}$ is within the range $\left[P_{i,j}^{min}, P_{i,j}^{max} \right]$ if not correct it
21:         **end for**
22:         evaluate the particle P_i using $OF(.)$ function
23:         **if** $OF(P_i)$ is better than $OF(Pbest_i)$ **then**
24:             assign the particle P_i to the particle *Pbest$_i$*
25:         **end if**
26:         select the best particle among Ne nearest neighbor of particle P_i and assign it to particle T
27:         **if** $OF(T)$ is better than $OF(Lbest_i)$ **then**
28:             assign the particle T to the particle *Lbest$_i$*
29:         **end if**
30:     **end for**
31:     select the best particle from swarm P and assign it to particle T
32:     **if** $OF(T)$ is better than $OF(Gbest)$ **then**
33:         assign the particle T to the particle *Gbest*
34:     **end if**
35: **end while**
36: return the *Gbest* as a result; select Re the best particles among *Lbest* particles and return these particles as a result

---

In step 5, we create the $D - th$ dimensional particle *Gbest*. This step is only executed in the GPSO algorithm. In the particle *Gbest* the best global solution will be stored. Next (step 6), we create a $D - th$ dimensional particle *Pbest$_i$* for each particle P_i. At the start the values in the particle *Pbest$_i$* are the same as they are in particle P_i (see step 7). In the next generations of the PSO algorithm, in the particle *Pbest$_i$*, the best particle which was found until now for particle P_i is stored. In step 8, we create a $D - dimensional$ particle *Lbest$_i$* for each particle P_i. Step 8 is only executed in the LPSO algorithm. In the *Lbest$_i$* particle, the best particle which was found for P_i particle until now among its Ne neighbors is stored. At the start the values stored in *Lbest$_i$* particle are equal to zero.

In step 9, we create a $D-th$ dimensional velocity vector V_i for each particle P_i. At the start, each $j - th$ value in vector V_i is equal to zero. In step 10, we evaluate the particle P_i using the previously defined objective function $OF(.)$. If the main goal of the algorithm is a minimization of the objective function then the best particle P_i is the particle with the lowest value of objective function $OF(.)$. If the main goal of the algorithm is a maximization of the objective function then the best particle P_i is the particle with the highest value of the objective function $OF(.)$. In step 12, we select and assign the best (from the whole swarm) particle P_i to the particle *Gbest* (only in the GPSO algorithm) or we select and assign to particle *Lbest$_i$* the best particle among Ne nearest neighbors of particle P_i (only in the LPSO algorithm). In step 13, the main loop of the algorithm is started. This loop is executed while the termination condition is not met. In the practical applications of the PSO algorithm we can mention three main termination criteria: maximal number of generations, maximal time of PSO algorithm operations, and PSO algorithm convergence. In step 16, we update the values from velocity vector V_i for each $j - th$ dimension. In the GPSO algorithm we use the equation from step 17, and in the LSPO algorithm we apply the equation from step 18. In both steps (17 and 18), the r_1 and r_2 are the randomly chosen values from the range $[0, 1)$. The r_1 and r_2 values are different for each $V_{i,j}$ value. In step 19, we update all the $j-th$ decision variable values in each $i-th$ particle. In step 20, we check if newly created value $P_{i,j}$ is within the range $[P_{i,j}^{min}, P_{i,j}^{max}]$. If the value of the $j - th$ decision variable for $i-th$ particle is higher than the $P_{i,j}^{max}$ ($P_{i,j} > P_{i,j}^{max}$) then the value $P_{i,j}^{max}$ is assigned to the $j - th$ decision variable in $i - th$ particle ($P_{i,j} = P_{i,j}^{max}$). If the value of the $j - th$ decision variable for $i - th$ particle is lower than the $P_{i,j}^{min}$ ($P_{i,j} < P_{i,j}^{min}$) then the value $P_{i,j}^{min}$ is assigned to the $j - th$ decision variable in $i - th$ particle ($P_{i,j} = P_{i,j}^{min}$). In step 22, we evaluate the particle P_i using objective function $OF(.)$. In step 23, we check whether the value of objective function $OF(.)$ for particle P_i is better than the value of objective function $OF(.)$ for particle *Pbest$_i$*. If yes, then we assign the particle P_i to its corresponding particle *Pbest$_i$* (step 24). In step 26, we select the best particle among the Ne nearest neighbor of particle P_i and then we assign this particle to temporary particle T. In step 27, we check whether the objective function $OF(.)$ for particle T is better than the objective function $OF(.)$ for particle $Lbest_i$. If yes, we assign the particle T to the particle $Lbest_i$. The steps 26-29 are only executed for the LPSO algorithm. In step 31, we select the best particle from the whole swarm P and next, we assign this particle to temporary particle T. In step 32, we check whether the objective function $OF(.)$ for particle T is better than the objective function $OF(.)$ for particle $Gbest$. If yes, we assign the particle T to the particle $Gbest$. The steps 31-34 are only executed for the GPSO algorithm. In step 36, we return the particle $Gbest$ as a result (only in the GPSO algorithm) or we select Re, the best particles among all $Lbest_i$ particles ($i \in [1, N]$) and return these particles as a result.

20.3 Modifications of the PSO algorithm

20.3.1 Velocity clamping

In the original PSO algorithm, the velocity value quickly explodes to the large values. One of the solutions for this problem is the introduction of the limited step sizes as shown in equation 20.1

$$V_{i,j} = \begin{cases} V_{i,j} & \text{if } |V_{i,j}| < V_j^{max} \\ V_j^{max} & \text{if } |V_{i,j}| \geq V_j^{max} \end{cases} \tag{20.1}$$

where: the V_j^{max} is a maximal value of velocity for $j - th$ decision variable. The V_j^{max} value can be equal by the all PSO generations or can be computed using formula 20.2 [4] or formula 20.5 [5].

$$V_j^{max} = \left(1 - \left(\frac{t}{n_t}\right)^h\right) \cdot V_j^{max} \tag{20.2}$$

where: t is a current number of PSO algorithm generation, n_t is an assumed maximal number of generations, h is a positive constant which can be chosen by trial-and-error.

$$V_j^{max} = \delta \cdot \left(P_j^{max} - P_j^{min}\right) \tag{20.3}$$

where: P_j^{max} and P_j^{min} are respectively the maximum and minimum values of the search domain for the $j - th$ decision variable, $\delta \in (0, 1]$ is a constant which can be chosen by trial-and-error.

20.3.2 Inertia weight

Inertia weight was developed to control the exploration and exploitation properties of the PSO algorithm. When we use the inertia weight factor the formula for velocity update is as follows for the GPSO algorithm:

$$V_{i,j} = \omega \cdot V_{i,j} + c_1 \cdot r_1 \cdot (Pbest_{i,j} - P_{i,j}) + c_2 \cdot r_2 \cdot (Gbest_j - P_{i,j}) \tag{20.4}$$

and for the LPSO algorithm:

$$V_{i,j} = w \cdot V_{i,j} + c_1 \cdot r_1 \cdot (Pbest_{i,j} - P_{i,j}) + c_2 \cdot r_2 \cdot (Lbest_{i,j} - P_{i,j}) \quad (20.5)$$

where: w is a factor called an inertia weight which significantly affects the convergence and exploration-exploitation trade-off in PSO. The typical value for the inertia weight factor is from the range $[0.4, 0.9]$ [3]. When $w \geq 1$ then velocities increase over time (swarm diverges) and particles fail to change direction toward more promising regions. When the $0 < w < 1$ then particles decelerate and more promising regions can be found. Of course the convergence of the PSO algorithm is also dependent on the values of c_1 and c_2. In the exploration and exploitation trade-off, we can say that the large values of w favor exploration, and the small values of w promote exploitation. It is hard to determine which value of w is the best value because it is highly dependent on the problem which is being solved. Sometimes, the decreasing inertia weight is used in PSO using the following formula.

$$w = \frac{(t^{max} - t)^n}{(t^{max})^n} \cdot (w_{max} - w_{min}) + w_{min} \quad (20.6)$$

where: $w \in [w_{min}, w_{max}]$, t^{max} is a maximal number of generations, t is a current number of generations.

20.3.3   Constriction coefficient

The constriction coefficient was developed to ensure convergence to a stable point without the need for velocity clamping. The new equation for velocity $V_{i,j}$ value (for the global version of the PSO algorithm) is as follows.

$$V_{i,j} = \gamma \cdot [V_{i,j} + \phi_1 \cdot (Pbest_{i,j} - P_{i,j}) + \phi_2 \cdot (Gbest_j - P_{i,j})] \quad (20.7)$$

where: γ is a constriction coefficient ($\gamma \in [0, 1]$) and is defined as follows.

$$\gamma = \frac{2 \cdot \kappa}{|2 - \phi - \sqrt{\phi \cdot (\phi - 4)}|} \quad (20.8)$$

and $\phi = \phi_1 + \phi_2$, $\phi_1 = c_1 \cdot r_1$, $\phi_2 = c_2 \cdot r_2$, the κ parameter controls the exploration ($\kappa \approx 0 \rightarrow$ fast convergence) and exploitation ($\kappa \approx 1 \rightarrow$ slow convergence) properties of the algorithm. If $\phi > 4$ and $\kappa \in [0, 1]$ then the swarm is guaranteed to converge to a stable point. Typically, the constriction factor is set to be $\gamma = 4.1$ and $c_1 = c_2 = 2.05$ [6].

20.3.4   Acceleration coefficients c_1 and c_2

The coefficient c_1 and c_2 are named as acceleration coefficients and they are responsible for the cognition part of the PSO algorithm (coefficient c_1) and the social part of the PSO algorithm (coefficient c_2). If $c_1 > 0$ and $c_2 = 0$ then we have a cognition only model of the PSO algorithm. Of course, if $c_1 = 0$ and $c_2 > 0$ then we have a social only model of the PSO algorithm. When $c_2 > c_1$ then the PSO algorithm is more suitable for the solving of unimodal problems. On the other hand when $c_1 < c_2$ then the algorithm PSO is more suitable for solving multimodal problems. In addition, we can introduce the adaptive changes of acceleration coefficient values using a formula.

$$c_1 = \left(c_1^{min} - c_1^{max} \right) \cdot \frac{t}{t^{max}} + c_1^{max} \tag{20.9}$$

$$c_2 = \left(c_2^{max} - c_1^{min} \right) \cdot \frac{t}{t^{max}} + c_2^{min} \tag{20.10}$$

where: $c_1 \in \left[c_1^{min}, c_1^{max} \right]$, $c_2 \in \left[c_2^{min}, c_2^{max} \right]$, t is a current number of generations ($t \in [1, t^{max}]$), t^{max} is a maximal assumed number of generations.

20.4 Application of PSO algorithm for IIR digital filter design

20.4.1 Problem description

In general the design of digital filters with non-standard amplitude characteristics is a very complex problem. The objective function describing this problem is a multimodal function, therefore the gradient optimization techniques can easily be stuck at local optima. Of course, digital filters with standard amplitude characteristics can be designed using existing standard approximations as for example with Butterworth, Cauer or Chebyshev. However, this problem becomes more complicated if the digital filter is supposed to possess non-standard amplitude characteristics (as for example in phase or amplitude equalizers). Then the standard approximations are not useful for us. The transfer function $H(z)$ in z domain for the IIR (Infinite Impulse Response) digital filter is given by formula 20.11.

$$H(z) = \frac{b_0 + b_1 \cdot z^{-1} + b_2 \cdot z^{-2} + \dots + b_{n-1} \cdot z^{-(n-1)} + b_n \cdot z^{-n}}{1 - \left(a_1 \cdot z^{-1} + a_2 \cdot z^{-2} + \dots + a_{n-1} \cdot z^{-(n-1)} + a_n \cdot z^{-n} \right)} \tag{20.11}$$

where: $b_0, ..., b_n$ and $a_1, ..., a_n$ are the transfer function coefficients, n is a filter order (positive integer value). The main objective for the PSO algorithm is to find such a set of transfer function coefficients ($b_0, ..., b_n$ and $a_1, ..., a_n$) which ensure that the digital filter will be stable (all poles of the transfer function $H(z)$ must be located inside the unitary circle in the z plane), and its amplitude characteristics will fulfill all design assumptions.

20.4.2   How can the PSO algorithm be used for this problem?

We want to create an optimization method which leads to a gradual improvement of the designed filter parameters. We would like to obtain the digital filter with the lowest deviation of amplitude characteristics from the idealized one. The proposed method is based on the local version of the PSO algorithm and it consists of eight steps which are as follows. In the first step, the initial swarm P which consists of N particles P_i (each particle is a D-dimensional vector; the dimension depends on the filter order n; if filter order is n then the number of dimensions D is equal to $2 \cdot n + 1$) is randomly created. The designed filter coefficients are coded in the particle P_i as follows.

$$P_i = \{b_0\ b_1\ b_2\ ...\ b_{n-1}\ b_n\ a_1\ a_2\ ...\ a_{n-1}\ a_n\}$$

Next, we create the $Pbest_i$, $Lbest_i$ and V_i vectors for each $i - th$ particle P_i and one vector $Gbest$ for whole swarm. Vector $Pbest_i$ determines the best position of the particle P_i in the search space among all positions obtained so far. Vector $Lbest_i$ determines the best position of another particle among NN nearest neighbors of particle P_i found until now. Vector V_i represents the value of particle P_i velocity. Vector $Gbest$ determines the global best solution found by the PSO algorithm. At the start vector P_i is assigned to the $Pbest_i$ vector for $i - th$ particle, vectors $Lbest_i$, V_i and $Gbest$ are equal to zero. Additionally, in the case when, during the run of algorithm, a return to the first step from the seventh step occurs then the vector $Gbest$ is assigned to the randomly chosen particle P_i from the swarm.

In the second step, each particle P_i is evaluated using objective function $OF(.)$. In order to compute the value of $OF(.)$ for particle P_i, the FFT (Fast Fourier Transform) is performed separately for the filter coefficients from nominator (coefficients b_k) and denominator (coefficients a_k). When we have FFT results, we can compute the amplitude characteristics $H(f)$ [dB] for the digital filter represented by filter coefficients which are stored in particle P_i as follows.

$$H\left(f\right) = 20 \cdot log_{10} \left(\sqrt{H_{real}(f)^2 + H_{imaginary}(f)^2} \right) \tag{20.12}$$

If we have the amplitude characteristics we can compute the objective function $OF(.)$ for particle P_i using the following formula.

$$OF(.) = \sum_{g=1}^{G} Error(f_g) + \sum_{m=1}^{M} Stability_m \tag{20.13}$$

where:

$$Error(f_g) = \begin{cases} |H(f_g) - C_1(f_g)| & \text{if } H(f_g) > C_1(f_g) \\ |H(f_g) - C_2(f_g)| & \text{if } H(f_g) < C_2(f_g) \\ 0 & \text{if } H(f_g) \in [C_2(f_g), C_1(f_g)] \end{cases} \tag{20.14}$$

$$Error(f_g) = \begin{cases} (|z_s| - 1) \cdot p + p & \text{if } |z_s| \geq 1 \\ 0 & \text{if } |z_s| < 1 \end{cases} \tag{20.15}$$

where: M is the number of poles of the transfer function 20.11, G is the number of output samples from FFT transform divided by 2 (we have assumed $G = 256$), p is a value of the penalty (we have assumed $p = 10^5$), f_g is a $g - th$ value of normalized frequency ($f_g \in [0, 1]$; Nyquist frequency = 1), $|z_s|$ is the module of the $s - th$ pole of the transfer function in the z plane, $C_1(f_g)$ is an upper constraint of the amplitude characteristics for frequency f_g, $C_2(f_g)$ is a lower constraint of the amplitude characteristics for frequency f_g, $H(f_g)$ is a value of amplitude characteristics for frequency f_g.

The value of objective function $OF(.)$ is higher when the sum of the absolute values of deviations between constraints and amplitude characteristics $H(f)$ for particle P_i is higher (first part of objective function 20.13). Additionally, the value of $OF(.)$ increases when the digital filter represented by particle P_i is not stable (second part of objective function 20.13). The PSO algorithm minimizes objective function $OF(.)$.

Next, when the value of the computed $OF(.)$ function for particle P_i is lower than the value written down in vector $Pbest_i$ then the values from particle P_i are assigned to the vector $Pbest_i$.

In the third step, for each particle P_i the best particle $Lbest_i$ is selected among NN nearest neighbors of the particle P_i. The similarity between particles was determined by the values of the $OF(.)$. The two particles are more similar to each other when their values of $OF(.)$ are close to each other.

In the fourth step, the vector V_i is computed for each particle P_i using the following formula:

$$V_{i,j} = \gamma \cdot V_{i,j} + \gamma \cdot c_1 \cdot r_1 \cdot (Pbest_{i,j} - P_{i,j}) + \gamma \cdot c_2 \cdot r_2 \cdot (Lbest_{i,j} - P_{i,j}) \tag{20.16}$$

where: j is the number of the decision variable ($j \in [1, D]$), γ is a scaling coefficient (we have assumed $\gamma = 0.729$), c_1 and c_2 are the learning coefficients (usually $c_1 = c_2$, we have assumed $c_1 = c_2 = 0.3$), r_1 and r_2 are random real numbers with uniform distribution selected separately for each $j - th$ dimension ($r_1 \in [0, 1]$ and $r_2 \in [0, 1]$).

In the fifth step, the new particle P_i is created using the following formula:

$$P_{i,j} = P_{i,j} + r_3 \cdot V_{i,j} \tag{20.17}$$

where: r_3 is a random real number selected separately for each $j - th$ dimension ($r_3 \in [0, 1]$).

In the sixth step, one randomly created particle is inserted into the place of the worst particle in the whole swarm. The worst particle means the particle P_i with the highest value of the $OF(.)$ function. Due to this the convergence of the presented algorithm is improved.

In the seventh step, we check whether the particle P_i for which the value of $OF(P_i) = 0$ is found. If such a particle is not found then the ninth step of the algorithm is executed. If such a particle P_i is found then this particle is assigned to *Gbest* particle and the $OF(.)$ function is modified by decreasing the allowed values of deviations of amplitude characteristics for the designed digital filter.

In the eighth step, the algorithm jumps to the first step.

In the ninth step, the algorithm termination criterion is checked. The algorithm is stopped when the absolute value of the difference between $OF(.)$ function for the particles *Gbest* in generation t_1 and in generation $t_1 + t$ is lower than ϵ. As a final result of the algorithm operation, we obtain the values stored in the current *Gbest* particle. If the termination criterion is not fulfilled then the algorithm jumps to the second step.

20.4.3   Description of experiments

The method presented was tested by the design of the $10 - th$ order ($n = 10, D = 2 \cdot n + 1 = 21$) IIR digital filter with linearly falling amplitude characteristics. The parameters of amplitude characteristics are as follows. The value of attenuation is equal to 0 [dB] for normalized frequency $f = 0$ and the value of attenuation is equal to 40 [dB] for normalized frequency $f = 1$. At the start of the algorithm, we have assumed that the maximal deviation value between amplitude characteristics represented by particle P_i and the digital filter design assumptions for any normalized frequency value cannot be higher than 10 [dB]. During algorithm operation, the maximal deviation value is decreased with 1 [dB] when a digital filter which fulfills all design assumptions is found. After achieving the maximal deviation value equal to 1 [dB], the value of this parameter was decreasing by 0.1 [dB] when the digital filter which fulfilled all design assumptions was found. Designing a stable digital filter whose amplitude characteristics fulfill all design assumptions with the lowest maximal deviation value from the ideal amplitude characteristics is the main goal of the presented algorithm. The values of the algorithm parameters are as follows. The number of particles $N = 100$, dimension of solution space $D = 21$, nearest neighbor parameter $NN = 3$, $t = 500$ generations and $\epsilon = 0.0001$. The initial values for particle P_i in each $j - th$ dimension were randomly chosen from the range $[-1, 1]$.

20.4.4   Results obtained

The IIR digital filter which fulfills all design assumptions was found after 19332 generations. This filter is stable (all poles of the transfer function are located in the unitary circle in the z plane). The maximal value of the amplitude characteristics' deviation from the ideal one does not exceed 0.3 [dB]. In Table 20.1 the number of generations which are required to design the IIR digital filter with assumed design assumptions (having deviations between ideal amplitude characteristics and amplitude characteristics generated by designed filter lower than maximal acceptable deviations) are shown.

TABLE 20.1
The number of generations required to obtain a digital filter with deviations of amplitude characteristics is lower than the prescribed maximal deviations from the ideal amplitude characteristics.

Maximal deviations [dB]								
10	9	8	7	6	5	4	3	2
15	21	32	50	98	172	230	304	446

Maximal deviations [dB]								
1	0.9	0.8	0.7	0.6	0.5	0.4	0.3	0.2
601	810	1050	1170	1321	2012	2415	8585	–

We can see that only 15 generations are required to obtain a digital filter with deviations of attenuation lower than 10 [dB], 570 generations are required for deviations not higher than 1 [dB], and 8585 generations are required for deviations lower than 0.3 [dB]. The $10 - th$ order IIR digital filter cannot be found for deviations of attenuation lower than 0.2 [dB].

20.5 Conclusions

In this paper the PSO algorithm, in a local and global version, was presented in detail. Some modifications of the PSO algorithm were demonstrated and discussed. Finally, the exemplary application of the PSO algorithm to a real-world problem was shown. The PSO algorithm was used for IIR digital filter design with non-standard amplitude characteristics. As shown in Section 20.4, it is possible to design digital filters with non-standard amplitude characteristics using the PSO algorithm. The designed digital filter fulfills all design assumptions and is a stable filter (all poles of transfer function are located in a unitary circle in the z plane). Using the PSO algorithm we can obtain higher automation of the digital filter design process, because expert knowledge concerning the filter design is not required. Also, it is worth noting that swarm based digital filter design is fundamentally different from the traditional filter design process. The swarm digital filter design process possesses fewer constraints than the design process which is based on human (designer) knowledge and experience [9]. The human designer is not only limited by the technology, but also by their own habits, intelligence and so on. The application of the PSO algorithm to digital filter design allows these limitations to be avoided and provides access to new possibilities.

References

1. J. Kennedy, R.C. Eberhart. "Particle swarm optimization" in *Proc. of IEEE International Conference on Neural Networks*, 1995, pp. 1942-1948.

2. A. Colorni, M. Dorigo, V. Maniezzo. "Distributed optimization by ant colonies" in *Proc. of the European Conf. on Artificial Life*, 1991, pp. 134-142.

3. J. Xin, G. Chen, Y. Hai. "A particle swarm optimizer with multi-stage linearly-decreasing inertia weight" in *Computational Sciences and Optimization*, vol. 1, 2009, pp. 505-508.

4. H. Fan. "A modification to particle swarm optimization algorithm". *Engineering Computations*, vol. 19(8), pp. 970-989, 2002.

5. I.K. Gupta. "A review on particle swarm optimization". *International Journal of Advances Research in Computer Science and Software Engineering*, vol. 5(4), pp. 618-623, 2015.

6. Y. Yang, J. Wen, X. Chen. "Improvements on particle swarm optimization algorithm for velocity calibration in microseismic monitoring". *Earthquake Science*, vol. 28(4), pp. 263-273, 2015.

7. A. Slowik, H. Kwasnicka. "Nature inspired methods and their industry applications – swarm intelligence algorithms". *IEEE Transactions on Industrial Informatics*, vol. 14(3), pp. 1004-1015, 2018.

8. M.H. Yar, V. Rahmati, H.R.D. Oskouei. "A survey on evolutionary computation: methods and their applications in engineering". *Modern Applied Science*, vol. 10(11), pp. 131-139, 2016.

9. A. Slowik, M. Bialko. "Design and optimization of IIR digital filters with non-standard characteristics using particle swarm optimization" in *Proc. of 14th IEEE International Conference on Electronics, Circuits and Systems, ICECS 2007*, pp. 162-165, 2007.

21

Salp Swarm Algorithm: Modification and Application

Essam H. Houssein
Faculty of Computers and Information
Minia University, Minia, Egypt

Ibrahim E. Mohamed
Faculty of Computers and Information
South Valley University, Luxor, Egypt

Aboul Ella Hassanien
Faculty of Computers and Information
Cairo University, Cairo, Egypt

CONTENTS

21.1 Introduction

Artificial Intelligence (AI) is the smartness displayed by machineries. It can be described as "the study and design of intelligent agents" [1], in which associate intelligent agents represent systems that identify their setting and take actions to exploit their goals. Now general methodologies of AI contain conventional statistical approaches [2], conventional symbolic AI, and Computational Intelligence (CI) [3]. It's a set of nature-inspired computing paradigms to capture information and make sense of it for which conventional methodologies are inefficient or useless and it includes Evolutionary Computation (EC), Artificial Neural Network and fuzzy logic [4].

Swarm Intelligence (SI) is highly adapted to a group of mobile agents that directly or indirectly communicate with each other, and collectively solve a set of basic problems that cannot be solved if the agents are operating independently [5, 6]. SI is in a pioneer tributary in computer science, bio-signals [7, 8] a complex discipline that expresses a set of nature-inspired mathematical models that are inspired by the collective behavior of natural or artificial decentralized and self-organized systems cum habits of organisms like plants, animals, fish, birds, ants and other elements in our ecosystem that employ the intuitive intelligence of the entire swarm/herd to solve some complex problems for a single agent [9, 10, 11]. The characteristics of swarm-based techniques are:

- They are population-based.

- Their searches are done using multiple agents.

- The agents that frame the population are typically homogenized.

- The collective behaviors of the system arise from each individual interaction with the other and with their environment.

- The agents are always moving randomly in a haphazard way.

- The agents' actions, principally movements are responsive to the environment.

- There is no centralized control, leaders' performances are solely the standard for their emergence in each iteration [12].

In 2017, Mirjalili et al. [13] suggested a recent meta-heuristic, the Salp Swarm Algorithm (SSA), deeply influenced by the swarming behaviour of deep-sea salps. SSA aims to develop a new optimizer based on populations by attempting to mimic salps' swarming conduct in the natural environment.

21.2   Salp Swarm Algorithm (SSA) in brief

SSA is an SI algorithm devised for continuous problem optimization. Compared to some existing algorithmic techniques, this algorithm seems to have a better or equivalent performance [14]. SSA is a stochastic algorithm in which, to start the optimization process, the initial population is formed by creating a set of initial random solutions, and then improving these solutions over the time in two stages, exploring and exploiting. In first stage, the promising regions are discovered by exploring the search space, while in exploitation, we hope to find better solutions than the current ones by searching the neighbourhood of specific solutions.

21.2.1   Inspiration analysis

Salps fit into the gelatinous salpidae family which are barrel-shaped and zoo plankton that form large swarms. They move slowly forwards through the sea as each zooid rhythmically contracts. This flow, hyped by muscle action, concurrently provides chemosensory details, food, exchange of respiratory gas, removal of solid and dissolved waste, sperm dispersal and propulsion by jet. To move forward, the body pumps water as propulsion [15]. SSA is the first technique to imitate the behaviour of salps in nature. The salps are marine organisms living in oceans and seas. They resemble jellyfish in their tissues and movement towards food sources [15]. Salps form groups (swarms) called salp chains; a leader and a set of followers are contained in each salp chain (please see the Figure 21.1). The leader salp attacks directly the target (feeding source), whereas all followers start moving directly or indirectly to ward the rest of the salps (and leader).

21.2.2   Mathematical model for salp chains

The swarming behaviours [16] and population of salp [17] are seldom in the literature to be mathematically modelled. Furthermore, to solve optimization problems, swarms of various animals (such as bees and ants) are commonly designed and utilized as a mathematical model, while it is rare to find mathematical pattern physical processes (like salp swarms) to solve various optimization issues. Through the next sub section, the standard model of salp chains in the review is proposed [13] to solve different problems with the optimizing process. Mathematically, the salp chains are divided into two groups by random division of the population (salps): leader and followers. The first salp in the series of salps is called the leader, whereas the remaining salps are regarded as followers. Through the given name of both types of these salps, the leader directs swarms and the remaining of these series follow each other (and the leader either explicitly or implicitly).

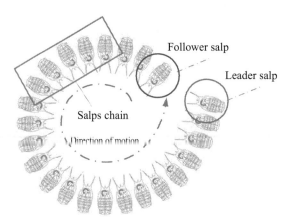

FIGURE 21.1
Demonstration of salp's series.

Given M is a counter for variables in a particular problem, like the other SI-based methodologies, the salps' location is denoted in a M-dimensional search space. Therefore, the population of salps X is composed of N swarms with M dimension. It could therefore be identified by a $N \times M$ matrix, as outlined in the equation below:

$$X_i = \begin{bmatrix} x_1^1 & x_2^1 & \cdots & x_M^1 \\ x_1^2 & x_2^2 & \cdots & x_M^2 \\ \vdots & \vdots & \ddots & \vdots \\ x_1^N & x_2^N & \cdots & x_M^N \end{bmatrix} \tag{21.1}$$

A feeding source termed F is also thought to be the target of the swarm in the search space. The position of the leader is updated by the following equation:

$$X_j^1 = \begin{cases} F_j + c_1((ub_j - lb_j)c_2 + lb_j) & c_3 \geq 0 \\ F_j - c_1((ub_j - lb_j)c_2 + lb_j) & c_3 < 0 \end{cases} \tag{21.2}$$

Where X_j^1 and F_j denote the positions of leaders and feeding source in the jth dimension, respectively. The ub_j and lb_j indicate the upper (superior) and lower (inferior) bounds of the jth dimension. c_2 and c_3 are two random floats from the closed interval $[0, 1]$. In actuality, they guide the next location in the jth dimension toward the $+\infty$ or $-\infty$ besides determining the step size. Equation 21.2 indicates that the leader only updates its location with respect to the feeding source. The coefficient c_1, the most effective parameter in SSA, gradually decreases over the course of iterations to balance exploration and exploitation, and is defined as follows:

$$c_1 = 2e^{-(\frac{4l}{L})^2} \tag{21.3}$$

where l and L represent the current iteration and maximum number of iterations, respectively. To update the position of the followers, the next equation is used (Newton's motion law):

$$X_j^i = \frac{X_j^i + X_j^{i-1}}{2} \tag{21.4}$$

where $i \geq 2$ and X_j^i is the location of the ith follower at the jth dimension. In SSA, the followers move toward the leader, whereas the leader moves toward the feeding source. During the process, the feeding source location can be changed and consequently the leader will move towards the new feeding source location. The flowchart of SSA is demonstrated in Fig. 21.2.

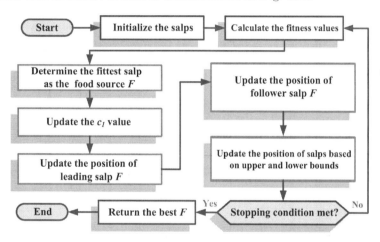

FIGURE 21.2
The flowchart of SSA.

21.3 Modifications of SSA

There are many techniques of SSA in the literature [18, 19, 20]. To categorize them, a certain classification scheme is required. This part is about enhancing SSA by using chaotic, robust, simplex, weight factor, adaptive mutation and other methods.

21.3.1 Fuzzy logic

In [21], Majhi et al. proposed an Automobile Insurance Fraud Detection System (AIFDS) that uses a hybrid fuzzy clustering technique using SSA (SSA-FCM) for outliers detection and removal. The statistical tests revealed that the use of clustering provides better accuracy. The proposed fuzzy clustering was applied for under-sampling of majority class samples of the automobile insurance data set for enhancing the effectiveness of the classifiers. The SSA helps in obtaining the optimal cluster center in the SSA-FCM. The SSA-FCM calculates the distance of the data-points from the cluster centers based on which suspicious classes are detected.

21.3.2 Robust

Suppose we have an application to an engineering optimization problem; multiple adjustment parameters are not desirable to solve a problem. In a limited calculation time, without trial and error for adjustment parameters, it is necessary to get as good a solution as possible. We therefore believe that meta-heuristics should have the robustness capability to ensure the performance of searching for pre-adjusted parameters against predetermined structural variation of problems to be solved.

Tolba et al. [22] developed a novel methodology that relies on SSA to find and optimize the size of Renewable Distribution Generators (RDGs) and Shunt Capacitor Banks (SCBs) on Radial Distributed Networks (RDNs). The approach offers various objectives, functions and different constraint conditions to enhance the voltage level, properly reduce energy loss and annual operating costs.

21.3.3 Simplex

The simplex methods are strategies of stochastic variants, which maximize population diversity and improve the algorithm's local search capability. This approach helps to accomplish a better trade off between the swarm algorithm's exploration and exploitation capabilities and makes the swarm algorithm more robust and faster. The simplex method [23] has the strong ability to avoid the local optimum and enhance the ability of searching the global optimum.

21.3.4 Weight factor and adaptive mutation

A weighting factor is usually used for calculating a weighted mean, to give less (or more) importance to group members. For balancing between global exploration and local exploitation, dynamic weight factor is included in updating the formulation of the place of population.

21.3.5 Levy flight

Levy flight is a specific random walk category that distributes the step size according to the tails of the heavy power law. Sometimes, the large step helps an approach perform a global search. It would be helpful to use the Levy flight trajectory [24, 25] to gain a better trade-off between exploring and exploiting the algorithm and based on the local optimum avoidance, it has positive points.

21.3.6 Binary

Different types of optimization problems cannot be solved by meta-heuristics. A binary optimization problem has diverse decision factors that are the components of the interval $[0, 1]$. In the term, 0 or 1 represents a digital meaning of each decision variable of the binary problems, such as Features Selection (FS) problem [26], unit commitment problem [27, 28], and Knapsack Problems (KP) [29, 30].

Using a modified Arctan transformation, Rizk-Allah et al. [31] suggested a novel binary approach of the SSA called BSSA with the objective of transforming the continuous space to binary space. To enhance the exploration and exploitation capabilities, multiplicity and mobility are two advantages with regard to the modification transfer function.

21.3.7 Chaotic

SSA can approximate an optimum solution with high convergence, but SSA is not yet beneficial in searching the optimum solution which affects the algorithm performance. Therefore, to decrease this impact and to enhance its potential and effectiveness, Chaotic SSA (CSSA) was proposed by Sayed et al. [32] by merging between SSA and the chaos algorithm. Chaos, a novel numerical approach, has recently been used to improve the execution of meta-heuristic approaches. Chaos is described as simulation for self-motivated conduct of a non-linear system [33]. The population of meta-heuristic methods has the same advantages including scalable approach, simplicity, and reduced computation time. However, these approaches have two intrinsic weaknesses; low convergence rate and recession in local optima [34].

To solve the graph colouring problem, Meraihi et al. [35] proposed a new Chaotic Binary SSA (CBSSA). First, the Binary SSA (BSSA) was gained from the standard SSA where the S-Shaped transfer function is used. Second, a common chaotic map, called a logistic map was used. Using the well-known DIMACS benchmark instances, the performance of the proposed approach was stronger in comparison with various relevant colouring methods. The experimental results verified the performance and strength of the proposed CBSSA approach compared to the previously mentioned algorithms.

In [36], five chaotic maps were utilized for diagnosing and designing different feature selection methods based on SSA for data classification. Ateya et al. [37] introduced the latency and cost aware Software-Defined Networking (SDN) controller placement problem. To minimize the deployment cost and the latency, to achieve the optimum amount of controllers and also the ideal allocation of switches to controllers, a CSSA method was constructed. By introducing chaotic maps, optimizer efficiency was enhanced and local optima were prevented. The method has been evaluated for different real functionalities from the topology of the zoo. The effect of variation of different network parameters on the performance was checked. In terms of reliability and execution time, simulation outcomes proved that the introduced algorithm outperforms a GT-based approach in addition to meta-heuristic methodologies.

To get over the potential shortcomings of native SSA, Zhang et al. [38] improved an SSA-based optimizer. The designed variant was defined as a Chaos-induced and Mutation-driven SSA (CMSSA) that simultaneously combines two approaches. First, to boost the exploitation of the algorithm, the basic SSA was used to introduce a chaotic exploitative mechanism with "shrinking" mode. Then, to get full benefit from the reliable diversification of Cauchy mutation and the strong intensification of Gaussian mutation, a combined mutation scheme was adapted. The statistical tests on a representative benchmark showed the effectiveness of the proposed method in solving optimization and engineering design problems by alleviating the precocious convergence of SSA.

21.3.8   Multi-Objective Problems (MOPS)

This part of our publication presents the basics of MOPS [39, 40]. The target of MOPS is assumed to minimize or maximize incompatible objectives functions [41, 42]. On the contrary, to improve individual target problems, including multiple objective and MOPS objectives contradict each other. A difficult task of MOPS is to find an optimal solution to optimize objectives of each function concurrently. Therefore, balancing should be done between all objectives of each function to achieve an optimum solution collection.

To find out an appropriate solution for a virtual machine locating problem, Alresheedi et al. [43] combined the SSA and sine cosine algorithm (SCA) with the means of improving MOPS techniques (MOSSASCA). The main purposes of the proposed MOSSASCA are to minimize quality of services infringements, to reduce power consumption, and maximize average time before an agent shutdown in addition to minimize conflict between the three objectives. In SCA, to increase convergence speed and to avoid getting trapped on a local optimum solution, a local search technique is followed to increase the performance of SSA. Various virtual and physical machines were in a set of experiments to evaluate the performance of the combined algorithm. Well-known MOP methods were compared with the results of MOSSASCA. Results indicate a balance between achieving the three objectives.

21.4   Application of SSA for welded beam design problem

21.4.1   Problem description

The real issue of welded beam structure gives rise to an optimization problem with four parameters. The beam is made from steel and has a shaped cross-section. The beam is focused on supporting a force at the beam's tip of F=6000

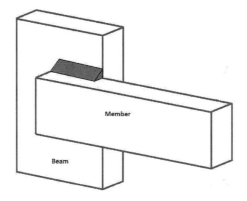

FIGURE 21.3
Welded beam design problem.

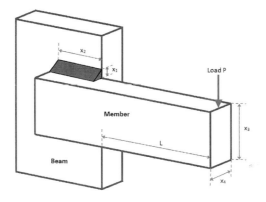

FIGURE 21.4
Parameters of the welded beam design problem.

lbs. The concern is to have four continuous factors of design: weld thickness h (x1), weld length l (x2), beam thickness t (x3) and beam width b (x4). The goal is to obtain minimum assembly costs. The τ, σ, δ and P-c care amounts are measured by treating the beam as a width "L" girder beam. The bending stress, buckling load, and weld stress are defined with notations as $\sigma(x), P_c(x)$, and $\tau(x)$. Both Fig. 21.3 and Fig. 21.4 illustrate the problem shape and its parameters.

21.4.2 How can SSA be used to optimize this problem?

To optimize the problem parameters of the welded beam design, we proposed a SSA-based method to achieve an optimal solution to the optimization problem. The details of our method are described follows:

Step 1, initialize the main problem factors like P = 6000 lb, L = 14 in, $\delta_{\max}$ = 0.25 in, E = 30 $\times 10^6$ psi,   G = 12 $\times$ 10^6 psi, $\tau_{\max}$ = 13600 *psi*, $\sigma_{\max}$ = 30000 *psi*.

Step 2, initialize the salps population randomly and the termination conditions shall be set. Step 3, calculate the fitness values on the basis of the objective function and the constraints (G1, G2, G3, G4, G5, G6, G7) are shown as follows:

Cost Function:

$\min_x f(x) = 1.10471x_1^2 x_2 + 0.04811x_3 x_4 (14 + x_2)$

Constraints

$G_1(x) = \tau_{\max} - \tau(x) \geq 0$

$G_2(x) = \sigma_{\max} - \sigma(x) \geq 0$

$G_3(x) = x_4 - x_1 \geq 0$

$G_4(x) = 5 - 1.10471x_1^2 x_2 + 0.04811x_3 x_4 (14.0 + x_2) \geq 0$ (21.5)

$G_5(x) = x_1 - 0.125 \geq 0$

$G_6(x) = P_c(x) - P \geq 0$

$G_7(x) = \delta \max = \delta(x) \geq 0$

Variables Boundaries:

$0.125 \leq x_1 \leq 5,$

$0.1 \leq x_2 \leq 10,$

$0.1 \leq x_3 \leq 10$

$0.125 \leq x_4 \leq 5$

The weld stress $\tau(x)$ has two components which are τ' (1st derivative of shear stress) and τ''(2nd derivative of shear stress). $\tau(x)$ is computed using the following equation:

$$\tau(x) = \sqrt{(\tau')^2 + \tau'' \cdot \tau' \frac{x_2}{2R} + (\tau'')^2}$$

where

$\tau' = P/(\sqrt{(2)} * x_1 * x_2),$

$\tau'' = (M * R)/J,$

$R = \sqrt{x_2^2/4 + ((x_1 + x_3)/2)^2},$

$J = 2 * (\sqrt{2} * x_1 * x_2 * ((x_2^2)/12 + ((x_1 + x_3)/2)^2)),$

$M = P * (L + x_2/2).$

The bar bending stress $\sigma(x)$ is calculated from the following equation:

$$\sigma(x) = \frac{6PL}{x_4 x_3^2}.$$

The bar buckling load is found from the following equation:

$$Pc(x) = \frac{4.013E\sqrt{x_3^2 x_4^6/36}}{L^2}\left(1 - \frac{x_3}{2L}\sqrt{\frac{E}{4G}}\right).$$

The bar displacement is computed using the following equation:

$$\delta(x) = \frac{4PL^3}{Ex_4 X_3^3}.$$

Step 4, determine the minimum fitness value as the source food F. Step 5, update the leader's position by equation 21.2 and followers by equation 21.4. Step 6, update the new salp positions to the upper and lower limits. Step 7, the conditions of termination are checked, the method is stopped when conditions are reached or jump to step 3 if the conditions of termination are not met. Finally, the best fitness value and positions are displayed where the minimum cost is achieved.

21.4.3 Result obtained

The algorithm was tested with 10000 iterations, 60 search agents and 4 design variables. The result is compared with several techniques like Gravitational Search Algorithm (GSA), modified Particle Swarm Optimization algorithm (CPSO), Genetic algorithm and Simplex (GA) algorithm as shown in Table 21.1.

21.5 Conclusion

The SSA algorithm has been presented in detail in this paper. Some SSA algorithm modifications have been illustrated and explained. Finally, the SSA algorithm's exemplary application to a real-world issue has been shown. The SSA

TABLE 21.1

A summary of comparison results for beam problem.

Algorithm	x_1	x_2	x_3	x_4	Fitness value
SSA	0.2057	3.4714	9.0366	0.2057	1.72491
GSA	0.18213	3.8569790	10.0	0.2023760	1.879950
CPSO	0.20237	3.5442140	9.04821	0.2057230	1.731480
GA	0.18290	4.04830	9.36660	0.20590	1.8242
Simplex	0.27920	5.62560	7.75120	0.27960	2.530730

algorithm was used for the design of welded beams with special attributes of design variables. As shown in Section 21.4, the SSA algorithm can be used to optimize welded beam design with characteristic design variables. Using the SSA algorithm we can achieve higher automation of the welded beam design process because there is no need for expert knowledge of welded beam design. It should also be noted that the swarm-based welded beam design differs essentially from the traditional welded beam design process based on individual intervention. The application of the SSA algorithm to the welded beams design problem enables the avoidance of errors with high efficiency and obtains minimum assembly making costs.

References

1. S. Russell, P. Norvig, Peter. *Artificial Intelligence: A Modern Approach*. Prentice Hall, 1995.

2. B.L. Agarwal. *Basic Statistics*. New Age International, 2006.

3. K.E. Voges, N. Pope. "Computational intelligence applications in business: A cross-section of the field" in *Business Applications and Computational Intelligence*, pp. 1-18, IGI Global, 2006.

4. Y. Zhang, S. Wang, G. Ji. "A comprehensive survey on particle swarm optimization algorithm and its applications". *Mathematical Problems in Engineering*, vol. 2015, Article ID 931256, 38 pages, 2015, doi: https://doi.org/10.1155/2015/931256.

5. A. Hamad, E.H. Houssein, A.E. Hassanien, A.A. Fahmy. "Hybrid grasshopper optimization algorithm and support vector machines for automatic seizure detection in eeg signals" in *International Conference on Advanced Machine Learning Technologies and Applications*, pp. 82-91, 2018.

6. M.M. Ahmed, E.H. Houssein, A.E. Hassanien, A. Taha, E. Hassanien. "Maximizing lifetime of wireless sensor networks based on whale optimization algorithm" in *International Conference on Advanced Intelligent Systems and Informatics*, pp. 724-733, 2017.

7. E.H. Houssein, A. Hamad, A.E. Hassanien, A.A. Fahmy. "Epileptic detection based on whale optimization enhanced support vector machine". *Journal of Information and Optimization Sciences*, vol. 40(3), pp. 699-723, 2019.

8. A. Hamad, E.H. Houssein, A.E. Hassanien, A.A. Fahmy. "A hybrid eeg signals classification approach based on grey wolf optimizer enhanced svms for epileptic detection" in *International Conference on Advanced Intelligent Systems and Informatics*, pp. 108-117, 2017.

9. V. Pandiri, A. Singh. "Swarm intelligence approaches for multidepot salesmen problems with load balancing". *Applied Intelligence*, vol. 44(4), pp. 849-861, 2016.

10. A.A. Ewees, M.A. Elaziz, E.H. Houssein. "Improved grasshopper optimization algorithm using opposition-based learning". *Expert Systems with Applications*, vol. 112, pp. 156-172, 2018.

11. A.G. Hussien, E.H. Houssein, A.E. Hassanien. "A binary whale optimization algorithm with hyperbolic tangent fitness function for feature selection" in *Eighth International Conference on Intelligent Computing and Information Systems (ICICIS)*, pp. 166-172, 2017.

12. R.S. Parpinelli, H.S. Lopes. "New inspirations in swarm intelligence: a survey" in *International Journal of Bio-Inspired Computation*, vol. 3(1), pp. 1-16, 2011.

13. S. Mirjalili, A.H. Gandomi, S.Z. Mirjalili, S. Saremi, H. Faris, S.M. Mirjalili. "Salp swarm algorithm: A bio-inspired optimizer for engineering design problems". *Advances in Engineering Software*, vol. 114, pp. 163-191, 2017.

14. H. Faris, S. Mirjalili, I. Aljarah, M. Mafarja, A.A. Heidari. "Salp swarm algorithm: Theory, literature review, and application in extreme learning machines" in *Nature-Inspired Optimizers*, pp. 185-199, Springer, 2020.

15. L.P. Madin. "Aspects of jet propulsion in salps". *Canadian Journal of Zoology*, vol. 68(4), pp. 765-777, 1990.

16. V. Andersen, P. Nival. "A model of the population dynamics of salps in coastal waters of the Ligurian Sea". *Journal of Plankton Research*, vol. 8(6), pp. 1091-1110, 1986.

17. N. Henschke, J.A. Smith, J.D. Everett, I.M. Suthers. "Population drivers of a thalia democratica swarm: insights from population modelling". *Journal of Plankton Research*, vol. 37(5), 1074-1087, 2015.

18. A.G. Hussien, A.E. Hassanien, E.H. Houssein. "Swarming behaviour of salps algorithm for predicting chemical compound activities" in *Eighth International Conference on Intelligent Computing and Information Systems (ICICIS)*, pp. 315-320, 2017.

19. H.M. Kanoosh, E.H. Houssein, M.M. Selim. "Salp swarm algorithm for node localization in wireless sensor networks". *Journal of Computer Networks and Communications*, 2019.

20. A.G. Hussien, A.E. Hassanien, E.H. Houssein, S. Bhattacharyya, M. Amin. "S-shaped binary whale optimization algorithm for feature selection" in *Recent Trends in Signal and Image Processing*, pp. 79-87, Springer, 2019.

21. S.K. Majhi, S. Bhatachharya, R. Pradhan, S. Biswal. "Fuzzy clustering using salp swarm algorithm for automobile insurance fraud detection". *Journal of Intelligent & Fuzzy Systems*, vol. 36(3), pp. 2333-2344, 2019.

22. M. Tolba, H. Rezk, A. Diab, M. Al-Dhaifallah. "A novel robust methodology based salp swarm algorithm for allocation and capacity of renewable distributed generators on distribution grids". *Energico*, vol. 11(10), pp. 25 56, 2018.

23. X.-S. Yang. *Engineering Optimization: An Introduction with Meta-heuristic Applications*. John Wiley & Sons, 2010.

24. X.-S. Yang, S. Deb. "Cuckoo search via Lévy flights" in *World Congress on Nature & Biologically Inspired Computing (NaBIC)*, pp. 210-214, 2009.

25. A.F. Kamaruzaman, A.M. Zain, S.M. Yusuf, A. Udin. "Levy flight algorithm for optimization problems-a literature review". *Applied Mechanics and Materials*, vol. 421, pp. 496-501, 2013.

26. H. Faris, M.M. Mafarja, A.A. Heidari, I. Aljarah, A. Ala'M, S. Mirjalili, H. Fujita. "An efficient binary salp swarm algorithm with crossover scheme for feature selection problems". *Knowledge-Based Systems*, vol. 154, pp. 43-67, 2018.

27. L.K. Panwar, S. Reddy, A. Verma, B.K. Panigrahi, R. Kumar. "Binary grey wolf optimizer for large scale unit commitment problem". *Swarm and Evolutionary Computation*, vol. 38, pp. 251-266, 2018.

28. Y.-K. Wu, H.-Y. Chang, S.M. Chang. "Analysis and comparison for the unit commitment problem in a large-scale power system by using three meta-heuristic algorithms". *Energy Procedia*, vol. 141, pp. 423-427, 2017.

29. Y. He, X. Wang. "Group theory-based optimization algorithm for solving knapsack problems". *Knowledge-Based Systems*, 2018.

30. E. Ulker, V. Tongur. "Migrating birds optimization (mbo) algorithm to solve knapsack problem". *Procedia Computer Science*, vol. 111, pp. 71-76, 2017.

31. R.M. Rizk-Allah, A.E. Hassanien, M. Elhoseny, M. Gunasekaran. "A new binary salp swarm algorithm: development and application for optimization tasks". *Neural Computing and Applications*, pp. 1-23, 2018.

32. G.I. Sayed, G. Khoriba, M.H. Haggag. "A novel chaotic salp swarm algorithm for global optimization and feature selection". *Applied Intelligence*, vol. 48(10), pp. 3462-3481, 2018.

33. L. dos Santos Coelho, V.C. Mariani. "Use of chaotic sequences in a biologically inspired algorithm for engineering design optimization". *Expert Systems with Applications*, vol. 34(3), pp. 1905-1913, 2008.

34. K.-L. Du, M.N.S. Swamy. "Particle swarm optimization" in *Search and Optimization by Metaheuristics*, pp. 153-173, Springer, 2016.

35. Y. Meraihi, A. Ramdane-Cherif, M. Mahseur, D. Achelia. "A chaotic binary salp swarm algorithm for solving the graph coloring problem" in *International Symposium on Modelling and Implementation of Complex Systems*, pp. 106-118, Springer, 2018.

36. A.E. Hegazy, M.A. Makhlouf, G.S. El-Tawel. "Feature selection using chaotic salp swarm algorithm for data classification". *Arabian Journal for Science and Engineering*, vol. 44(4), pp. 3801-3816, 2019.

37. A.A. Ateya, A. Muthanna, A. Vybornova, A.D. Algarni, A. Abuarqoub, Y. Koucheryavy, A. Koucheryavy. "Chaotic salp swarm algorithm for sdn multi-controller networks". *Engineering Science and Technology*, pp. 31243-31261, 2019.

38. Q. Zhang, H. Chen, A.A. Heidari, X. Zhao, Y. Xu, P. Wang, Y. Li, C. Li. "Chaos-induced and mutation-driven schemes boosting salp chains-inspired optimizers". *IEEE Access*, vol. 7, pp. 31243-31261, 2019.

39. S.Z. Mirjalili, S. Mirjalili, S. Saremi, H. Faris, I. Aljarah. "Grasshopper optimization algorithm for multi-objective optimization problems". *Applied Intelligence*, vol. 48(4), pp. 805-820, 2018.

40. A. Tharwat, E.H. Houssein, M.M. Ahmed, A.E. Hassanien, T. Gabel. "Mogoa algorithm for constrained and unconstrained multi-objective optimization problems". *Applied Intelligence*, pp. 48(8), pp. 2268-2283, 2018.

41. A. Zhou, B.-Y. Qu, H. Li, S.-Z. Zhao, P.N. Suganthan, Q. Zhang. "Multiobjective evolutionary algorithms: A survey of the state of the art". *Swarm and Evolutionary Computation*, vol. 1(1), pp. 32-49, 2011.

42. B.Y. Qu, Y.S. Zhu, Y.C. Jiao, M.Y.Wu, P.N. Suganthan, J.J. Liang. "A survey on multi-objective evolutionary algorithms for the solution of the environmental/economic dispatch problems". *Swarm and Evolutionary Computation*, vol. 38, pp. 1-11, 2018.

43. S.S. Alresheedi, S. Lu, M.A. Elaziz, A.A. Ewees. "Improved multiobjective salp swarm optimization for virtual machine placement in cloud computing". *Human-centric Computing and Information Sciences*, vol. 9(1), pp. 15-38, 2019.

22

Social Spider Optimization – Modifications and Applications

Ahmed F. Ali
Department of Computer Science
Suez Canal University, Ismaillia, Egypt

Mohamed A. Tawhid
Department of Mathematics and Statistics, Faculty of Science
Thompson Rivers University, Kamloops, Canada

CONTENTS

22.1 Introduction

The social spider optimization (SSO) is a resent swarm intelligence algorithm proposed by Cuevas et al. [1]. The SSO algorithm simulates the social behavior of the social spiders and their movements on the communal web. The SSO algorithm starts by generating random solutions (spiders) which contain female and male spiders. The number of the females is greater than the number of the males in the colony (population). Each spider transmits its information through the communal web by encoding it as small vibrations. The weight of each spider (fitness function) is presented by calculating its fitness function value. The female spiders update their positions according to the attraction toward or the repulsion from other males. There are two types of male spiders according to their weights. The spider male with the weight greater than the median weight of the other males in the colony is the dominant male, while the other males with weight less than the median weight of other males are non-dominant males. The dominant male is responsible for mating the females in his mating range. The males with the highest weight have a big chance of influencing a new offspring. This chapter is structured as follows. In Section 22.2, we give a brief introduction to the SSO algorithm and how it works. We present some modifications of the SSO algorithm in the female and male cooperative operators in Section 22.3. In Section 22.4, we solve an economic load dispatch problem by a modification of the SSO algorithm. Finally, we summarize the conclusion in Section 22.5.

22.2   Original SSO algorithm in brief

In this section, we highlight the main steps of the SSO algorithm. These steps are shown in Algorithm 24.

22.2.1   Description of the original SSO algorithm

The SSO algorithm starts by initializing the values of the number of solutions N in the population size S, threshold PF and maximum number of iterations max_{itr} as shown in step 1. The number of females N_f and males N_m solutions are assigned in steps 2 as shown in Equations 22.1–22.2.

$$N_f = floor[(0.9 - rand(0,1) \cdot 0.25) \cdot N] \qquad (22.1)$$

$$N_m = N - N_f \qquad (22.2)$$

The counter of the initial iteration t is initialized in step 3. In steps 4–7, the initial female solutions are randomly generated as shown in Equation 22.3.

$$f_{i,j}^0 = p_j^{low} + rand(0,1) \cdot (p_j^{high} - p_j^{low}) \qquad (22.3)$$
$$i = 1, 2, \ldots, N_f; j = 1, 2, \ldots, n$$

The initial male solutions are randomly generated as shown in Equation 22.4 in steps 9–13.

$$m_{k,j}^0 = p_j^{low} + rand(0,1) \cdot (p_j^{high} - p_j^{low}) \tag{22.4}$$
$$k = 1, 2, \ldots, N_m; j = 1, 2, \ldots, n$$

Each solution in the population is evaluated by calculating its weight (fitness function) w_i by assigning the best $best_s$ and worst $worst_s$ solutions as shown in Equation 22.5 in steps 15–17.

$$w_i = \frac{J(s_i) - worst_s}{best_s - worst_s} \tag{22.5}$$

In step 19, the vibrations (transmitted information) between the solution i and the best solution b (s_b) in the population can be defined as follows.

$$Vibb_i = w_b \cdot e^{-d_{i,b}^2} \tag{22.6}$$

while the vibrations (transmitted information) between the solution i and the nearest female solution f (s_f) can be defined as

$$Vibf_i = w_f \cdot e^{-d_{i,f}^2} \tag{22.7}$$

The female spiders (solutions) update their positions in steps 20–25, where α, β, δ and rand are random numbers in [0,1], whereas t is the number of iterations as shown in Equation 22.8.

$$f_i^{t+1} = \begin{cases} f_i^t + \alpha \cdot Vibc_i.(s_c - f_i^t) + \beta \cdot Vibb_i \cdot (s_b - f_i^t) \\ \qquad\qquad +\delta \cdot (rand - 0.5) \quad \text{at } PF \\ f_i^t - \alpha \cdot Vibc_i.(s_c - f_i^t) - \beta \cdot Vibb_i \cdot (s_b - f_i^t) \\ \qquad\qquad +\delta \cdot (rand - 0.5) \quad \text{at } 1 - PF \end{cases} \tag{22.8}$$

where α, β, δ and rand are random numbers in [0,1], whereas t is the number of iterations.

The dominant spider with a weight value above the median value of the other males in the population is calculated in step 26. In steps 27–34, the male spiders update their positions as shown in Equation 22.9

$$m_i^{t+1} = \begin{cases} m_i^t + \alpha \cdot Vibf_i \cdot (s_f - m_i^t) + \delta \cdot (rand - 0.5) & \text{if } w_{N_f+i} > w_{N_f+m} \\ m_i^t + \alpha \cdot \left(\frac{\sum_{h=1}^{N_m} m_h^t \cdot w_{N_f+h}}{\sum_{h=1}^{N_m} w_{N_f+h}} - m_i^t \right) \end{cases} \tag{22.9}$$

where the solution s_f represents the nearest female solution to the male solution i.

The range r (range of mating) is calculated in step 35.

In steps 36–46, the spider with a heavier weight has a big chance to influence the new product. The influence probability Ps_i of each solution is assigned by the roulette wheel selection method as follows.

$$Ps_i = \frac{w_i}{\sum_{j \in T^t} w_j} \tag{22.10}$$

The iteration counter is increasing in step 47. Finally, the overall best solution is submitted in step 49.

Algorithm 24 Social spider optimization algorithm.

1: Set the initial value of total number of solutions N in the population size S, threshold PF, and maximum number of iterations Max_{itr}
2: Set the number of female spiders N_f and number of males spiders N_m
3: Set $t := 0$ ▷ **Counter initialization**
4: **for** $(i = 1; i < N_f + 1; i + +)$ **do**
5:    **for** $(j = 1; j < n + 1; j + +)$ **do**
6:      $f_{i,j}^t = p_j^{low} + rand(0,1) \cdot (p_j^{high} - p_j^{low})$
7:    **end for**
8: **end for**   ▷ **Initialize randomly the female spider**
9: **for** $(k = 1; k < N_m + 1; k + +)$ **do**
10:    **for** $(j = 1; j < n + 1; j + +)$ **do**
11:      $m_{k,j}^t = p_j^{low} + rand(0,1) \cdot (P_j^{high} - p_j^{low})$
12:    **end for**
13: **end for**   ▷ **Initialize randomly the male spider**
14: **repeat**
15:    **for** $(i = 1; i < N + 1; i + +)$ **do**
16:      $w_i = \frac{J(s_i) - worst_s}{best_s - worst_s}$
17:    **end for**   ▷ **Evaluate the weight (fitness function) of each spider**
18:    **for** $(i = 1; i < N_f + 1; i + +)$ **do**
19:      Calculate the vibrations of the best local and best global solutions $Vibc_i$ and $Vibb_i$
20:      **if** $(r_m < PF)$ **then**
21:        $f_i^{t+1} = f_i^t + \alpha \cdot Vibc_i \cdot (s_c - f_i^t) + \beta \cdot Vibb_i \cdot (s_b - f_i^t) + \delta \cdot (rand - 0.5)$
22:      **else**
23:        $f_i^{t+1} = f_i^t - \alpha \cdot Vibc_i \cdot (s_c - f_i^t) - \beta \cdot Vibb_i \cdot (s_b - f_i^t) + \delta \cdot (rand - 0.5)$
24:      **end if**
25:    **end for**
26:    Find the median male individual $(w_{N_f + m})$ from M
27:    **for** $(i = 1; i < N_m + 1; i + +)$ **do**
28:      Calculate $Vibf_i$
29:      **if** $(w_{N_f i} > w_{N_f + m})$ **then**
30:        $m_i^{t+1} = m_i^t + \alpha \cdot Vibf_i \cdot (s_f - m_i^t) + \delta \cdot (rand - 0.5)$
31:      **else**
32:        $m_i^{t+1} = m_i^t + \alpha \cdot \left(\frac{\sum_{h=1}^{N_m} m_h^t \cdot w_{N_f + h}}{\sum_{h=1}^{N_m} w_{N_f + h}} - m_i^t \right)$
33:      **end if**
34:    **end for**
35:    Calculate the radius of mating r, where $r = \frac{\sum_{j=1}^{n} (p_i^{high} - p_j^{low})}{2 \cdot n}$ ▷ **Perform the mating operation**
36:    **for** $(i = 1; i < N_m + 1; i + +)$ **do**
37:      **if** $(m_i \in D)$ **then**
38:        Find E^i
39:        **if** E^i is not empty **then**
40:          Form s_{new} using the roulette method

```
41:                if w_new > w_wo then
42:                    Set s_wo = s_new
43:                end if
44:            end if
45:        end if
46:    end for
47:    t = t + 1                          ▷ Iteration counter is increasing
48: until (t ≥ Max_itr)                   ▷ Termination criteria are satisfied
49: Produce the best solution.
```

22.3 Modifications of the SSO algorithm

It is known that the search process in a population-based meta-heuristic is categorized in two essential phases, namely, exploration and exploitation. A random behavior in the exploration phase is to figure out the search space as much as possible. On the other hand, the exploitation toward the promising regions is the essential objective of the latter phase. It is a difficult task to find a suitable balance between these two phases because of the stochastic nature of population-based meta-heuristic algorithms. Researchers show that chaotic maps are able to improve both phases. The literature indicates that combining the chaos theory into population-based algorithms is one of the efficient and effective methods for fostering both exploration and exploitation [8], [9], [10], [11], [12].

This work combined four chaotic maps (namely, Chebyshev, circle, Gauss/-mouse, and iterative) into the population-based meta-heuristic algorithm called Social Spider Optimization algorithm. We present the effect of replacing the random parameters α, β, δ and rand in Equations 22.8–22.9 by the chaotic maps $C1, C2, C3$, and $C4$ (Chebyshev, circle, Gauss/mouse, and iterative), respectively. The proposed algorithm is called Chaotic Social Spider Optimization (CSSO) algorithm as shown in the following subsections.

22.3.1 Chaotic maps

The chaotic maps show a random behavior although they have no random variables. The mathematical forms of these maps are shown in Table 22.1.

TABLE 22.1
Chaotic maps.

No Name	Chaotic map	Range
C1 Chebyshev [5]	$x_{i+1} = \cos(i\cos^{-1}(x_i))$	(-1,1)
C2 Circle [6]	$x_{i+1} = \mod(x_i + b - (\frac{a}{2\pi})\sin(2\pi x_k), 1), a = 0.5$ and $b = 0.2$	(0,1)
C3 Gauss/mouse [2]	$x_{i+1} = \begin{cases} 1 & x_i = 0 \\ \frac{1}{\mod(x_i,1)} & \text{otherwise} \end{cases}$	(0,1)
C4 Iterative [7]	$x_{i+1} = \sin(\frac{a\pi}{x_i}), a = 0.7$	(-1,1)

We invoke some of these maps in the female and male spiders position update equations instead of using the standard random parameters to increase the diversity of the search and avoid trapping in local minima. The initial point for the four chaotic maps is random number between $[0,1]$. We use the initial point $x^0 = 0.7$ as suggested in [4].

22.3.2   Chaotic female cooperative operator

The female spiders update their positions as shown in Equation 22.11 by replacing the random α, β, δ and rand parameters by four chaotic maps.

$$
f_i^{t+1} = \begin{cases} f_i^t + C1 \cdot Vibc_i \cdot (s_c - f_i^t) + C2 \cdot Vibb_i \cdot (s_b - f_i^t) \\ \qquad\qquad\qquad\qquad + C3 \cdot (C4 - 0.5) \quad \text{at } PF \\ f_i^t - C1 \cdot Vibc_i \cdot (s_c - f_i^t) - C2 \cdot Vibb_i \cdot (s_b - f_i^t) \\ \qquad\qquad\qquad\qquad + C3 \cdot (C4 - 0.5) \quad \text{at } 1 - PF \end{cases}
\tag{22.11}
$$

where $C1, C2, C3$, and $C4$ are chaotic maps and t is the number of iterations.

22.3.3   Chaotic male cooperative operator

The position of the male spider can be calculated as shown in Equation 22.12.

$$
m_i^{t+1} = \begin{cases} m_i^t + C1 \cdot Vibf_i \cdot (s_f - m_i^t) + C3 \cdot (C3 - 0.5) \quad \text{if } w_{N_f+i} > w_{N_f+m} \\ m_i^t + C1 \cdot \left(\dfrac{\sum_{h=1}^{N_m} m_h^t \cdot w_{N_f+h}}{\sum_{h=1}^{N_m} w_{N_f+h}} - m_i^t \right) \end{cases}
\tag{22.12}
$$

where the solution s_f represents the nearest female solution to the male solution i, and $C1, C2$ are chaotic maps.

22.4   Application of SSO algorithm for an economic load dispatch problem

In this section, we apply the modified SSO algorithm for solving the economic load dispatch problem (ELD)

22.4.1   Economic load dispatch problem

The economic load dispatch problem (ELD) [3] is a continuous optimization problem. In this problem, we try to find the optimal combination of power generation where the total production cost of the system is minimized. The convex and non-convex cost function of a generator can be defined as follows.

$$
F_i(P_i) = a_i P_i^2 + b_i P_i + c_i
\tag{22.13}
$$

$$F_i(P_i) = a_i P_i^2 + b_i P_i + c_i + |d_i sin[e_i \cdot (P_i^{min} - P_i)]| \qquad (22.14)$$

where P_i is the active power output, $F_i(P_i)$ is the generation cost, P_i^{min} is the minimum output limit of the generator and the a_i, b_i, c_i, d_i, e_i are the cost coefficients of the generator. The fuel cost of all generators can be defined in Equation 22.15.

$$MinF(P) = \sum_i^{Ng} a_i P_i^2 + b_i P_i + c_i + |d_i sin[e_i \cdot (P_i^{min} - P_i)]| \qquad (22.15)$$

where Ng is the number of generating units.

22.4.2   Problem Constraints

The ELD problem is a constrained optimization problem. The total power generated should cover the power demand and the active power losses as shown in Equation 22.16.

$$\sum_{i=1}^{Ng} P_i = P_D + P_{loss} \qquad (22.16)$$

where P_D is the total demand load and P_{loss} is the total transmission losses computed and it can be calculated as shown in Equation 22.17.

$$P_{loss} = \sum_{i=1}^{Ng} \sum_{j=1}^{Ng} P_i B_{ij} P_j \qquad (22.17)$$

where B_{ij} is the loss coefficient matrix.

The output power of each generator should be within a lower and an upper limit as shown in Equation 22.18.

$$P_i^{min} \leq P_i \leq P_i^{max} \qquad (22.18)$$

where P_i^{min} and P_i^{max} are the lower and the upper limit of the *ith* generators.

22.4.3   Penalty function

In order to transform the constrained optimization problem to an unconstrained optimization problem, we use the penalty function technique. We can handle the constraints of the ELD problem as shown in Equation 22.19.

$$\text{Min}(Q(P_i)) = \quad F_c(P_i) + pen \cdot G[h_k(P_i)] \qquad (22.19)$$
$$\text{Subject to}: \quad g_j(P_i) \leq 0, j = 1, \ldots, J$$

where *pen* is the penalty factor and $G[h_k(P_i)]$ is the penalty function calculated as shown in Equation 22.20.

$$G[h_k(P_i)] = \varepsilon^2 + \omega^2 \qquad (22.20)$$

where ε and ω are equality and inequality constraint violation penalties, respectively and can be calculated as follows.

$$\varepsilon = \left| P_D + P_{loss} - \sum_{j=1}^{Ng} P_j \right| \tag{22.21}$$

and

$$\omega = \begin{cases} |P_i^{min} - P_i| & P_i^{min} > P_i \\ 0 & P_i^{min} < P_i < P_i^{max} \\ |P_i - P_i^{max}| & P_i^{max} < P_i \end{cases} \tag{22.22}$$

22.4.4 How can the SSO algorithm be used for an economic load dispatch problem?

In this subsection, we apply the proposed CSSO algorithm to solve an economic load dispatch problem. The first step is the initialization of all parameters in the algorithm such as the population size S and maximum number of iterations Max_{itr} which are set to be 20 and 500 respectively. The number of female N_f and male N_m spiders are assigned. In step 2, the iteration counter t is initialized. In steps 4–13, the initial population is randomly generated where the problem dimension D is set to be 6 which is the length of the input data in Table 22.3 where the lower and upper bounds of the generated data are set to $L = \{10, 10, 35, 35, 130, 125\}$, $U = \{125, 150, 225, 210, 325, 315\}$. In steps 15-17, the objective function of each solution is evaluated by calculating its fitness function OF(.). The best $(best_s)$ and worst $(worst_s)$ solutions in the population are assigned; then the weight w_i of each solution is calculated. In steps 18-25, the vibrations of the best local and best global solutions $Vibc_i$ and $Vibb_i$ as shown in Equation 22.12 are calculated. In step 26, the median male individual (w_{N_f+m}) is calculated. In steps 27–34, the male solutions are updated after calculating the radius of mating. The overall process is repeated until the maximum number of iterations is reached.

22.4.5 Description of experiments

We test the proposed CSSO algorithm on six-generator test systems for a total demand of 700 MW and 800 MW. We show the parameters of the proposed CSSO in Table 22.2. To simplify the problem, we set the values of parameters d, e in Equation 22.14 to zero. The limits of the generator active power and the coefficients of the fuel cost are given in Tables 22.3–22.4, respectively. We

TABLE 22.2

Parameter setting.

Parameters	Definitions	Values
n	Population size	20
l	Attractive length scale	1.5
f	Intensity of attraction	0.5
PF	Probability threshold	0.7
Max_{itr}	Maximum number of iterations	500

TABLE 22.3

Limits of the generator active power.

Generator	1	2	3	4	5	6
$P_{m}in(MW)$	10	10	35	35	130	125
$P_{m}ax(MW)$	125	150	225	210	325	315

TABLE 22.4

The coefficients of the fuel cost.

No.	a	b	c
1	0.15240	38.53973	756.79886
2	0.10587	46.15916	451.32513
3	0.02803	40.39655	1049.9977
4	0.03546	38.30553	1243.5311
5	0.02111	36.32782	1658.5596
6	0.01799	38.27041	1356.6592

set the number of the population size to 20 and the maximum number of iterations Max_{itr} to 500.

22.4.6 Results obtained

In Figures 22.1–22.2, we show the performance of the proposed CSSO algorithm at total system demand at 700, 800 MW by plotting the number of iterations versus the cost($/h). Also, to test the efficiency of the proposed CSSO algorithm, we compare it with the standard SSO algorithm on six-generator test systems at total demand of 700 MW and 800 MW. We apply the same termination criteria by setting the error tolerance to 0.01 MW or when the maximum number of iterations Max_{itr} reaches 500. We report the best, average (Avg) and standard deviation (Std) results in Tables 22.5–22.6. The results in Tables 22.5–22.6 show that the proposed CSSO algorithm is a promising algorithm and its performance is better than the standard SSO algorithm.

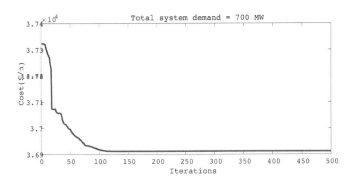

FIGURE 22.1
Total system demand at 700 MW.

TABLE 22.5
Comparison between CSSO and SSO total system demand = 700 MW.

Algorithm		P1(MW)	P2(MW)	P3(MW)	P4(MW)	P5(MW)	P6(MW)	Loss(MW)	Cost($/h)
	Best	28.0745	10.0572	119.987	117.7735	231.1342	212.397	19.4241	36911.82
SSO	Avg	28.3961	10.2698	119.162	119.0377	230.2975	212.255	19.4099	36912.75
	Std	0.6923	0.2655	2.2236	1.7081	2.9535	3.79	0.05938	1.0014
	Best	28.1475	10.0377	119.715	118.048	231.0216	212.417	19.4304	36911.18
CSSO	Avg	28.3622	10.0211	119.0855	118.582	231.1235	212.709	19.4225	36911.75
	Std	0.1477	0.0115	0.4114	0.38235	0.08133	0.2066	0.0132	0.9145

TABLE 22.6
Comparison between CSSO and SSO total system demand = 800 MW.

Algorithm		P1(MW)	P2(MW)	P3(MW)	P4(MW)	P5(MW)	P6(MW)	Loss(MW)	Cost($/h)
	Best	32.46756	14.34413	141.9085	135.7281	257.7262	243.1414	25.3342	41895.74
SSO	Avg	32.58655	14.49139	141.7118	136.2047	257.3585	242.9544	25.3224	41896.15
	Std	0.38255	0.49511	0.97066	0.88625	1.2138	1.3818	0.037027	0.25815
	Best	32.4951	14.34415	141.9048	135.7519	257.7254	243.1414	25.3341	41895.53
CSSO	Avg	32.5745	14.4824	141.5423	135.8414	257.6575	243.0017	25.3384	41895.71
	Std	0.09235	0.09821	0.2135	0.05445	0.9441	0.9462	0.0772	0.09919

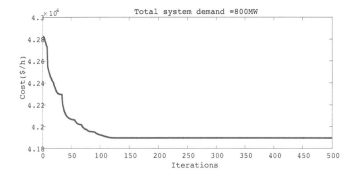

FIGURE 22.2
Total system demand at 800 MW.

22.5 Conclusions

In this chapter, we present a social spider optimization (SSO) algorithm. We apply some modifications to the standard SSO algorithm. The modified SSO algorithm is called the Chaotic Social Spider Optimization (CSSO) algorithm. We solve an economic load dispatch problem by CSSO. The chaotic maps are able to enhance both exploitation and exploration of SSO. Also, we test the modified SSO algorithm on six-generator test systems for a total demand of 700MW and 800MW and compare with the standard SSO algorithm. The results show that the CSSO algorithm is a promising algorithm and can solve an economic load dispatch problem in reasonable time.

References

1. E. Cuevas, M. Cienfuegos, D. Zaldívar and M. Pérez-Cisneros. "A swarm optimization algorithm inspired in the behavior of the social-spider". *Expert Systems with Applications*, vol. 40, no. 16, pp. 6374–6384, 2013.

2. V. Jothiprakash, R. Arunkumar. "Optimization of hydropower reservoir using evolutionary algorithms coupled with chaos". *Water Resour Manag*, vol. 27, no. 7, pp. 1963–1979, 2013.

3. F.E.R.C. Staff. "Economic Dispatch: Concepts, Practices and Issues". Presentation to the Joint Board for the Study of Economic Dispatch, RC Staff, Palm Springs, California, November 13, 2005.

4. S. Saremi,S. Mirjalili, and A. Lewis. "Grasshopper optimisation algorithm: Theory and application". *Advances in Engineering Software*, vol. 105, pp. 30–47, 2017.

5. N. Wang, L. Liu. "Genetic algorithm in chaos". *OR Trans*, vol. 5, pp. 1–10, 2001.

6. LJ. Yang, TL. Chen. "Application of chaos in genetic algorithms". *Commun Theor Phys* vol. 38, pp. 168–172, 2002.

7. G. Zhenyu, C. Bo, Y. Min, C. Binggang. "Self-Adaptive Chaos Differential Evolution. In: Jiao L., Wang L., Gao X., Liu J., Wu F. (eds) Advances in Natural Computation. ICNC 2006. Lecture Notes in Computer Science, vol. 4221, pp. 972–975, Springer, 2006.

8. D. Yang, G. Li, G. Cheng. "On the efficiency of chaos optimization algorithms for global optimization". Chaos, Solitions & Fractals, vol. 34 pp. 136–1375, 2007.

9. G. Gharooni-fard, F. Moein-darbari, H. Deldari, A. Morvaridi. "Scheduling of scientific workflows using a chaos-genetic algorithm". *Procedia Computer Science*, vol. 1, pp. 1445–1454, 2010.

10. A.H. Gandomi, G.J. Yun, X.S. Yang, S. Talatahari. "Chaos-enhanced accelerated particle swarm algorithm". *Communications in Nonlinear Science and Numerical Simulation*, vol. 18, no. 2, pp. 327–340, 2013.

11. B. Alatas. "Chaotic bee colony algorithms for global numerical optimization". *Expert Systems with Applications*, vol. 37, pp. 5682–5687, 2010.

12. J. Mingjun, T. Huanwen. "Application of chaos in simulated annealing". *Chaos, Solitons & Fractals*, vol. 21, pp. 933–941, 2004.

23

Stochastic Diffusion Search: Modifications and Application

Mohammad Majid al-Rifaie

School of Computing and Mathematical Sciences, University of Greenwich
Old Royal Naval College, London, United Kingdom

J. Mark Bishop

Department of Computing
Goldsmiths, University of London, United Kingdom

CONTENTS

23.1 Introduction

Stochastic Diffusion Search (SDS) [1] introduced a new probabilistic approach for solving best-fit pattern recognition and matching problems. SDS, as a multi-agent population-based global search and optimisation algorithm proposed in 1989, is a distributed mode of computation utilising interaction between simple agents [2].

SDS has been applied to a wide range of applications and problems, including but not limited to computer vision and medical imaging [3, 32, 5, 6, 7, 8] optimisation [9, 10, 11, 12] machine learning [13, 14, 15, 16, 17, 18] computational creativity [19, 20, 21, 22, 23] digital arts [24, 25, 26, 27, 28] and other theoretical or practical domains [29, 30, 31, 32, 33, 34, 35, 36, 37].

This paper first provides a brief description of the algorithm, and then a set of suggested variations and modifications is presented. These include an introduction to various recruitment strategies which govern the diffusion of information amongst the population. In addition to providing details about the initialisation of agents' hypotheses and their criteria for the algorithm termination, the concept of partial function evaluation is also expanded. The paper is then concluded with a real-world application of SDS for identifying metastasis in bone scans and how the algorithm's features are adopted to deal with this particular problem.

23.2   SDS algorithm

SDS algorithm commences a search or optimisation by initialising its population (e.g. agents). In any SDS search, each agent maintains a hypothesis, h, defining a possible problem solution. After initialisation two phases are followed (see the high-level description of SDS in Algorithm 25):

- Test Phase

- Diffusion Phase

In the Test Phase, SDS checks whether the agent hypothesis is successful or not by performing a *partial function evaluation, pFE*, which is some function of the agent's hypothesis; $pFE = f(h)$; subsequently, this evaluation returns a domain independent boolean value. Later in the iteration, contingent on the strategy employed, successful hypotheses diffuse across the population and in this way information on potentially good solutions spreads throughout the entire population of agents.

In the Diffusion Phase, each agent recruits another agent for interaction and potential communication of the hypothesis. Various flavours of the communication strategies which are deployed in the Diffusion Phase are explained in the next section.

Algorithm 25 SDS Algorithm.

```
1   Initialising agents()
2   While (stopping condition is not met)
3      Testing hypotheses()
4         Determining agents' activities (active/inactive)
5      Diffusing hypotheses()
6         Exchanging of information
7   End While
```

23.3 Further modifications and adjustments

In this section, further relevant details related to SDS are discussed, including different recruitment strategies, initialisation of agents, termination criteria and partial function evaluation.

In SDS, similar to other optimisation algorithms, the goal is finding the best solution based on the criteria specified in the objective function. The collection of all candidate solutions (hypotheses) forms the search space and each point in the search space is represented by an objective value, from which the objective function is formed. In the minimisation mode, for example, the lower the objective value is, the better the result.

One of the issues related to SDS is the mechanism behind allocating resources to ensure that while potential areas of the problem space are exploited, exploration is not ignored. For this purpose, different recruitment methods, where one agent recruits another one, are investigated:

23.3.1 Recruitment Strategies

Three recruitment strategies are proposed in [38]: active, passive and dual. These strategies are used in the Diffusion Phase of SDS. Each agent can be in either one of the following states: It is *active* if the agent is successful in the Test Phase; an agent is *inactive* if it is not successful; and it is *engaged* if it is invloved in a communication with another agent.

The standard SDS algorithm [1] uses the passive recruitment mode, which will be described next along with other recruitment modes.

23.3.1.1 Passive recruitment mode

In the *passive recruitment mode*, if the agent is not active, another agent is randomly selected and if the randomly selected agent is active, the hypothesis of the active agent is communicated (or *diffused*) to the inactive one. Otherwise a new random hypothesis is generated for the inactive agent and there will be no flow of information (see Algorithm 26).

Algorithm 26 Passive Recruitment Mode.

```
1    For ag = 1 to No_of_agents
2        If ( !ag.activity() )
3            r_ag = pick a random agent()
4            If ( r_ag.activity() )
5                ag.setHypothesis( r_ag.getHypothesis() )
6            Else
7                ag.setHypothesis( randomHypothesis() )
8            End If/Else
9        End If
10   End For
```

23.3.1.2   Active recruitment mode

In the *active recruitment mode*, active agents are in charge of communication with other agents. An active agent randomly selects another agent. If the randomly selected agent is neither active nor engaged in communication with another active agent, then the hypothesis of the active agent is communicated to the inactive one and the agent is flagged as engaged. The same process is repeated for the rest of the active agents. However if an agent is neither active nor engaged, a new random hypothesis is generated for it (see Algorithm 27).

Algorithm 27 Active Recruitment Mode.

```
1    For ag = 1 to No_of_agents
2        If ( ag.activity() )
3            r_ag = pick a random agent()
4            If ( !r_ag.activity() AND !r_ag.getEngaged() )
5                r_ag.setHypothesis( ag.getHypothesis() )
6                r_ag.setEngaged(true)
7            End If
8        End If
9    End For
10
11   For ag = 1 to No_of_agents
12       If ( !ag.activity() AND !ag.getEngaged() )
13           ag.setHypothesis( randomHypothesis() )
14       End If
15   End For
```

23.3.1.3   Dual recruitment mode

In *dual recruitment mode*, both active and inactive agents randomly select other agents. When an agent is active, another agent is randomly selected. If the randomly selected agent is neither active nor engaged, then the hypothesis of the active agent is shared with the inactive one and the inactive agent is flagged as engaged. Also, if there is an agent which is neither active nor engaged, it selects another agent randomly. If the newly selected agent is active, there will be a flow of information from the active agent to the inactive one and the inactive agent is flagged as engaged. Nevertheless, if there remains an agent that is neither active nor engaged, a new random hypothesis is chosen for it.

Algorithm 28 Dual Recruitment Mode.

```
1   For ag = 1 to No_of_agents
2       If ( ag.activity() )
3           r_ag = pick a random agent()
4           If ( !r_ag.activity() AND !r_ag.getEngaged() )
5               r_ag.setHypothesis( ag.getHypothesis() )
6               r_ag.setEngaged(true)
7           End If
8       Else
9           r_ag = pick a random agent ()
10          If ( r_ag . activity () AND ! ag . getEngaged () )
11              ag . setHypothesis ( r_ag . getHypothesis () )
12              ag . setEngaged ( true )
13          End If
14      End If/Else
15  End For
16
17  For ag = 1 to No_of_agents
18      If ( !ag.activity() AND !ag.getEngaged() )
19          ag.setHypothesis( randomHypothesis() )
20      End If
21  End For
```

23.3.1.4 Context sensitive mechanism

Comparing the above-mentioned recruitment modes, it is theoretically determined in [38] that robustness and greediness decrease in the active recruitment mode. Conversely, these two properties are increased in dual recruitment strategy. Although, the greediness of dual recruitment mode results in decreasing the robustness of the algorithm, the use of *Context Sensitive Mechanism* limits this decrease [38, 39]. In other words, the use of context sensitive mechanism biases the search towards global exploration. In the context sensitive mechanism if an active agent randomly chooses another active agent that maintains the same hypothesis, the selecting agent is set inactive and adopts a random hypothesis. This mechanism frees up some of the resources in order to have a wider exploration throughout the search space as well as preventing cluster size from overgrowing, while ensuring the formation of large clusters in case there exists a perfect match or good sub-optimal solutions (see Algorithm 29).

Algorithm 29 Context Sensitive Mechanism.

```
1   If ( ag.activity() )
2       r_ag = pick a random agent ()
3       If ( r_ag.activity() AND
4               ag.getHypothsis() == r_ag.getHypothsis() )
5           ag.setActivity ( false )
6           ag.setHypotheis ( randomHypothsis() )
7       End If
8   End If
```

23.3.1.5   Context free mechanism

In *Context Free Mechanism,* which is another recruitment mechanism, the performance is similar to context sensitive mechanism, where each active agent randomly chooses another agent. However, if the selected agent is active (irrespective of having the same hypothesis or not), the selecting agent becomes inactive and picks a new random hypothesis. By the same token, this mechanism ensures that even if one or more good solutions exist, about half of the agents explore the problem space and investigate other possible solutions (see Algorithm 30).

Algorithm 30 Context Free Mechanism.

```
1    If ( ag.activity() )
2        r_ag = pick a random agent ()
3        If ( r_ag.activity() )
4            ag.setActivity ( false )
5            ag.setHypotheis ( randomHypothsis() )
6        End If
7    End If
```

23.3.2   Initialisation and termination

Although normally agents are uniformly distributed throughout the search space, if the search space is of a specific type, or knowledge exists about it *a priori*, it is possible to use a more intelligent (than uniform random distribution) startup by biasing the initialisation of the agents.

If there is a pre-defined pattern to find in the search space, the goal will be locating the best match or, if it does not exist, its best instantiation in the search space [40]. Similarly, in a situation which lacks a pre-defined pattern, the goal will be finding the best pattern in accordance with the objective function.

In both cases, it is necessary to have a termination strategy. In one method[1], SDS terminates the process when a statistical equilibrium state is reached, which means that the threshold of the number of active agents is exceeded and the population maintained the same state for a specified number of iterations. In [41], four broad types of halting criteria are introduced:

1. No stopping criterion, where the user interrupts the course of action of the search or optimisation and is usually preferred when dealing with *dynamically changing* problem spaces or when there is no pre-defined pattern to look for

2. Time-based criterion, in which passing a pre-set duration of time is the termination point of the algorithm

3. Activity-based criterion, which is a problem-dependent halting criterion, and it is the most prevalent form in the SDS algorithm.

---

[1]Ibid

The termination of the process is decided upon through monitoring the overall activity of the agents (e.g. reaching a certain user defined activity level, reaching a stable population state after a sudden increase in their activities)

4. Cluster-based criterion that keeps track of the formation of stable clusters.

Determining the termination criteria without having a fixed a priori threshold as a prerequisite is a possible approach (e.g. Quorum sensing, which is a system of stimulus and response correlated to population density, is used in some social insects in search for nest sites or many species of bacteria to coordinate gene expression based on the density of their local population [42]). By the same analogy and as stated in [41], it is possible to implement the termination criterion as a random sampling process: for example, a cluster-based termination procedure may consider monitoring the hypotheses of a subset of the population until the same hypothesis is encountered more than once. This provides partial evidence of the formation of a cluster. The size of the sample taken from the population can then be increased.

The two most common termination strategies in SDS (introduced in [40]) are the following:

• Weak halting criterion is the ratio of the active agents to the total number of agents. In this criterion, cluster sizes are not the main concern.

• Strong halting criterion investigates the number of active agents that form the largest cluster of agents all adopting the same hypothesis.

Therefore, the choice of the halting mechanism is based on whether to favour the active agents in the whole of the agent populations (weak halting mechanism), which is similar to the activity-based criterion, or to consider the largest cluster of active agents (strong halting mechanism), which is similar to the cluster-based criterion.

23.3.3 Partial function evaluation

One of the concerns associated with many optimisation algorithms (e.g. Genetic Algorithm [43], Particle Swarm Optimisation [44] etc.) is the repetitive evaluation of computationally expensive fitness functions. In some applications, such as tracking a rapidly moving object, the repetitive function evaluation significantly increases the computational cost of the algorithm. Therefore, in addition to reducing the number of function evaluations, other measures should be taken in order to reduce the computations carried out during the evaluation of each possible solution as part of the optimisation or search processes.

The commonly used benchmarks for evaluating the performance of swarm intelligence algorithms are typically small in terms of their objective functions computational costs [45, 46], which is often not the case in real-world applications. Examples of costly evaluation functions are seismic data interpretation [46], selection of sites for the transmission infrastructure of wireless communication networks and radio wave propagation calculations of one site [37] etc.

Costly functions have been investigated under different conditions [47] and the following two broad approaches have been proposed to reduce the cost of function evaluations:

• The first is to estimate the fitness by taking into account the fitness of the neighbouring elements, the former generations or the fitness of the same element through statistical techniques introduced in [48, 49].

• In the second approach, the costly fitness function is substituted with a cheaper, approximate fitness function.

When agents are about to converge, the original fitness function can be used for evaluation to check the validity of the convergence [47].

Many fitness functions are decomposable to components that can be evaluated separately. In partial evaluation of the fitness function in SDS, the evaluation of one or more of the components may provide partial information and means for guiding the optimisation.

The partial function evaluation of SDS allows the algorithm to absorb certain types of noise in the objective function without affecting the convergence time or the size and stability of the clusters.

Additionally, noise, which does not alter the averaged probabilities of the test score (probability of producing active agents during the test phase, averaged over all component functions) but increases the variance in the evaluation of component functions, has no effect on the resource allocation process of SDS [50]. However, if the value of the test score changes as a result of noise presence, the resource allocation process may be influenced either:

• positively if the value of the test score increases.

• or negatively if the value of the test score decreases.

23.4 Application: Identifying metastasis in bone scans

This section discusses the application of SDS as a tool to identify metastasis in bone scans. This work has been originally published in [6, 7]. SDS is adapted for this particular purpose and its performance is investigated by running SDS agents on sample bone scans (see Fig. 23.1-Top) whose statuses have been determined by the experts.

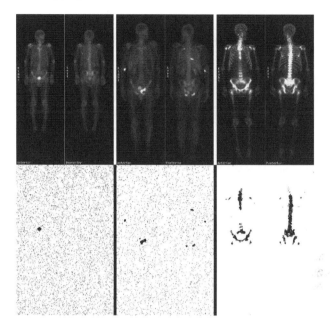

FIGURE 23.1
Bone Scans. Top: Typically 2-6 hours after intravenous administration of technetium-99m-labelled diphosphonates. Brighter areas indicate a higher radiotracer uptake. Bottom: The scans are processed using Stochastic Diffusion Search algorithm. Left: Healthy; middle: partially affected; right: metastatic disease spread.

23.4.1 Experiment setup

Each scan, as shown in Fig. 23.1 (Top) are processed by the SDS agents which are responsible for locating the affected area(s). According to the description provided by the experts, Fig. 23.1 (Top-middle and right) are the areas of metastasis.

The reproducibility and the accuracy of the SDS algorithm can be utilised in developing a standardised system to interpret the bone scans preventing operator errors and discrepancies. This technology can be employed as an adjunct by radiologists to assess the various parts of the bone scan making the diagnosis of the lesions more thorough and less time consuming. Additionally this technique can be effectively used to develop programs for teaching and training medical students and junior doctors.

The number of agents used in this experiment is 10,000 and the algorithm is run for 10 iterations (i.e. 10 cycles of test and diffusion phases). At the beginning of the process, all the agents are initialised randomly throughout the search space. In other words, each agent randomly picks a pixel from the image of the bone scan (i.e. one pixel from 460×690). During the test phase of SDS

o	o	o
o	x	o
o	o	o

FIGURE 23.2
Agent's Neighbours in Test Phase. The symbol x represents the position of the agent and the o's represent the neighbours used during the test phase.

o	o	o	o	o
o	o	o	o	o
o	o	x	o	o
o	o	o	o	o
o	o	o	o	o

FIGURE 23.3
Diffusion Area. The symbol x represents the position of the active agent and the o's represent the accessible places during the diffusion phase.

algorithm, each agent's status should be determined. The method used here to set the activity of the agents is to find the average of the colour intensity[2] ($avgIn$) of each agent and its neighbours (see Fig. 23.2). If $avgIn > 180$ the agent is flagged active, otherwise inactive[3].

Note that in this application, standard SDS which uses passive recruitment mode, is employed.

During the diffusion phase, each inactive agent randomly selects another agent from the population; if the selected agent is active, the selecting agent adopts the hypothesis (i.e. location) of the active agent and the information sharing takes place. The strategy used for information sharing is to randomly pick an area surrounding the active agent (see Fig. 23.3). Active agents also check their position by continuously picking a random pixel in the neighbourhood; this way, an area which does not have a good enough potential is discarded from one iteration to the next.

23.4.2   Results

As shown in Fig. 23.1 (Bottom), areas with higher potential of metastasis are identified. Other than urinary bladder activity, faint renal activity, and minimal soft-tissue activity which are normally present in the scan (Fig. 23.1 Bottom-left), the existence of multiple, randomly distributed areas of increased uptake of varying size, shape, and intensity are highly suggestive

---

[2]Colour intensity signifies the brightness of each pixel, which is a spectrum between 0 and 255.

[3]The value of 180 is problem-dependent and could be adjusted to increase or decrease the sensitivity of the system.

of bone metastases (Fig. 23.1 Bottom-middle). Additionally as stated before, when the metastatic process is distributed, almost all of the radiotracer congregates in the skeleton, with little or no activity in the soft tissues or urinary tract (Fig. 23.1 Bottom-right).

In order to show the behaviour of the algorithm in each iteration, the position of each agent during the process of metastasis identification is illustrated in Fig. 23.4. In this figure, the three original bone scans referred to earlier are used as input to the system, and as the figure shows, successful agents diffuse their positions across the population and this way, information on potentially good solutions spreads throughout the entire population of agents. This process is caused through the recruitment strategy, where each agent recruits another agent for interaction and potential communication of the promising areas.

In order to decide whether the activity of the agents when presented with different types of bone scans (e.g. not affected, affected and highly affected), would be a distinctive indicator, a statisical analysis is performed. TukeyHSD Test [51] is used to highlight whether there is a significant difference between the activity of the agents when processing the bone scans. Table 23.1 (a) shows the activity rate of the populations over each iteration. Three different samples are used for this analysis: Samples 1, 2 and 3 refer to the scans

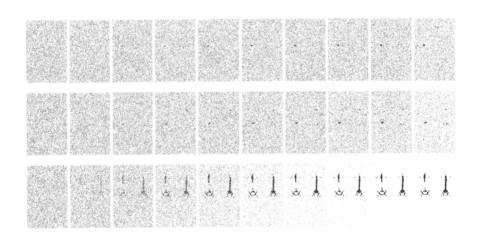

FIGURE 23.4
SDS algorithm processing bone scans in 10 iterations. Each row shows the behaviour of the agents when presented with one bone scan. Each bone scan is processed by 10,000 agents (illustrated as black dots) and through communication, agents explore different areas of the bone scan to identify potential areas of metastasis. The leftmost figures in each row show the location of the agents on the first iteration, and the rightmost ones represent the last iteration. Top: Healthy; middle: partially affected; bottom: metastatic disease spread.

TABLE 23.1

Activity Status of Agents.

Itr	Sample 1	Sample 2	Sample 3
0	0±0	0±0	0±0
1	5±2	17±4	277±16
2	15±4	47±9	763±37
3	33±8	100±18	1602±76
4	66±18	201±31	2991±137
5	129±33	379±51	4992±188
6	245±62	697±84	7260±198
7	461±110	1250±141	8947±123
8	852±201	2201±230	9583±51
9	1557±351	3650±330	9708±22

(a) Mean ±standard deviation of the number of active agents in each iteration is shown (rounded to the nearest number).

Itr	s1 − s2	s1 − s3	s2 − s3
0	−	−	−
1	o − X	o − X	o − X
2	o − X	o − X	o − X
3	o − X	o − X	o − X
4	o − X	o − X	o − X
5	o − X	o − X	o − X
6	o − X	o − X	o − X
7	o − X	o − X	o − X
8	o − X	o − X	o − X
9	o − X	o − X	o − X

(b) Based on TukeyHSD Test, if the difference between each pair of samples is significant, the pairs are marked (o − X shows that the right sample has significantly more active agents than the left one). This test uses 95% family-wise confidence level. The aim is to show that agents dealing with scans which have different levels of metastasis exhibit significantly different behaviour.

in Fig. 23.1 (left to right). Table 23.1 (b) shows that other than the first iteration where the agents are just initialised, different bone scans would result in significantly different activity rates. This could be used as an indicator, highlighting the difference between various scans and whether they are healthy, partially affected or the metastasis is spread.

23.4.3   Concluding remarks

The results of applying SDS to detect areas of metastasis in this experiment demonstrate the suitability of the approach. A statistical analysis further investigates the behaviour of the agents in the population and the outcome highlights that the algorithm exhibits a statistically significant difference when applied to bone scans for healthy, partially affected or heavily affected individuals. Finally, it is important to note that the presented technique could be effectively utilised as an adjunct to the experts' eyes of a specialist.

23.5 Conclusion

This paper provides the principal topics on the standard SDS algorithm followed by details of various flavours to SDS; these include, recruitment strategies, initialisation of agents and termination of the process as well as partial function evaluation. The paper then provides a real-world application of SDS where it is used to identify metastasis in bone scans. In this medical imaging problem, a set of clinically pre-determined bone scans is presented to SDS algorithm which are then processed and the outcomes are analysed. While this application is designed to be an adjunct to a clinician, the process can be expanded and further complexity can be added to identify different types of tissues and affected organs.

References

1. J.M. Bishop. "Stochastic Searching Networks" in *Proc. of 1st IEE Conf. on Artificial Neural Networks*, pp. 329-331, 1989.

2. K. de-Meyer, J.M. Bishop, S.J. Nasuto. *Stochastic diffusion: Using recruitment for search* in Proc. of Symposium on Evolvability and Interaction, pp. 60-65, The University of London, UK, 2003.

3. J.M. Bishop, P. Torr. "The Stochastic Search Network" in *Neural Networks for Images, Speech and Natural Language*, pp. 370-387, Chapman & Hall, New York, 1992.

4. E. Grech-Cini. "Locating Facial Features". PhD Thesis, University of Reading, Reading, UK, 1995.

5. M.M. al-Rifaie, A. Aber, D.J. Hemanth. "Deploying swarm intelligence in medical imaging identifying metastasis, micro-calcifications and brain image segmentation". *Systems Biology, IET*, vol. 9(6), pp. 234-244, 2015.

6. M.M. al-Rifaie, A. Aber. "Identifying Metastasis in Bone Scans with Stochastic Diffusion Search" in *Information Technology in Medicine and Education (ITME)*, 2012.

7. M.M. al-Rifaie, A. Aber, A.M. Oudah. "Utilising Stochastic Diffusion Search to identify metastasis in bone scans and microcalcications on mammographs" in *Proc. of IEEE International Conference on Bioinformatics and Biomedicine Workshops*, pp. 280-287, 2012.

8. M.M. al-Rifaie, A. Aber, A.M. Oudah. "Ants intelligence framework; identifying traces of cancer" in The House of Commons, UK Parliment. SET for BRITAIN 2013. Poster exhibitions in Biological and Biomedical Science, 2013

9. M.M. al-Rifaie, M.J. Bishop, T. Blackwell. "An investigation into the merger of stochastic diffusion search and particle swarm optimisation" in *Proc. of the 13th Annual Conference on Genetic and Evolutionary Computation, GECCO 2011*, pp. 37-44, 2011.

10. M.M. al-Rifaie, J.M. Bishop, T. Blackwell. "Information sharing impact of stochastic diffusion search on differential evolution algorithm". *Memetic Computing*, vol. 4(4), pp. 327-338, 2012.

11. M.M. al-Rifaie, M. Bishop, T. Blackwell. "Resource Allocation and Dispensation Impact of Stochastic Diffusion Search on Differential Evolution Algorithm" in *Nature Inspired Cooperative Strategies for Optimization* (NICSO 2011). Studies in Computational Intelligence, vol. 387, pp. 21-40, Springer, 2012.

12. M.G.H. Omran, A. Salman. "Probabilistic stochastic diffusion search" in *Swarm Intelligence*, pp. 300-307, Springer, 2012.

13. M.M. al-Rifaie, M. Y.-K. Matthew and M. d'Inverno. "Investigating Swarm Intelligence for Performance Prediction" in *Proc. of the 9th International Conference on Educational Data Mining*, pp. 264-269, 2016.

14. H. Alhakbani, M.M. al-Rifaie. "Feature Selection Using Stochastic Diffusion Search" in *Proc. of the Genetic and Evolutionary Computation Conference*, pp. 385-392, 2017.

15. M.M. al-Rifaie, D. Joyce, S. Shergill, M. Bishop. "Investigating stochastic diffusion search in data clustering" in *SAI Intelligent Systems Conference (IntelliSys)*, pp. 187-194, 2015.

16. H.A. Alhakbani, M.M. al-Rifaie. "Exploring Feature-Level Duplications on Imbalanced Data Using Stochastic Diffusion Search" in *Multi-Agent Systems and Agreement Technologies*, pp. 305-313, Springer, 2016.

17. I. Aleksander, T.J. Stonham. "Computers and Digital Techniques 2(1)" in *Lecture Notes in Artificial Intelligence*, vol. 1562, pp. 29-40, Springer, 1979.

18. H.A. Alhakbani, M.M. al-Rifaie. "A Swarm Intelligence Approach in Undersampling Majority Class" in *Proc. of 10th International Conference, ANTS 2016*, pp. 225-232, 2016.

19. M.M. al-Rifaie, J.M. Bishop, S. Caines. "Creativity and autonomy in swarm intelligence systems". *Cognitive Computation*, vol. 4(3), pp. 320-331, 2012.

20. M.M. al-Rifaie, J.M. Bishop. "Weak and Strong Computational Creativity" in *Computational Creativity Research: Towards Creative Machines. Atlantis Thinking Machines*, vol. 7, pp. 37-49, Atlantis Press, 2015.

21. M.M. al-Rifaie, F.F. Leymarie, W. Latham, M. Bishop. "Swarmic autopoiesis and computational creativity". *Connection Science*, vol. 29(4), pp. 276-294, 2017.

22. J.M. Bishop, M.M. al-Rifaie. "Autopoiesis, creativity and dance". *Connection Science*, vol. 29(1), pp. 21-35, 2017.

23. M.M. al-Rifaie, A. Cropley, D. Cropley, M. Bishop. "On evil and computational creativity". *Connection Science*, vol. 28(1), pp. 171-193, 2016.

24. M.M. al-Rifaie, A. Aber, M. Bishop. "Cooperation of Nature and Physiologically Inspired Mechanisms in Visualisation" in *Biologically-Inspired Computing for the Arts: Scientific Data through Graphics*, IGI Global, United States, 2012.

25. M.M. al-Rifaie, M. Bishop. "Swarmic Paintings and Colour Attention". *Lecture Notes in Computer Science*, vol. 7834, pp. 97-108, Springer, 2013.

26. M.M. al-Rifaie, M. Bishop. "Swarmic Sketches and Attention Mechanism". *Lecture Notes in Computer Science*, vol. 7834, pp. 85-96, Springer, 2013.

27. A.M. al-Rifaie, M.M. al-Rifaie. "Generative Music with Stochastic Diffusion Search". *Lecture Notes in Computer Science*, vol. 9027, pp. 1-14, Springer, 2015.

28. M.M. al-Rifaie, W. Latham, F.F. Leymarie, M. Bishop. "Swarmic Autopoiesis: Decoding de Kooning" in *Proc. of 3rd Symposium on Computational Creativity, AISB 2016*, 2016.

29. M.A.J. Javid, M.M. al-Rifaie, R. Zimmer. "Detecting Symmetry in Cellular Automata Generated Patterns Using Swarm Intelligence" in *Proc. of 3rd International Conference on the Theory and Practice of Natural Computing (TPNC 2014)*, pp. 83-94, 2014.

30. F.M. al-Rifaie, M.M. al-Rifaie. "Investigating Stochastic Diffusion Search in DNA Sequence Assembly Problem" in *Proc. of SAI Intelligent Systems Conference (IntelliSys)*, pp. 625-631, 2015.

31. M.A.J. Javid, W. Alghamdi, A. Ursyn, R. Zimmer, M.M. al-Rifaie. "Swarmic approach for symmetry detection of cellular automata behaviour". *Soft Computing*, vol. 21(19), pp. 5585-5599, 2017.

32. E. Grech-Cini. "Locating Facial Features". PhD Thesis, University of Reading, Reading, UK, 1995.

33. P.D. Beattie, J.M. Bishop. "Self-localisation in the SENARIO autonomous wheelchair". *Journal of Intellingent and Robotic Systems*, vol. 22, pp. 255-267, 1998.

34. A.K. Nircan. "Stochastic Diffusion Search and Voting Methods". PhD Thesis, Bogaziki University, 2006.

35. K. de-Meyer, M. Bishop, S. Nasuto. "Small World Effects in Lattice Stochastic Diffusion Search". *Lecture Notes in Computer Science*, vol. 2415, pp. 147-152, Springer, 2002.

36. T. Tanay, J.M. Bishop, S.J. Nasuto, E.B. Roesch, M.C. Spencer. "Stochastic Diffusion Search Applied to Trees: a Swarm Intelligence Heuristic Performing Monte-Carlo Tree Search" in *Proc. of AISB 2013*, 2013.

37. R.M. Whitaker, S. Hurley. "An Agent Based Approach to Site Selection for Wireless Networks" in *Proc. of 1st IEE Conf. on Artificial Neural Networks*, 2002.

38. D.R. Myatt, S.J. Nasuto, J.M. Bishop. *Artificial Life X Proceedings of the Tenth International Conference on the Simulation and Synthesis of Living Systems*, Bloomington, USA, 2006.

39. S.J. Nasuto. "Resource Allocation Analysis of the Stochastic Diffusion Search". PhD Thesis, University of Reading, Reading, UK.

40. S.J. Nasuto, J.M. Bishop. "Convergence analysis of stochastic diffusion search". *Parallel Algorithms and Applications*, vol. 14(2), 1999.

41. K. de-Meyer. "Foundations of Stochastic Diffusion Search". PhD Thesis, University of Reading, Reading, UK, 2003.

42. M.B. Miller, B.L. Bassler. "Quorum sensing in bacteria". *Annual Reviews in Microbiology*, vol. 55(1), pp. 165-199, 2001.

43. D.E. Goldberg. *Genetic Algorithms in Search, Optimization and Machine Learning*, Addison-Wesley Longman Publishing Co., 1989.

44. J. Kennedy, R.C. Eberhart. "Particle swarm optimization" in *Proc. of the IEEE International Conference on Neural Networks*, pp. 1942-1948, 1995.

45. J. Digalakis, K. Margaritis. "An experimental study of benchmarking functions for evolutionary algorithms". *Int. J. Comput. Math.* vol. 79, pp. 403-416, 2002.

46. D. Whitley, S. Rana, J. Dzubera, K.E. Mathias. "Evaluating evolutionary algorithms". *Artificial Intelligence*, vol. 85(1-2), pp. 245-276, 1996.

47. Y. Jin. "A comprehensive survey of fitness approximation in evolutionary computation". *Soft Computing*, vol. 9, pp. 3-12, 2005.

48. J. Branke, C. Schmidt, H. Schmeck. "Efficient Fitness Estimation in Noisy Environments" in *Proc. of Genetic and Evolutionary Computation Conference*, 2001.

49. M.A. el-Beltagy, A. J. Keane. "Evolutionary Optimization for Computationally Expensive Problems using Gaussian Processes" in *Proc. of Int. Conf. on Artificial Intelligence'01*, pp. 708-714, 2001.

50. K. de-Meyer, S.J. Nasuto, J.M. Bishop. "Stochastic Diffusion Optimisation: the Application of Partial Function Evaluation and Stochastic recruitment in Swarm Intelligence Optimisation" in *Swarm Intelligence and Data Mining*, Springer, 2006.

51. R.G. Miller. *Simultaneous Statistical Inference*, Springer, 1981.

24

Whale Optimization Algorithm – Modifications and Applications

Ali R. Kashani

Department of Civil Engineering
University of Memphis, Memphis, Tennessee, United States

Charles V. Camp

Department of Civil Engineering
University of Memphis, Memphis, Tennessee, United States

Moein Armanfar

Department of Civil Engineering
Arak University, Arak, Iran

Adam Slowik

Department of Electronics and Computer Science
Koszalin University of Technology, Koszalin, Poland

CONTENTS

24.1   Introduction

The whale optimization algorithm (WOA) algorithm as a nature-inspired metaheuristic optimization algorithm imitates the social behavior of humpback whales [1]. WOA is based on a mathematical model of their hunting behaviors. To be more exact, humpback whales utilize an exploration procedure called the bubble net feeding method. Humpback whales desire to hunt a school of krill or small fishes near the surface. This foraging is accomplished by creating distinctive bubbles along a circle or '9'-shaped path. To accomplish this task, whales utilize two different tactics: upward-spirals and double-loops. The main procedure of a WOA can be summarized in three main steps: encircling prey; bubble-net attacking method; and search for prey. The first feature, encircling the prey, pushes all the search agents to move toward the best-found solution (leader). Next, the bubble-net attacking method simulates whales' movement path to approach the prey. To be more precise, whales move on a shrinking circular route along a spiral-shape path. This behavior provides the exploitation ability in the WOA. Another important characteristic of a metaheuristic algorithm is exploration; WOA uses search for prey to explore the search space. In this step, the position of the i-th search agent will be updated using one of the randomly selected search agents. The remainder of this chapter is organized accordingly. In Section 24.2 a summary of the fundamental of WOA algorithm is presented. In Section 24.3 some modifications of WOA are described. Finally, in Section 24.4 the application of a WOA to the optimization of shallow foundation is discussed.

24.2   Original WOA algorithm in brief

In this section, the fundamentals of the original WOA are discussed using a step-by-step pseudo-code. As presented in Algorithm 31, the WOA conducts an optimization procedure for a D-dimensional objective function $OF(.)$. In step 2, the boundary limitations for the j-th variable where $j = \{1, ..., number\ of\ decision\ variables\}$ is defined for the i-th agent as $[X_{i,j}^{min}, X_{i,j}^{max}]$. Next, parameters for the algorithm are defined, such as number of search agents ($SearchAgents\_no$) and maximum number of iterations (MI). In the next step, a population of $SearchAgents\_no$ search agents is generated randomly. Each agent is defined by a D-dimensional vector accordingly:

$$X_i = \{x_{i,1}, x_{i,2}, ..., x_{i,D}\} \tag{24.1}$$

where $x_{i,1}$ to $x_{i,D}$ are decision variables varying between $X_{i,j}^{min}$ and $X_{i,j}^{max}$.

As a fundamental in a WOA, search agents try to search the space around a leader (agent representing the current best solution). There are two strategies for selecting the leader: the overall best solution (for exploitation); and a randomly selected agent (for exploration). Therefore, in step 5, the global best solution and its associated X_i vector are attributed to the leader. After that, the current iteration counter is initialized as zero. Now, the main loop of the WOA is started. In step 9, each i-th search agent position is updated through an iterative loop where the value of α is updated following a decreasing pattern. Next, two coefficient vectors, A and C, are updated in steps 11 and 12, respectively. An important feature of the hunting behavior of humpback whales is the way they get close to the prey. Humpback whales move along a shrinking circle (steps 16 and 19) and spiral-shaped path (step 22), simultaneously. In the WOA, one of these movements is chosen based on a probability of 50%. In step 13, a random number p is generated between 0 and 1 and if $p < 0.5$ the search for global optimum solution uses a shrinking circle; otherwise, the spiral-shaped path strategy is applied. If the shrinking circle is selected, and $|A| < 1$ then the best-found solution is chosen as a leader and used to update position; otherwise, an agent is selected randomly and used as the leader. In step 22, if $p > 0.5$ the position of the i-th agent is updated following a spiral-shaped path. In step 25, estimate the $OF(.)$ for all the search agents. Next, the overall global best solution is updated. Similar to other optimization algorithms, the best-found solution and its associated X_i vector are proposed as the final result when the termination criteria are satisfied.

Algorithm 31 Pseudo-code of the original WOA.

1: determine the $D - th$ dimensional objective function $OF(.)$
2: determine the range of variability for each $j - th$ dimension $\left[X_{i,j}^{min}, X_{i,j}^{max}\right]$
3: determine the WOA algorithm parameter values such as $SearchAgents\_no$ – number of search agents, MI – maximum iteration
4: randomly create positions X_i for $SearchAgentsA\_no$ (each agent is a D-dimensional vector)
5: find the best search agent and call it leader
6: $Iter = 0$
7: **while** termination condition not met (here is reaching MI) **do**
8:     **for** each i-th search agent **do**
9:         update α value to decrease from 2 to 0 using $\alpha = 2 - Iter \times \left(\frac{2}{MI}\right)$
10:         $A = 2\alpha \times rand - \alpha$
11:         $C = 2 \times rand$
12:         determine p as a random number between 0 and 1
13:         **if** p<0.5 **then**
14:             **if** $| A |$<1 **then**
15:                 update the position of i-th agent using: $X\left(Iter\right) = X^*\left(Iter - 1\right) - A \cdot \mid C \cdot X^*\left(Iter - 1\right) - X\left(Iter - 1\right) \mid$
16:             **else**
17:                 randomly select one of the search agents as a leader

18:             update the position of the i-th agent using: $X(Iter) = X_{rand} - A \cdot |C \cdot X_{rand} - X(Iter - 1)|$

19:             **end if**

20:         **else**

21:             update the position of i-th agent using $X(Iter) = D' \cdot exp(bl) \cdot cos(2\pi l) + X^*(Iter - 1)$

22:         **end if**

23:     **end for**

24:     calculate $OF'(.)$

25:     update best-found solution

26:     Iter=Iter+1

27: **end while**

28: post-processing the results

24.3   Modifications of WOA algorithm

24.3.1   Chaotic WOA

In a study by Kaur and Arora [2] a chaotic version of the WOA (CWOA) was developed to improve its performance and convergence speed. Like many other metaheuristic algorithms, there are some stochastic parameters in a WOA which affect the algorithm's performance. It has been proved using chaotic maps to control the randomness of those components can be considerably advantageous. Kaur and Arora [2] proposed to use 10 different chaotic maps to control parameter p in line 13 of the Algorithm 31. Based on this approach, an initial random number is selected for the chaotic map. Afterward, the value of p will be updated in each generation via a given chaotic map. Table 24.1 lists the chaotic maps utilized by Kaur and Arora [2].

In Table 24.1, a is a control parameter and CP_{i+1} is a pseudo random produced value using chaotic maps. CP_i varies between 0 and 1 where in the CWOA an initial point of 0.7 has been chosen.

In this effort, 20 well-known benchmark functions were studied to assess the effectiveness of the proposed modification. Results illustrated that the convergence rate of WOA has been improved by applying chaos maps. Statistical testing and convergence histories of modified WOA by chaotic maps determined superiority of chaotic WOA over the original WOA. Kaur and Arora [2] demonstrated that the Tent map was the best chaos mapping strategy.

24.3.2   Levy-flight WOA

Metaheuristic optimization algorithms explore the solution space based on exploration and exploitation. Incorporating a Levy-flight distribution to algorithms has been proved to be effective in this regard. Abdel-Basset et al. [3]

TABLE 24.1

Different chaos maps.

ID	Name	Formulation
1.	Logistic	$CP_{i+1} = aCP_i(1 - CP_i)$
2.	Cubic	$CP_{i+1} = aCP_i(1 - CP_i^2)$
3.	Sine	$CP_{i+1} = \frac{a}{4}sin(\pi CP_i)$
4.	Sinusoidal	$CP_{i+1} = aCP_i^2 sin(\pi CP_i)$
5.	Singer	$CP_{i+1} = \mu(7.86CP_i - 23.31CP_i^2 + 28.75CP_i^3 - 13.302875CP_i^4), \mu = 1.07$
6.	Circle	$CP_{i+1} = mod\left(CP_i + b - \left(\frac{a}{2\pi}\right)sin(2\pi CP_i), 1\right)$ $a = 0.5$ and $b = 0.2$
7.	Iterative	$CP_{i+1} = sin\left(\frac{a\pi}{CP_i}\right)$
8.	Tent	$CP_{i+1} = \begin{cases} \frac{CP_i}{0.7} & CP_i < 0.7 \\ \frac{10}{3}(1 - CP_i) & CP_i \geq 0.7 \end{cases}$
9.	Piecewise	$CP_{i+1} = \begin{cases} \frac{CP_i}{p} & 0 \leq CP_i < p \\ \frac{CP_i - p}{0.5 - p} & p \leq CP_i < 0.5 \\ \frac{1 - p - CP_i}{0.5 - p} & 0.5 \leq CP_i < 1 - p \\ \frac{1 - CP_i}{p} & 1 - p \leq CP_i < 1 \end{cases}$
10.	Gauss/mouse	$CP_{i+1} = \begin{cases} 1 & CP_i = 0 \\ \frac{1}{mod(CP_i, 1)} & otherwise \end{cases}$

developed an improved version of the WOA based on incorporating Levy-flight to WOA as well as using a chaos theory for the random parameter. In this paper, the authors tried to provide broader movement domain for the search agents by adjusting the parameter C in line 12 of Algorithm 31. To this end, this parameter is replaced with an isotropic Levy step as Equation (24.2):

$$Levy = \frac{\lambda\Gamma(\lambda)\,sin\left(\frac{\pi\lambda}{2}\right)}{\pi} \times \frac{1}{s^{1+\lambda}}, s >> s_0 > 0 \tag{24.2}$$

$$s = \frac{U}{|V|^{\lambda-1}}, U \sim N\left(0, \sigma_u^2\right), V \sim N\left(0, \sigma_v^2\right) \tag{24.3}$$

$$\sigma_u^2 = \left[\frac{\Gamma(1+\lambda)}{\lambda\Gamma\left(\frac{1+\lambda}{2}\right)} \times \frac{sin\left(\frac{\pi\lambda}{2}\right)}{2^{\frac{(\lambda-1)}{2}}}\right]^{\frac{1}{\lambda}}, \sigma_v^2 = 1 \tag{24.4}$$

where *Levy* is a step size, $\Gamma(\lambda)$ is the standard gamma function with large steps $s > 0$ which is drawn according to Magneta algorithm [4], U and V are samples produced by a Gaussian normal distribution with zero mean value and σ_u^2 and σ_v^2 variance values.

In addition, a logistic chaotic map (presented as ID 1 in Table 24.1) is selected to adjust parameter p.

Another modification in this algorithm is enlisting a mutation operator in case of observing no improvement in the solution. This mutation operator works based on three functions: swap, displacement, and reversion. The first function swaps two individuals randomly. The second function replaces a random subset of the individuals and inserts it in another random section of the population. The third function reserves a random subset of the individuals.

The results demonstrated that this algorithm worked more efficiently as complexity of the optimization problems increased.

24.3.3   Binary WOA

There are many optimization problems with discrete solution spaces. In these problems, it is necessary to modify an algorithm's suitability for a binary search space. Kumar and Kumar [5] proposed a binary version of a WOA based on changing three main phases of the standard WOA, shrinking and encircling, bubble-net attacking method, and prey search. In this modified WOA, the position of the search agents during the shrinking and encircling prey phases are updated accordingly:

$$X\left(Iter\right) = \begin{cases} complement\left(X\left(Iter-1\right)\right) & \text{if } rand < C_{step} \\ X\left(Iter-1\right) & \text{otherwise} \end{cases} \tag{24.5}$$

where C_{step} is the step size given as

$$C_{step} = \frac{1}{1 + e^{-10(A \times D - 0.5)}} \tag{24.6}$$

In Equation (24.6), A is the coefficient vector in line 11 of Algorithm 31 and D is as follows (line 16 Algorithm 31):

$$D = \left| C \times X^*\left(Iter-1\right) - X\left(Iter-1\right) \right| \tag{24.7}$$

The second modification was applied to the bubble-net behavior of whales as follows:

$$X\left(Iter\right) = \begin{cases} complement\left(X\left(Iter-1\right)\right) & \text{if } rand < C_{step'} \\ X\left(Iter-1\right) & \text{otherwise} \end{cases} \tag{24.8}$$

where $C_{step'}$ is as follows:

$$C_{step'} = \frac{1}{1 + e^{-10(A \times D' - 0.5)}} \tag{24.9}$$

And D' is proposed as:

$$D' = X^*\left(Iter-1\right) - X\left(Iter-1\right) \tag{24.10}$$

The third modification considered in the searching for prey phase is based on the following

$$X\left(Iter\right) = \begin{cases} complement\left(X\left(Iter-1\right)\right) & \text{if } rand < C_{step''} \\ X\left(Iter-1\right) & \text{otherwise} \end{cases} \tag{24.11}$$

where $C_{step''}$ is proposed to be as follows:

$$C_{step''} = \frac{1}{1 + e^{-10(A \times D'' - 0.5)}} \tag{24.12}$$

In the Equation (24.13), D'' is defined by the following formulation:

$$D'' = |C \times X_{rand} - X(Iter - 1)| \tag{24.13}$$

where X_{rand} is a randomly selected whale.

Kumar and Kumar [5] compared their proposed binary WOA algorithm with six other optimization algorithms for several benchmark test functions. In their paper, Kumar and Kumar [5] claimed that their improved algorithm demonstrated better performance based on the obtained results.

24.3.4 Improved WOA

To achieve a better performance with a WOA, several studies have been conducted with different strategies. Sun et al. [6] attempted to improve WOA by adopting three modifications: considering a quadratic interpolation, employing a Levy flight, and using cosine function for updating α in each iteration. Quadratic interpolation (QI) finds the minimum point of the quadratic curve passing through three selected points in the solution space [6]. Following this strategy, quadratic crossover produces a new solution using the best search (X^*) agent and two other solution vectors $(Y$ and $Z)$ using the equation below:

$$x_i = 0.5 \times \frac{\left(y_i^2 - z_i^2\right) \times f(X^*) + \left(z_i^2 - x_i^{*2}\right) \times f(Y) + \left(x_i^{*2} - y_i^2\right) \times f(Z)}{(y_i - z_i) \times f(X^*) + (z_i - x_i^*) \times f(Y) + (x_i^* - y_i) \times f(Z)} \tag{24.14}$$

where i represents i-th dimension of vectors X^*, Y and Z.

This crossover operator used in the exploration phase with a probability threshold of 0.6. To be more exact, a probability decision value will be produced randomly. Then, the previously mentioned quadratic crossover will be utilized to produce new solutions for the probability values of less than 0.6. Otherwise, the spiral-shaped path as the original WOA algorithm methodology will be utilized.

Another improvement is to use a Levy-flight approach. In this modified WOA, a Levy-flight approach replaces the shrinking encircling mechanism (line 15 to 20 in Algorithm 31) to help the algorithm avoid a local minimum. To this end, a simple power-law vision of the Levy distribution is utilized as follows:

$$L(s) \sim |s|^{-1-\beta}, 0 < \beta \leq 2, s = \frac{\mu}{|v|^{\frac{1}{\beta}}} \tag{24.15}$$

where β is an index, s is the step length of Levy flight and μ and v are given by the following equations:

$$\mu \sim N\left(0, \sigma_u^2\right), v \sim N\left(0, \sigma_v^2\right), \sigma_\mu = \left[\frac{\Gamma\left(1+\beta\right) \cdot sin\left(\pi \cdot \frac{\beta}{2}\right)}{\Gamma\left(\frac{1+\beta}{2}\right) \cdot \beta \cdot 2^{\frac{(\beta-1)}{2}}}\right], \sigma_v = 1$$

(24.16)

Based on this method, the position of whales would be updated using the following equation:

$$X\left(Iter\right) = X\left(Iter - 1\right) + \frac{1}{\sqrt{Iter - 1}} \times sign\left(rand - 0.5\right) \oplus Levy \quad (24.17)$$

where $\oplus$ denotes entry-wise multiplications, $sign$ is a sign function with three possible outputs of -1, 0, and 1. In the above equation $Levy$ can be calculated as follows:

$$Levy = random\left(size\left(D\right)\right) \oplus L\left(\beta\right) \sim 0.01\left(\frac{\mu}{|v|^{\frac{1}{\beta}}}\right)\left(X_i - X^*\right) \quad (24.18)$$

where $size\left(D\right)$ is the scale of the problem, and X_i is the i-th solution vector. As the third modification, a nonlinear controlling pattern was proposed for parameter a as follows:

$$a = 2cos\left(\frac{Iter}{MI}\right) \quad (24.19)$$

In this modified version of WOA, the parameter a is updated using Equation (24.18) and then used to compute the A value. Sun et al. [6] demonstrated improved solutions with their proposed modified WOA for several benchmark functions. It noted in this paper that these modifications prevented WOA from premature convergence and helped the algorithm to escape from the local minima.

24.4 Application of WOA algorithm for optimum design of shallow foundation

24.4.1 Problem description

Shallow footings are critical geotechnical structures that play an important role in behavior of the whole foundation system. To help maintain the serviceability of footings, building codes provide many limitations and rules for their design and analysis. The main challenge for an engineer is to meet all the code's regulations while attempting to find an optimum design. Metaheuristic optimization algorithms are very useful in obtaining a footing design that meets code requirements and minimizes cost. To this end, the design procedure should be defined in the form of an objective function and input design variables. As demonstrated in Figure 24.1, the design variables are length of footing (X_1), width (X_2), thickness (X_3), depth (X_4), bar number along the longer direction (R_1), the number of bars in long direction (R_2), bar number along the shorter direction (R_3), number of bars in short direction (R_4), bar number of dowels (R_5). Regularly, a shallow foundation may be designed using these mentioned design variables. In this study, this method is simulated for solving the examples 1 and 2 in Section 24.4. Gandomi and Kashani [7] proposed two additional input parameters (E_x and E_y) to determine the location of the column at the top of shallow foundation. It can be helpful to reach more uniform tension distribution under the shallow footing by which a more cost-effective design results. The effect of this modification is examined through the third numerical example in Section 24.4.

The objective function is based on the cost of the shallow footing given as [7, 8]:

$$f_{cost} = C_e V_e + C_f A_f + \xi C_r M_r + \frac{f'_c}{f'_{cmin}} C_c V_c + C_b V_b \qquad (24.20)$$

where C_e is the unit cost of excavation, C_f is the unit cost of the framework, C_r is the unit cost of reinforcement, C_c is the unit cost of concrete, C_b is the unit cost of backfill, respectively. Table 24.2 lists the cost values for this problem. ξ is a scale factor that gives the reinforcing steel term a magnitude comparable to that of the other terms, f'_{cmin} is the minimum allowable strength of concrete, V_e is volume of excavation, A_f is area of framework, M_r is the mass of reinforcement, V_c is volume of concrete, and V_b is volume of compacted backfill.

As mentioned earlier, some structural and geotechnical conditions are required to guarantee the strength and stability of the foundation. These limitations have been applied to the design procedure as constraints tabulated in Table 24.3 [8].

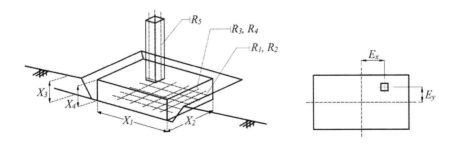

FIGURE 24.1

Schematic view of a shallow foundation with design variables.

TABLE 24.2

Unit cost values.

Input parameter	Unit	Symbol	Value
Excavation	$/m^3$	C_e	25.16
Concrete framework	$/m^2$	C_f	51.97
Reinforcement	$/kg$	C_r	2.16
Concrete	$/m^3$	C_c	173.96
Compacted backfill	$/m^3$	C_b	3.07

The parameters utilized in Table 24.3 are: FS_B, factor of safety against bearing capacity; FS_{min}, the minimum factor of safety for bearing capacity; δ_{max}, the maximum allowable settlement; V_n, the shear capacity of reinforced concrete foundation; V_u, the ultimate shear force; M_n, the moment capacity of foundation; M_u, the ultimate moment; A_s, steel reinforcement area; $A_{s,min}$, minimum reinforcement area; ϵ_s, tension steel strain; b_{col}, width of the column; cover, the concrete cover; n_{bar}, number of bars in either long or short direction; ω, either X_1 in long direction or X_2 in short direction; d, diameter of rebars and d_{dowel} the bend diameter of the hook dowels. To handle the mentioned constraints a static penalty function approach proposed by Homaifar et al. [9] is utilized. In this way, the objective function would be imposed by a factored violations term following the below equation:

$$fitness_i\left(X\right) = f_i\left(X\right) + \sum_{j=1}^{m} R_{k,j}\phi_j^2\left(X\right) \tag{24.21}$$

where $R_{i,j}$ are the penalty coefficients used, ϕ_j is amount of violation, m is the number of constraints, $f\left(X\right)$ is the unpenalized objective function, and $k = 1, 2, ..., l$, where l is the number of levels of violation defined by the user.

24.4.2   How can WOA algorithm be used for this problem?

In this sub-section, the automation of the design of shallow footings using the WOA algorithm is presented. First, *SearchAgents_no* search agents are randomly generated within the permitted bound constraints. There are 11 decision variables in the problem; each agent is an 11-dimensional vector. A penalty function proposed by Homaifar et al. [9] is used to reflect any violation of the constraints defined by standard codes to the objective function. In the next step, the best solution will be found by evaluating all the agents' objective function values. Now, the values of α, A, C, and p are initialized. For every i-th search agent, when $p < 0.5$ and $|A| < 1$, the best search agent is chosen as the leader; otherwise, a randomly selected search agent is the leader. When $p > 0.5$, the spiral updating path method is applied. Finally, the penalized objective function is evaluated through a minimization procedure by WOA.

TABLE 24.3

Geotechnical and structural constraints.

Constraint	Function
$g_1(x)$	$\frac{FS_B}{FS_{min}} - 1 > 0$
$g_2(x)$	$\frac{\delta_{max}}{\delta} - 1 > 0$
$g_{[3-5]}(x)$	$\frac{V_n}{V_u} - 1 > 0$
$g_{[6-7]}(x)$	$\frac{M_n}{M_u} - 1 > 0$
$g_{[8-10]}(x)$	$\frac{A_s}{A_{s,min}} - 1 > 0$
$g_{[11-12]}(x)$	$\frac{\epsilon_s}{0.005} - 1 > 0$
$g_{13}(x)$	$\frac{\left(\frac{X_1}{2} - \frac{b_{col}}{2} - cover\right)}{l_{d,short}} - 1 > 0$
$g_{14}(x)$	$\frac{\left(\frac{X_2}{2} - \frac{b_{col}}{2} - cover\right)}{l_{d,long}} - 1 > 0$
$g_{15}(x)$	$\frac{H}{2cover + d_{b,long} + d_{b,short} + \frac{d_{bend}}{2} + d_{dowel} + l_{d,dowel}} - 1 > 0$
$g_{[16-17]}(x)$	$\frac{H}{2cover + n_{bars}d_b + (n_{bars} - 1)s_{min}} - 1 > 0$
$g_{[18-19]}(x)$	$\frac{2cover + d_b + (n_{bars} - 1)s_{max}}{\omega} - 1 > 0$
$g_{20}(x)$	$\frac{P_{bearing}}{P_u} - 1 > 0$
$g_{21}(x)$	$\frac{D_{max}}{B} - 1 > 0$
$g_{22}(x)$	$\frac{B}{D_{min}} - 1 > 0$

This procedure will be repeated iteratively until satisfying the termination criteria.

24.4.3 Description of experiments

In this section, a numerical example is presented to design a low-cost shallow footing using the WOA. Table 24.4 lists combinations of dead and live vertical loads and moments applied to a foundation situated on a cohesionless soil. A column with a cross-sectional area of 400 mm × 400 mm and reinforced with 8 numbers of reinforcement with diameters of 16 mm is selected to direct the effective loads and moments onto the footing. Table 24.5 lists the soil parameters for this design. Over excavation of the length and the width of the footing are equal to 0.3 m. The yield strength of steel reinforcements is 400 MPa and the compressive strength of concrete is 21 MPa. The unit weight of concrete is 23.5 kN/m^3 and the concrete cover is 7 cm. The depth to the bottom of the footing from the ground surface (D) is 0.5 m. Table 24.6 lists the domains for the decision variables defined in Sub-section 24.4.1. There are several inequality constraints to control design of the footing. The geotechnical related constraints guarantee bearing capacity of the soil foundation and prescribe allowable settlement. However, structural constraints estimate the footing strength for bearing capacity of concrete, flexural moment, one-way and two-way shear forces. Finally, the geometrical criteria are related to the

TABLE 24.4

Loads and moments combinations for the case study.

Loading type	Dead	Live
Uniaxial load	650 kN	350 kN
Moment	400 $kN \cdot m$	150 $kN \cdot m$

TABLE 24.5

Utilized soil parameters for the case study.

Input parameters	Unit	Symbol	Values
Internal friction angle of base soil	°	ϕ	30
Unit weight of base soil	kN/m^3	γ_s	18
Poisson's ratio	-	μ_s	0.3
Elasticity modulus of soil	kPa	E_s	10.500
Maximum allowable settlement	mm	δ	25
Factor of safety for bearing capacity	-	$SF_{Bdesign}$	3

TABLE 24.6

Design variables permitted domain.

Design variables	X_1	X_2	X_3	X_4	R_1	R_2
Unit	cm	cm	cm	cm	$-$	$-$
Lower bound	400	400	0	300	2	3
Upper bound	4000	4000	3000	3000	20	18

Design variables	R_3	R_4	R_5	E_x	E_y
Unit	$-$	$-$	$-$	cm	cm
Lower bound	2	3	4	-2000	-2000
Upper bound	20	18	20	2000	2000

location of the column at the top of foundation and the depth of the shallow foundation.

24.4.4   Results obtained

Table 24.7 lists the final design results for three different shallow footing designs obtained with the WOA. For the current simulations, the maximum number of generations and search agents are 1000 and 50, respectively. Over 101 runs, Table 24.7 lists values for the best and worst designs, and the mean and standard deviation (SD) for the entire set. For all cases, the total number of function evaluations is 50,000. In Case I, a shallow footing with a uniaxial load is designed. A footing design which satisfies all the geotechnical and structural limitations was found after 65 generations. For Case II, an additional external moment is added to the loadings in Case I. In this case, the WOA found valid solutions with no constraint violations after 45 generations

TABLE 24.7
Design cost values for numerical case studies.

Model	Best	Worst	Mean	SD	Median
Case I	43,449.51	50,867.28	44,099.64	1,896.90	43,548.68
Case II	82,913.69	86,744.27	84,424.89	905.71	84,722.97
Case III	49,218.44	113,860.89	62,590.00	21,403.45	50,576.35

and the best solution was obtained after 399 generations. Due to the additional applied moment in CASE II, the best design has a 95.63% increase in cost compared to Case I. In Case III, the position of the column on the top of the footing is changed dynamically using two additional input variables (E_x and E_y). This attempt was helpful to decrease the final cost about 27.17%. In this case, the WOA did not find a stable design until the 23rd generation and the best design was obtained after 348 iterations.

24.5 Conclusions

In this study, several whale optimization algorithms were presented in a detailed step-by-step description. Many important modifications to the general WOA were described and reviewed. To demonstrate the effectiveness of the algorithm, three shallow footings are designed using the WOA. To this end, we explained how WOA deals with a shallow footing optimization problem using a step-by-step algorithm. As can be seen in Section 24.4 the algorithm reached a valid and stable design after some initial iterations. The WOA provides an automated approach to the often tedious and monotonous trial-and-error procedure of designing a foundation that meets all the geotechnical stability and structural strength criteria. More importantly, the WOA significantly reduces the overall cost of the structure while meeting all design criteria.

References

1. S. Mirjalili, A. Lewis, A. "The whale optimization algorithm". *Advances in Engineering Software*, vol. 95, pp. 51-67, 2016.

2. G. Kaur, S. Arora. "Chaotic Whale Optimization Algorithm". *Journal of Computational Design and Engineering*, vol. 5(3), pp. 275-284, 2018.

3. M. Abdel-Basset, L. Abdle-Fatah, A.K. Sangaiah. "An improved Lévy based whale optimization algorithm for bandwidth-efficient virtual machine placement in cloud computing environment". *Cluster Computing*, pp. 1-16, 2018.

4. R. N. Mantegna. "Fast, accurate algorithm for numerical simulation of Levy stable stochastic processes". *Physical Review E*, vol. 49(5), pp. 4677-4683, 1994.

5. V. Kumar, D. Kumar. "Binary whale optimization algorithm and its application to unit commitment problem". *Neural Computing and Applications*, pp. 1-29, 2018.

6. Y. Sun, X. Wang, Y. Chen, Z. Liu. "A modified whale optimization algorithm for large-scale global optimization problems". *Expert Systems with Applications*, vol. 114, pp. 563-577, 2018.

7. A.H. Gandomi, A.R. Kashani. "Construction cost minimization of shallow foundation using recent swarm intelligence techniques". *IEEE Transactions on Industrial Informatics*, vol. 14(3), pp. 1099-1106, 2018.

8. C.V. Camp, A. Assadollahi. "CO2 and cost optimization of reinforced concrete footings using a hybrid big bang-big crunch algorithm". *Structural and Multidisciplinary Optimization*, vol. 48(2), pp. 411-426, 2013.

9. A. Homaifar, S.H.Y Lai, X. Qi. "Constrained optimization via Genetic Algorithms". *Simulation*, vol. 62(4), pp. 242- 254, 1994.

Index

Printed and bound by CPI Group (UK) Ltd, Croydon, CR0 4YY

21/10/2024

01777049-0017